Lecture Notes in Mathematics

Volume 2334

This series reports on new developments in all areas of mathematics and their applications - quickly, informally and at a high level. Mathematical texts analysing new developments in modelling and numerical simulation are welcome. The type of material considered for publication includes:

1. Research monographs
2. Lectures on a new field or presentations of a new angle in a classical field
3. Summer schools and intensive courses on topics of current research.

Texts which are out of print but still in demand may also be considered if they fall within these categories. The timeliness of a manuscript is sometimes more important than its form, which may be preliminary or tentative.

Titles from this series are indexed by Scopus, Web of Science, Mathematical Reviews, and zbMATH.

Dinh Dũng • Van Kien Nguyen •
Christoph Schwab • Jakob Zech

Analyticity and Sparsity in Uncertainty Quantification for PDEs with Gaussian Random Field Inputs

Springer

Dinh Dũng
Information Technology Institute
Vietnam National University
Hanoi, Vietnam

Van Kien Nguyen
Department of Mathematical Analysis
University of Transport and Communications
Hanoi, Vietnam

Christoph Schwab
Seminar for Applied Mathematics
ETH Zürich
Zürich, Switzerland

Jakob Zech
Interdisciplinary Center for Scientific Computing
Heidelberg University
Heidelberg, Germany

ISSN 0075-8434 ISSN 1617-9692 (electronic)
Lecture Notes in Mathematics
ISBN 978-3-031-38383-0 ISBN 978-3-031-38384-7 (eBook)
https://doi.org/10.1007/978-3-031-38384-7

Mathematics Subject Classification: 65D40, 65N21, 41A25, 41A46, 33C45, 35B30, 65N30, 65N15, 28C20, 35J57, 35J15

This work was supported by NAFOSTED (102.01-2020.03)

This Springer imprint is published by the registered company Springer Nature Switzerland AG
The registered company address is: Gewerbestrasse 11, 6330 Cham, Switzerland

Paper in this product is recyclable.

Preface

We establish sparsity and summability results for coefficient sequences of Wiener-Hermite polynomial chaos expansions of countably parametric solutions of linear elliptic and parabolic divergence-form partial differential equations with Gaussian random field inputs.

The novel proof technique developed here is based on analytic continuation of parametric solutions into the complex domain. It differs from previous works that used bootstrap arguments and induction on the differentiation order of solution derivatives with respect to the parameters. The present holomorphy-based argument allows a unified, "differentiation-free" proof of sparsity (expressed in terms of ℓ^p-summability or weighted ℓ^2-summability) of sequences of Wiener-Hermite coefficients in polynomial chaos expansions in various scales of function spaces. The analysis also implies corresponding analyticity and sparsity results for posterior densities in Bayesian inverse problems subject to Gaussian priors on uncertain inputs from function spaces.

Our results furthermore yield dimension-independent convergence rates of various *constructive* high-dimensional deterministic numerical approximation schemes such as single-level and multi-level versions of Hermite-Smolyak anisotropic sparse-grid interpolation and quadrature in both forward and inverse computational uncertainty quantification.

Hanoi, Vietnam	Dinh Dũng
Hanoi, Vietnam	Van Kien Nguyen
Zürich, Switzerland	Christoph Schwab
Heidelberg, Germany	Jakob Zech

Acknowledgement

The work of Dinh Dũng and Van Kien Nguyen is funded by Vietnam National Foundation for Science and Technology Development (NAFOSTED) under Grant No. 102.01-2020.03. A part of this work was done when Dinh Dũng was visiting the Forschungsinstitut für Mathematik (FIM) of the ETH Zürich invited by Christoph Schwab, and when Dinh Dũng and Van Kien Nguyen were at the Vietnam Institute for Advanced Study in Mathematics (VIASM). They would like to thank the FIM and VIASM for providing a fruitful research environment and working condition. Dinh Dũng thanks Christoph Schwab for invitation to visit the FIM and for his hospitality.

Contents

List of Symbols

$C^s(D)$	Space of s-Hölder continuous functions on D
Δ	Laplace operator
div	Divergence operator
D	Domain in $\mathbb{R}^d$
∂D	Boundary of the domain D
$\boldsymbol{e}_j$	The multiindex $(\delta_{ij})_{i\in\mathbb{N}} \in \mathbb{N}_0^\infty$
γ	Gaussian measure
γ_d	Gaussian measure on $\mathbb{R}^d$
∇	Gradient operator
$H(\gamma)$	Cameron-Martin space of the Gaussian measure γ
H_k	kth normalized probabilistic Hermite polynomial
$H^s(D)$	$W^{s,2}(D)$
$H_0^1(D)$	Space of functions $u \in W^{1,2}(D)$ such that $u\|_{\partial D} = 0$
$H_0^s(D)$	$H^s(D) \cap H_0^1(D)$
$H^{-1}(D)$	Dual space of $H_0^1(D)$
$\mathbf{I}_\Lambda$	Smolyak interpolation operator
$\mathbf{I}_{\mathbf{l}}^{\mathrm{ML}}$	Multilevel Smolyak interpolation operator
$\Im(z)$	Imaginary part of the complex number z
$\mathcal{K}_\varkappa^s(D)$	Space of functions $u : D \to \mathbb{C}$ such that $r_D^{\|\boldsymbol{\alpha}\|-\varkappa} D^{\boldsymbol{\alpha}} u \in L^2(D)$ for all $\|\boldsymbol{\alpha}\| \le s$
λ_d	Lebesgue measure on $\mathbb{R}^d$
Λ	Set of multiindices
$\hat{\mu}$	Fourier transform of the measure μ
$L^p(\Omega)$	Space of Lebesgue measurable, p-integrable functions on Ω
$L^p(\Omega, \mu)$	Space of μ-measurable, p-integrable functions on Ω
$L^p(\Omega, X; \mu)$	Space of functions $u : \Omega \to X$ such that $\\|u\\|_X \in L^p(\Omega, \mu)$
$L^\infty(\Omega)$	Space of Lebesgue measurable, essentially bounded functions on Ω
$\ell^p(I)$	Space of sequences $(y_j)_{j\in I}$ such that $(\sum_{j\in I} \|y_j\|^p)^{1/p} < \infty$
$\\|f\\|_X$	Norm of f in the space X
$\mathbf{Q}_\Lambda$	Smolyak quadrature operator

$\mathbf{Q}_{\mathbf{l}}^{\mathrm{ML}}$	Multilevel Smolyak quadrature operator
r_D	Smooth function $D \to \mathbb{R}_+$ which equals $\|\boldsymbol{x} - \boldsymbol{c}\|$ in the vicinity of each corner of D
$\Re(z)$	Real part of the complex number z
U	$\mathbb{R}^\infty$
$W^{s,q}(D)$	Sobolev spaces of integer order s and integrability q on D
$\mathcal{W}^s_\infty(D)$	Space of functions $u : D \to \mathbb{C}$ such that $r_D^{\|\boldsymbol{\alpha}\|} D^{\boldsymbol{\alpha}} u \in L^\infty(D)$ for all $\|\boldsymbol{\alpha}\| \le s$

List of Abbreviations

BIP	Bayesian inverse problems
BVP	boundary value problem
FE	finite element
FEM	finite element method
GM	Gaussian measure
GRF	Gaussian random fields
IBVP	initial boundary value problem
KL	Karhunen-Loève
MC	Monte Carlo
ONB	orthonormal basis
PC	polynomial chaos
PDE	partial differential equations
QMC	quasi-Monte Carlo
RKHS	reproducing kernel Hilbert space
RV	random variable
UQ	uncertainty quantification

Chapter 1
Introduction

Gaussian random fields (GRFs for short) play a fundamental role in the modelling of spatio-temporal phenomena subject to uncertainty. In several broad research areas, particularly, in spatial statistics, data assimilation, climate modelling and meteorology to name but a few, GRFs play a pivotal role in mathematical models of physical phenomena with distributed, uncertain input data. Accordingly, there is an extensive literature devoted to mathematical, statistical and computational aspects of GRFs. We mention only [3, 75, 87] and the references there for mathematical foundations, and [56, 89] and the references there for a statistical perspective on GRFs.

In recent years, the area of *computational uncertainty quantification* (UQ for short) has emerged at the interface of the fields of applied mathematics, numerical analysis, scientific computing, computational statistics and data assimilation. Here, a key topic is the mathematical and numerical analysis of partial differential equations (PDEs for short) with random field inputs, and in particular with GRF inputs. The mathematical analysis of PDEs with GRF inputs addresses questions of well-posedness, pathwise and L^p-integrability and regularity in scales of Sobolev and Besov spaces of random solution ensembles of such PDEs. The numerical analysis focuses on questions of efficient numerical simulation methods of GRF inputs (see, e.g., [12, 14, 28, 29, 50, 57, 89, 100] and the references there), and the numerical approximation of corresponding PDE solution ensembles, which arise for GRF inputs. This concerns in particular the efficient representation of such solution ensembles (see [8, 9, 43, 52, 71]), and the numerical quadrature of corresponding solution fields (see, e.g., [31, 43, 51, 59, 66, 67, 71, 83, 98] and the references there). Applications include for instance subsurface flow models (see, e.g., [50, 58]) but also other PDE models for media with uncertain properties (see, e.g., [76] for electromagnetics). The careful analysis of efficient computational sampling of solution families of PDEs subject to GRF inputs is also a key ingredient in numerical data assimilation, e.g., in Bayesian inverse problems (BIPs for short); we refer to the

D. Dũng et al., *Analyticity and Sparsity in Uncertainty Quantification for PDEs with Gaussian Random Field Inputs*, Lecture Notes in Mathematics 2334,
https://doi.org/10.1007/978-3-031-38384-7_1

surveys [47, 48] and the references therein for a mathematical formulation of BIPs for PDEs subject to Gaussian prior measures and function space inputs.

In the past few years there have been considerable developments in the analysis and numerical simulation of PDEs with random field input subject to Gaussian measures (GMs for short). The method of choice in many applications for the numerical treatment of GMs is Monte-Carlo (MC for short) sampling. The (mean-square) convergence rate $1/2$ in terms of the number of MC samples is assured under rather mild conditions (existence of MC samples, and of finite second moments). We refer to, e.g., [30, 36, 73, 106] and the references there for a discussion of MC methods in this context. Given the high cost of MC sampling, recent years have seen the advent of numerical techniques which afford higher convergence orders than $1/2$, also on infinite-dimensional integration domains. Like MC, these techniques are not prone to the so-called curse of dimensionality. Among them are Hermite-Smolyak sparse-grid interpolation (also referred to as "stochastic collocation"), see e.g. [43, 45, 52], and sparse-grid quadrature [31, 43, 45, 52, 63], and quasi-Monte Carlo (QMC for short) integration as developed in [59, 66, 67, 78, 83, 96] and the reference there.

The key condition which emerged as governing the convergence rates of numerical integration and interpolation methods for a function is *a sparsity of the coefficients of its Wiener-Hermite polynomial chaos (PC for short) expansion*, see, e.g., [12, 71]. Rather than counting the ratio of nonzero coefficients, the sparsity is quantified by *ℓ^p-summability and/or weighted ℓ^2-summability* of these coefficients. This observation forms the foundation for the current text.

1.1 An Example

To indicate some of the mathematical issues which are considered in this book, consider in the interval $D = (0, 1)$ and in a probability space $(\Omega, \mathcal{A}, \mathbb{P})$, a GRF $g : \Omega \times D \to \mathbb{R}$ which takes values in $L^\infty(D)$. That is to say, that the map $\omega \mapsto g(\omega, \cdot)$ is an element of the Banach space $L^\infty(D)$. *Formally*, at this stage, we represent realizations of the random element $g \in L^\infty(D)$ with a *representation system* $(\psi_j)_{j=1}^J \subset L^\infty(D)$ in *affine-parametric form*

$$g(\omega, x) = \sum_{j=1}^{J} y_j(\omega)\psi_j(x), \tag{1.1}$$

where the coefficients $(y_j)_{j=1}^J$ are assumed to be i.i.d. standard normal random variables (RVs for short) and J may be a finite number or infinity. Representations such as (1.1) are widely used both in the analysis and in the numerical simulation of random elements g taking values in a function space. The coefficients $y_j(\omega)$ being standard normal RVs, the sum $\sum_{j=1}^J y_j\psi_j(x)$ may be considered as a *parametric deterministic map* $g : \mathbb{R}^J \to L^\infty(D)$. The random element $g(\omega, x)$ in (1.1) can then be obtained by evaluating this deterministic map in random coordinates, i.e., by sampling it in Gaussian random vectors $(y_j(\omega))_{j=1}^J \in \mathbb{R}^J$.

Gaussian random elements as inputs for PDEs appear in particular, in coefficients of diffusion equations. Consider, for illustration, in D, and for given $f \in L^2(D)$, the boundary value problem: find a random function $u : \Omega \to V$ with $V := \{w \in H^1(D) : w(0) = 0\}$ such that

$$f(x) + \frac{\mathrm{d}}{\mathrm{d}x}\left(a(x,\omega)\frac{\mathrm{d}}{\mathrm{d}x}u(x,\omega)\right) = 0 \quad \text{in} \quad D, \quad a(1,\omega)u'(1,\omega) = \bar{f}. \tag{1.2}$$

Here, $a(x, \omega) = \exp(g(x, \omega))$ with GRF $g : \Omega \to L^\infty(D)$, and $\bar{f} := F(1)$ with

$$F(x) := \int_0^x f(\xi)\,\mathrm{d}\xi \in V, \quad x \in D.$$

In order to dispense with summability and measurability issues, let us temporarily assume that the sum in (1.1) is finite, with $J \in \mathbb{N}$ terms. We find that a random solution u of the problem must satisfy

$$u'(x, \omega) = -\exp(-g(x, \omega))F(x), \quad x \in D, \omega \in \Omega.$$

Inserting (1.1), this is equivalent to the *parametric, deterministic family of solutions* $u(x, \boldsymbol{y}) : D \times \mathbb{R}^J \to \mathbb{R}$ given by

$$u'(x, \boldsymbol{y}) = -\exp(-g(x, \boldsymbol{y}))F(x), \quad x \in D, \boldsymbol{y} \in \mathbb{R}^J. \tag{1.3}$$

Hence

$$\|u'(\cdot, \boldsymbol{y})\|_{L^2(D)} = \|\exp(-g(\cdot, \boldsymbol{y}))F\|_{L^2(D)}, \quad \boldsymbol{y} \in \mathbb{R}^J,$$

which implies the (sharp) bounds

$$\|u'(\cdot, \boldsymbol{y})\|_{L^2(D)} \begin{cases} \geq \exp(-\|g(\cdot, \boldsymbol{y})\|_{L^\infty(D)})\|F\|_{L^2(D)} \\ \leq \exp(\|g(\cdot, \boldsymbol{y})\|_{L^\infty(D)})\|F\|_{L^2(D)}. \end{cases}$$

Due to the homogeneous Dirichlet condition at $x = 0$, up to an absolute constant the same bounds also hold for $\|u(\cdot, \boldsymbol{y})\|_V$.

It is evident from the explicit expression (1.3) and the upper and lower bounds, that for every parameter $\boldsymbol{y} \in \mathbb{R}^J$, the solution $u \in V$ exists. However, we can not, in general, expect uniform w.r.t. $\boldsymbol{y} \in \mathbb{R}^J$ a-priori estimates, also of the higher derivatives, for smoother functions $x \mapsto g(x, \boldsymbol{y})$ and $x \mapsto f(x)$. Therefore, the parametric problem (1.2) is nonuniformly elliptic, [28, 73]. In particular, also a-priori error bounds for various discretization schemes will contain this uniformity w.r.t. $\boldsymbol{y}$. The random solution will be recovered from (1.3) by inserting for the coordinates y_j samples of i.i.d. standard normal RVs.

This book focuses on developing a regularity theory for countably-parametric solution families $u(\cdot, \boldsymbol{y}) : \boldsymbol{y} \in \mathbb{R}^J$ with a particular emphasis on the case $J = \infty$. This allows for arbitrary Gaussian random fields $g(\cdot, \omega)$ in (1.2). Naturally, our results also cover the finite-parametric setting where the number J of random parameters is finite, but may be very large. Then, all constants in our error estimates are either independent of the parameter dimension J or their dependence on J is explicitly indicated. Previous works [8, 9, 59] addressed the ℓ^p-summability of the Wiener-Hermite PC expansion coefficients of solution families $\{u(\cdot, \boldsymbol{y}) : \boldsymbol{y} \in \mathbb{R}^\infty\} \subset V$ for the forward problem, based on moment bounds of derivatives of parametric solutions w.r.t. GM. Estimates for these coefficients and, in particular, for the summability, were obtained in [8, 9, 59, 70, 71]. In these references, all arguments were based on real-variable, bootstrapping arguments with respect to $\boldsymbol{y}$.

1.2 Contributions

We make the following contributions to the area computational UQ for PDEs with GRF inputs. First, we provide novel proofs of some of the sparsity results in [8, 9, 71] of the infinite-dimensional parametric forward solution map to PDEs with GRF inputs. The presently developed proof technique is based on holomorphic continuation and complex variable arguments in order to bound derivatives of parametric solutions, and their coefficients in Wiener-Hermite PC expansions. This is in line with similar arguments in the so-called "uniform case" in [32, 39]. There, the random parameters in the representation of the input random fields range in compact subsets of $\mathbb{R}$. Unlike in these references, in the present text due to the Gaussian setup the parameter domain $\mathbb{R}^\infty$ is not compact. This entails significant modifications of mathematical arguments as compared to those in [32, 39].

Contrary to the analysis in [8, 9, 59], where parametric regularity results were obtained by real-variable arguments combined with induction-based bootstrapping with respect to the derivative order, the present text develops derivative-free, complex variable arguments which allow directly to obtain bounds of the Wiener-Hermite PC expansion coefficients of the parametric solutions in scales of Sobolev and Besov spaces in the physical domain D in which the parametric PDE is posed. They also allow to treat in a unified manner parametric regularity of the solution map in several scales of Sobolev and Kondrat'ev spaces in the physical domain D which is the topic of Sect. 3.8, resulting in novel sparsity results for the solution operators to linear elliptic and parabolic PDEs with GRF inputs in scales of Sobolev and Besov spaces. We apply the quantified holomorphy of parametric solution families to PDEs with GRF inputs and preservation of holomorphy under composition, to problems of Bayesian PDE inversion conditional on noise observation data in Chap. 5, establishing in particular quantified parametric holomorphy of the corresponding Bayesian posterior.

We construct deterministic sparse-grid interpolation and quadrature methods for the parametric solution with convergence rate bounds that are free from the curse of dimensionality, and that afford possibly high convergence rates, given a sufficient sparsity in the Wiener-Hermite PC expansion of the parametric solutions. For sampling strategies in deterministic numerical quadrature, our findings show improved convergence rates, as compared to previous results in this area. Additionally, our novel sparsity results provided in scales of function spaces of varying spatial regularity enable us to construct apriori *multilevel versions of sparse-grid interpolation and quadrature*, with corresponding approximation rate bounds which are free from the curse of dimensionality, and explicit in terms of the overall number of degrees of freedom. Lastly, and in contrast to previous works, leveraging the preservation of holomorphy under compositions with holomorphic maps, our holomorphy-based arguments enable us to establish that our algorithms and bounds are applicable to posterior distributions in Bayesian inference problems involving GRF or Besov priors, as developed in [48, 94] and the references there.

1.3 Scope of Results

We prove quantified holomorphy of countably-parametric solution families of linear elliptic and parabolic PDEs. The parameter range equals $\mathbb{R}^\infty$, corresponding to countably-parametric representations of GRF input data, taking values in a separable locally convex space, in particular, Hilbert or Banach space of uncertain input data, endowed for example with a Gaussian product measure γ on $\mathbb{R}^\infty$.

The results established in this text and the related bounds on partial derivatives w.r.t. the parameters in Karhunen-Loève or Lévy-Cieselsky expansions of uncertain GRF inputs imply convergence rate bounds for several families of computational methods to numerically access these parametric solution maps. Importantly, we prove that in terms of $n \geq 1$, an integer measure of work and memory, an approximation accuracy $O(n^{-a})$ for some parameter $a > 0$ can be achieved, where the convergence rate a depends on the approximation process and on the amount of sparsity in the Wiener-Hermite PC expansion coefficients of the GRF under consideration. In the terminology of computational complexity, a prescribed numerical tolerance $\varepsilon > 0$ can be reached in work and memory of order $O(\varepsilon^{-1/a})$. *In particular, the convergence rate a and the constant hidden in the Landau $O(\cdot)$ symbol do not depend on the dimension of the space of active parameters involved in the approximations which we construct.* The approximations developed in the present text are *constructive and linear* and can be realized computationally by *deterministic algorithms* of so-called "stochastic collocation" or "sparse-grid" type. Error bounds are proved in L^2-type Bochner spaces with respect to the GM γ on the input data space of the PDE, in natural Hilbert or Banach spaces of solutions of the PDEs under consideration. Here, it is important to notice that the sparsity of the Wiener-Hermite PC expansion coefficients used in constructive linear approximation algorithms and in estimating convergence rates, takes the form of weighted ℓ^2-summability, but not ℓ^p-summability as in best n-term approximations [8, 9, 71]. Furthermore, ℓ^p-summability results are implied from the corresponding weighted ℓ^2-summability ones.

All approximation rates for deterministic sampling strategies in the present text are free from the so-called *curse of dimensionality*, a terminology coined apparently by R.E. Bellmann (see [17]). The rates are in fact only limited by the sparsity of the Wiener-Hermite PC expansion coefficients of the deterministic, countably-parametric solution families. In particular, dimension-independent convergence rates $> 1/2$ are possible, provided a sufficient Wiener-Hermite PC expansion coefficient sparsity, that the random inputs feature sufficient pathwise regularity, and the affine representation system (being a tight frame on space of admissible input realizations) are stable in a suitable smoothness scale of inputs.

1.4 Structure and Content of This Text

We briefly describe the structure and content of the present text.

In Chap. 2, we collect known facts from functional analysis and GM theory which are required throughout this text. In particular, we review constructions and results on GMs on separable Hilbert and Banach spaces. Special focus will be on constructions via countable products of univariate GMs on countable products of real lines. We also review assorted known results on convergence rates of Lagrangian finite elements for linear, second order, divergence-form elliptic PDEs in polytopal domains D with Lipschitz boundary ∂D.

In Chap. 3, we address the analyticity and sparsity for elliptic divergence-form PDEs with log-Gaussian coefficients. In Sect. 3.1, we introduce a model linear, second order elliptic divergence-form PDE with log-Gaussian coefficients, with variational solutions in the "energy space" $H_0^1(D)$. This equation was investigated with parametric input data in a number of references in recent years [8, 9, 32, 38, 39, 42, 59, 71, 83, 114]. It is considered in this work mainly to develop the holomorphic approach to establish our mathematical approach to parametric holomorphy and sparsity of Wiener-Hermite PC expansions of parametric solutions in a simple setting, and to facilitate comparisons with the mentioned previous works and results. We review known results on its well-posedness in Sect. 3.1, and Lipschitz continuous dependence on the input data in Sect. 3.2. We discuss regularity results for parametric coefficients in Sect. 3.3. Sections 3.4 and 3.5 describe uncertainty modelling by placing GMs on sets of admissible, countably parametric input data, i.e., formalizing mathematically aleatoric uncertainty in input data. Here, the Gaussian series introduced in Sect. 2.5 will be seen to take a key role in converting operator equations with GRF inputs to infinitely-parametric, deterministic operator equations. The Lipschitz continuous dependence of the solutions on input data from function spaces will imply strong measurability of corresponding random solutions, and render well-defined the *uncertainty propagation*, i.e., the push-forward of the GM on the input data. In Sect. 3.6, we connect the quantified holomorphy of the parametric, deterministic solution manifold $\{u(\boldsymbol{y}) : \boldsymbol{y} \in \mathbb{R}^\infty\}$ in the space $H_0^1(D)$ with a sparsity (weighted ℓ^2-summability and ℓ^p-summability) of the coefficients $(u_{\boldsymbol{\nu}})_{\boldsymbol{\nu}\in\mathcal{F}}$ of the ($H_0^1(D)$-valued) Wiener-Hermite PC expansion. With this methodology in place, we show in Sect. 3.7 how to obtain holomorphic regularity of the parametric solution family $\{u(\boldsymbol{y}) : \boldsymbol{y} \in \mathbb{R}^\infty\}$ in Sobolev spaces $H^s(D)$ of possibly high smoothness order $s \in \mathbb{N}$ and how to derive from here the corresponding sparsity. The argument is self-contained and provides parametric holomorphy for any differentiation order $s \in \mathbb{N}$ in a unified way, in domains D of sufficiently high regularity and for sufficiently high almost sure regularity of coefficient functions. In Sect. 3.8, we extend these results for linear second order elliptic differential operators in divergence form in a bounded polygonal domain $D \subset \mathbb{R}^2$. Here, corners are well-known to obstruct high almost sure pathwise regularity in the usual Sobolev and Besov spaces in D for both, PC coefficients and parametric solutions. Therefore, we develop summability of the Wiener-Hermite PC expansion coefficients $(u_{\boldsymbol{\nu}})_{\boldsymbol{\nu}\in\mathcal{F}}$ of the random solutions in terms of corner-weighted Sobolev spaces, originating with V.A. Kondrat'ev (see, e.g., [24, 61, 90] and the references there). In Sect. 3.9, we briefly recall some known related results [8–11, 32, 37–39, 71] on ℓ^p-summability and weighted ℓ^2-summability of the generalized PC expansion coefficients of solutions to parametric divergence-form elliptic PDEs, as well as applications to best n-term approximation.

In Chap. 4, we investigate sparsity of the Wiener-Hermite PC expansions coefficients of holomorphic functions. In Sect. 4.1, we introduce a concept of $(\boldsymbol{b}, \xi, \delta, X)$-holomorphy of parametric deterministic functions on the parameter domain $\mathbb{R}^\infty$ taking values in a separable Hilbert space X. This concept is fairly broad and covers a large range of parametric PDEs depending on log-Gaussian distributed data. In order to extend the results and the approach to bound Wiener-Hermite PC expansion coefficients via quantified holomorphy beyond the simple, second order diffusion equation introduced in Chap. 3, we address sparsity of the Wiener-Hermite PC expansions coefficients of $(\boldsymbol{b}, \xi, \delta, X)$-holomorphic functions. In Sect. 4.2, we show that composite functions of a certain type are $(\boldsymbol{b}, \xi, \delta, X)$-holomorphic under certain conditions. The significance of such functions is that they cover solution operators of a collection of linear, elliptic divergence-form PDEs in a unified way along with structurally similar PDEs with log-Gaussian random input data. This will allow to apply the ensuing results on convergence rates of deterministic collocation and quadrature algorithms to a wide range of PDEs with GRF inputs and functionals on their random solutions. In Sect. 4.3, we analyze some examples of holomorphic functions which are solutions to certain PDEs, including linear elliptic divergence-form PDEs with parametric diffusion coefficient, linear parabolic PDEs with parametric coefficient, linear elastostatics equations with log-Gaussian modulus of elasticity, Maxwell equations with log-Gaussian permittivity.

In Chap. 5, we apply the preceding abstract results on parametric holomorphy to establish quantified holomorphy of countably-parametric, posterior densities of corresponding BIPs where the uncertain input of the forward PDE is a countably-parametric GRF taking values in a separable Banach space of inputs. As an example, we analyze the BIP for the parametric diffusion coefficient of the diffusion equation with parametric log-Gaussian inputs.

In Chap. 6, we discuss deterministic interpolation and quadrature algorithms for approximation and numerical integration of $(\boldsymbol{b}, \xi, \delta, X)$-holomorphic functions. Such algorithms are necessary for the approximation of certain statistical quantities (expectations, statistical moments) of the parametric solutions with respect to a GM on the parameter space. The proposed algorithms are variants and generalizations of so-called "stochastic collocation" or "sparse-grid" type approximation, and proved to outperform sampling methods such as MC methods, under suitable sparsity conditions on coefficients of the Wiener-Hermite PC expansion of integrands. In the quadrature case, they are also known as "Smolyak quadrature" methods. Their common feature is (a) the deterministic nature of the algorithms, and (b) the possibility of achieving convergence rates $> 1/2$ independent of the dimension of parameters and therefore the curse of dimensionality is broken. They offer, in particular, the perspective of deterministic numerical approximations for GRFs under nonlinear pushforwards (being realized via the deterministic data-to-solution map of the PDE of interest). The decisive analytic property to be established are dimension-explicit estimates of individual Wiener-Hermite PC expansion coefficients of parametric solutions, and based on these, sharp summability estimates of norms of the coefficients of Wiener-Hermite PC expansion of parametric, deterministic solution families are given. In Sects. 6.1 and 6.2, we construct sparse-grid Smolyak-type interpolation and quadrature algorithms. In Sects. 6.3 and 6.4, we prove the convergence rates of interpolation and quadrature algorithms for $(\boldsymbol{b}, \xi, \delta, X)$-holomorphic functions.

Chapter 7 is devoted to multilevel interpolation and quadrature of parametric holomorphic functions. We construct deterministic interpolation and quadrature algorithms for generic $(\boldsymbol{b}, \xi, \delta, X)$-holomorphic functions. For linear second order elliptic divergence-form PDEs with log-Gaussian coefficients, the results on the weighted ℓ^2-summability of the Wiener-Hermite PC expansion coefficients of parametric, deterministic solution families with respect to corner-weighted Sobolev spaces on spatial domain D finally also allow to analyze methods for constructive, deterministic linear approximations of parametric solution families. Here, a truncation of Wiener-Hermite PC expansions is combined with approximating the Wiener-Hermite PC expansion coefficients in the norm of the "energy space" $H_0^1(D)$ of these solutions from finite-dimensional approximation spaces which are customary in the numerical approximation of solution instances. Importantly, *required approximation accuracies of the Wiener-Hermite PC expansion coefficients* $u_{\boldsymbol{\nu}}$ *will depend on the relative importance of* $u_{\boldsymbol{\nu}}$ *within the Wiener-Hermite PC expansion.* This observation gives rise to *multilevel approximations* where a prescribed overall accuracy in mean square w.r.t. the GM γ with respect to $H_0^1(D)$ will be achieved by a $\boldsymbol{\nu}$-dependent discretization level in the physical domain. Multilevel approximation and integration and the corresponding error estimates will be developed in this section in an abstract setting: Besides $(\boldsymbol{b}, \xi, \delta, X)$-holomorphy, it is neccessary to require an assumption on the discretization error in the physical domain in the form of stronger holomorphy of the approximation error in this discretization. A combined assumption for guaranteeing constructive multilevel approximations is formulated in Sect. 7.1. In Sect. 7.2 we introduce multilevel algorithms for interpolation and quadrature of $(\boldsymbol{b}, \xi, \delta, X)$-holomorphic functions, and discuss work models and choices of discretization levels. A key for the sparse-grid integration and interpolation approaches is to efficiently numerically allocate discretization levels to Wiener-Hermite PC expansion coefficients. We develop such an approach in Sect. 7.3. It is based on greedy searches and suitable thresholding of (suitable norms of) Wiener-Hermite PC expansion coefficients and on a-priori bounds for these quantities which are obtained by complex variable arguments. In Sects. 7.4 and 7.5, we establish convergence rate bounds of multilevel interpolation and quadrature algorithms for $(\boldsymbol{b}, \xi, \delta, X)$-holomorphic functions. In Sect. 7.6, we verify the abstract hypotheses of the sparse-grid multilevel approximations for the forward and inverse problems for concrete linear elliptic and parabolic PDEs on corner-weighted Sobolev spaces (Kondrat'ev spaces) with log-GRF inputs. In Sect. 7.7, we briefly recall some results from [43] (see also, [45] for some corrections) on linear multilevel (fully discrete) interpolation and quadrature in abstract Bochner spaces based on weighted ℓ^2-summabilities. These results are subsequently applied to parametric divergence-form elliptic PDEs and to parametric holomorphic functions.

1.5 Notation and Conventions

Additional to the real numbers $\mathbb{R}$, the complex numbers $\mathbb{C}$, and the positive integers $\mathbb{N}$, we set $\mathbb{R}_+ := \{x \in \mathbb{R} : x \geq 0\}$ and $\mathbb{N}_0 := \{0\} \cup \mathbb{N}$. We denote by $\mathbb{R}^\infty$ the set of all sequences $\boldsymbol{y} = (y_j)_{j\in\mathbb{N}}$ with $y_j \in \mathbb{R}$, and similarly define $\mathbb{C}^\infty$, $\mathbb{R}^\infty_+$ and $\mathbb{N}^\infty_0$. Both, $\mathbb{R}^\infty$ and $\mathbb{C}^\infty$, will be understood with the product topology from $\mathbb{R}$ and $\mathbb{C}$, respectively. For $\boldsymbol{\alpha}, \boldsymbol{\beta} \in \mathbb{N}^d_0$, $d \in \mathbb{N} \cup \{\infty\}$, the inequality $\boldsymbol{\beta} \leq \boldsymbol{\alpha}$ is understood component-wise, i.e., $\boldsymbol{\beta} \leq \boldsymbol{\alpha}$ if and only if $\beta_j \leq \alpha_j$ for all j.

Denote by $\mathcal{F}$ the countable set of all sequences of nonnegative integers $\boldsymbol{\nu} = (\nu_j)_{j\in\mathbb{N}}$ such that $\operatorname{supp}(\boldsymbol{\nu})$ is finite, where $\operatorname{supp}(\boldsymbol{\nu}) := \{j \in \mathbb{N} : \nu_j \neq 0\}$ denotes the "support" of the multi-index $\boldsymbol{\nu}$. Similarly, we define $\operatorname{supp}(\boldsymbol{\rho})$ of a sequence $\boldsymbol{\rho} \in \mathbb{R}^\infty_+$. For $\boldsymbol{\nu} \in \mathcal{F}$, and for a sequence $\boldsymbol{b} = (b_j)_{j\in\mathbb{N}}$ of positive real numbers, the quantities

$$\boldsymbol{\nu}! := \prod_{j\in\mathbb{N}} \nu_j!, \qquad |\boldsymbol{\nu}| := \sum_{j\in\mathbb{N}} \nu_j, \qquad \text{and} \qquad \boldsymbol{b}^{\boldsymbol{\nu}} := \prod_{j\in\mathbb{N}} b_j^{\nu_j}$$

are finite and well-defined.

For a multi-index $\boldsymbol{\alpha} \in \mathbb{N}^d_0$ and a function $u(\boldsymbol{x}, \boldsymbol{y})$ of $\boldsymbol{x} \in \mathbb{R}^d$ and parameter sequence $\boldsymbol{y} \in \mathbb{R}^\infty$ we use the notation $D^{\boldsymbol{\alpha}} u(\boldsymbol{x}, \boldsymbol{y})$ to indicate the partial derivatives taken with respect to $\boldsymbol{x}$. The partial derivative of order $\boldsymbol{\alpha} \in \mathbb{N}^\infty_0$ with respect to $\boldsymbol{y}$ *of finite total order* $|\boldsymbol{\alpha}| = \sum_{j\in\mathbb{N}} \alpha_j$ is denoted by $\partial^{\boldsymbol{\alpha}} u(\boldsymbol{x}, \boldsymbol{y})$. In order to simplify notation, we will systematically suppress the variable $\boldsymbol{x} \in D \subset \mathbb{R}^d$ in mathematical expressions, except when necessary. For example, instead $\int_D v(\boldsymbol{x})\,\mathrm{d}\boldsymbol{x}$ we will write $\int_D v\,\mathrm{d}\boldsymbol{x}$, etc. For a Banach space X, we denote by $X_{\mathbb{C}} := X + \mathrm{i}X$ the complexification of X. The space $X_{\mathbb{C}}$ is also a Banach space endowed with the (minimal, among several possible equivalent ones, see [91]) norm $\|x_1 + \mathrm{i}x_2\|_{X_{\mathbb{C}}} := \sup_{0\le t\le 2\pi} \|x_1 \cos t - x_2 \sin t\|_X$. The space X^∞ is defined in a similar way as $\mathbb{R}^\infty$.

By $\mathcal{L}(X, Y)$ we denote the vector space of bounded, linear operators between to Banach spaces X and Y. With $\mathcal{L}_{\mathrm{is}}(X, Y)$ we denote the subspace of boundedly invertible, linear operators from X to Y.

For a function space $X(D)$ defined on the domain D, if there is no ambiguity, when writing the norm of $x \in X(D)$ we will omit D, i.e., we write $\|x\|_X$ instead of $\|x\|_{X(D)}$.

For $0 < p \le \infty$ and a finite or countable index set J, we denote by $\ell^p(J)$ the quasi-normed space of all $\boldsymbol{y} = (y_j)_{j\in J}$ with $y_j \in \mathbb{R}$, equipped with the quasi-norm $\|\boldsymbol{y}\|_{\ell^p(J)} := \left(\sum_{j\in J} |y_j|^p\right)^{1/p}$ for $p < \infty$, and $\|\boldsymbol{y}\|_{\ell^\infty(J)} := \sup_{j\in J} |y_j|$. Sometimes, we make use of the abbreviation $\ell^p = \ell^p(J)$ in a particular context if there is no misunderstanding of the meaning. We denote by $(\boldsymbol{e}_j)_{j\in J}$ the standard basis of $\ell^2(J)$, i.e., $\boldsymbol{e}_j = (e_{j,i})_{i\in J}$ with $e_{j,i} = 1$ for $i = j$ and $e_{j,i} = 0$ for $i \ne j$.

Chapter 2
Preliminaries

A key technical ingredient in the analysis of numerical approximations of PDEs with GRF inputs from function spaces, and of numerical methods for their efficient numerical treatment are constructions and numerical approximations of GRFs on real Hilbert and Banach spaces. Due to their high relevance in many areas of science (theoretical physics, quantum field theory, spatial and high-dimensional statistics, etc.), a rich theory has been developed in the past decades and a large body of literature is available now. We recapitulate basic definitions and key results, in particular on GMs, that are necessary for the ensuing developments. We do not attempt to provide a comprehensive survey. We require the exposition on GMs on real-valued Hilbert and Banach spaces, as most PDEs of interest are formulated for real-valued inputs and solutions. However, we crucially use in the ensuing sections of this text analytic continuation of parametric representations to the complex parameter domain. This is required in order to bring to bear complex variable methods for derivative-free, sharp bounds on Hermite expansion coefficients of GRFs. Therefore, we develop in our presentation solvability, well-posedness and regularity for the PDEs that are subject to GRF inputs in Hilbert and Banach spaces of complex-valued fields.

The structure of this section is as follows. In Sect. 2.1, we recapitulate GMs on finite dimensional spaces, in particular on $\mathbb{R}^d$ and $\mathbb{C}^d$. In Sect. 2.2, we extend GMs to separable Banach spaces. Section 2.3 reviews the Cameron-Martin space. In Sect. 2.4 we recall a notion of Gaussian product measures on a Cartesian product of locally convex spaces. Section 2.5 is devoted to a summary of known representations of a GRF by a Gaussian series. A key object in these and more general spaces is the concept of *Parseval frame* which we introduce. For details, the reader may consult, for example, the books [3, 21, 87].

In Sect. 2.6 we recapitulate, from [6, 25, 54], (known) technical results on approximation properties of Lagrangian Finite Elements (FEs for short) in polygonal and polyhedral domains $D \subset \mathbb{R}^d$, on regular, simplicial partitions of D with local refinement towards corners (and, in space dimension $d = 3$, towards edges).

D. Dũng et al., *Analyticity and Sparsity in Uncertainty Quantification for PDEs with Gaussian Random Field Inputs*, Lecture Notes in Mathematics 2334,
https://doi.org/10.1007/978-3-031-38384-7_2

These will be used in Chap. 6 in conjunction with collocation approximations in the parameter space of the GRF to build deterministic numerical approximations of solutions in polygonal and in polyhedral domains.

2.1 Finite Dimensional Gaussian Measures

2.1.1 Univariate Gaussian Measures

In dimension $d = 1$, for every $\mu, \sigma \in \mathbb{R}$, there holds the well-known identity

$$\frac{1}{\sigma\sqrt{2\pi}} \int_{\mathbb{R}} \exp\left(-\frac{(y-\mu)^2}{2\sigma^2}\right) \mathrm{d}y = 1.$$

A Borel probability measure γ on $\mathbb{R}$ is *Gaussian* if it is either a Dirac measure δ_μ at $\mu \in \mathbb{R}$ or its density with respect to Lebesgue measure λ on $\mathbb{R}$ is given by

$$\frac{\mathrm{d}\gamma}{\mathrm{d}\lambda} = p(\cdot; \mu, \sigma^2), \quad p(\cdot; \mu, \sigma^2) := y \mapsto \frac{1}{\sigma\sqrt{2\pi}} \exp\left(-\frac{(y-\mu)^2}{2\sigma^2}\right).$$

We shall refer to μ as *mean*, and to σ^2 as *variance* of the GM γ. The case that $\gamma = \delta_\mu$ is understood to correspond to $\sigma = 0$. If $\sigma > 0$, we shall say that *the GM γ is nondegenerate*. Unless explicitly stated otherwise, we assume GMs to be nondegenerate.

For $\mu = 0$ and $\sigma = 1$, we shall refer to the GM γ_1 as *the standard GM on* $\mathbb{R}$. A GM with $\mu = 0$ is called *centered* (or also *symmetric*). For a GM γ on $\mathbb{R}$, there holds

$$\mu = \int_{\mathbb{R}} y \, \mathrm{d}\gamma(y), \quad \sigma^2 = \int_{\mathbb{R}} (y-\mu)^2 \, \mathrm{d}\gamma(y).$$

Let $(\Omega, \mathcal{A}, \mathbb{P})$ be a probability space with sample space Ω, σ-fields $\mathcal{A}$, and probability measure $\mathbb{P}$. A *Gaussian random variable* ("Gaussian RV" for short) $\eta : \Omega \to \mathbb{R}$ is a RV whose law is Gaussian, i.e., it admits a Gaussian distribution. If η is a Gaussian RV with mean μ and variance σ^2 we write $\eta \sim \mathcal{N}(\mu, \sigma^2)$.

Linear transformations of Gaussian RVs are Gaussian: every Gaussian RV η can be written as $\eta = \sigma\xi + \mu$, where ξ is a standard Gaussian RV, i.e., a Gaussian RV whose law is a standard GM on $\mathbb{R}$.

The Fourier transformation of a GM γ on $\mathbb{R}$ is defined, for every $\xi \in \mathbb{R}$, as

$$\hat{\gamma}_1(\xi) := \int_{\mathbb{R}} \exp(\mathrm{i}\xi y) \, \mathrm{d}\gamma(y) = \exp\left(\mathrm{i}\mu\xi - \frac{1}{2}\sigma^2\xi^2\right).$$

We denote by Φ the distribution function of γ_1. For the standard normal distribution

$$\Phi(t) = \int_{-\infty}^{t} p(s; 0, 1)\, \mathrm{d}s \qquad \forall t \in \mathbb{R}.$$

With the convention $\Phi^{-1}(0) := -\infty$, $\Phi^{-1}(1) := +\infty$, the inverse function Φ^{-1} of Φ is defined on $[0, 1]$.

2.1.2 Multivariate Gaussian Measures

Consider now a finite dimension $d > 1$. A Borel probability measure γ on $(\mathbb{R}^d, \mathcal{B}(\mathbb{R}^d))$ is called Gaussian if for every $f \in \mathcal{L}(\mathbb{R}^d, \mathbb{R})$ the measure $\gamma \circ f^{-1}$ is a GM on $\mathbb{R}$, where as usually, $\mathcal{B}(\mathbb{R}^d)$ denotes the σ-field on $\mathbb{R}^d$. Since d is finite, we may identify $\mathcal{L}(\mathbb{R}^d, \mathbb{R})$ with $\mathbb{R}^d$, and we denote the Euclidean inner product on $\mathbb{R}^d$ by $(\cdot, \cdot)$. The Fourier transform of a Borel measure γ on $\mathbb{R}^d$ is given by

$$\hat{\gamma} : \mathbb{R}^d \to \mathbb{C} : \hat{\gamma}(\boldsymbol{\xi}) = \int_{\mathbb{R}^d} \exp\left(\mathrm{i}(\boldsymbol{\xi}, \boldsymbol{y})\right) \mathrm{d}\gamma(\boldsymbol{y}).$$

For a GM γ on $\mathbb{R}^d$, the Fourier transform $\hat{\gamma}$ uniquely determines γ.

Proposition 2.1 ([21, Proposition 1.2.2]) *A Borel probability measure γ on $\mathbb{R}^d$ is Gaussian iff*

$$\hat{\gamma}(\boldsymbol{\xi}) = \exp\left(\mathrm{i}(\boldsymbol{\xi}, \boldsymbol{\mu}) - \frac{1}{2}(\boldsymbol{K}\boldsymbol{\xi}, \boldsymbol{\xi})\right), \quad \boldsymbol{\xi} \in \mathbb{R}^d.$$

Here, $\boldsymbol{\mu} \in \mathbb{R}^d$ and $\boldsymbol{K} \in \mathbb{R}^{d \times d}$ is a symmetric positive semidefinite matrix.

We shall say that a GM γ on $\mathbb{R}^d$ has a density with respect to Lebesgue measure λ on $\mathbb{R}^d$ iff the matrix $\boldsymbol{K}$ is nondegenerate. Then, this density is given by

$$\frac{\mathrm{d}\gamma}{\mathrm{d}\lambda}(\boldsymbol{x}) : \boldsymbol{x} \mapsto \frac{1}{\sqrt{(2\pi)^d \det \boldsymbol{K}}} \exp\left(-\frac{1}{2}(\boldsymbol{K}^{-1}(\boldsymbol{x} - \boldsymbol{\mu}), \boldsymbol{x} - \boldsymbol{\mu})\right).$$

Furthermore,

$$\boldsymbol{\mu} = \int_{\mathbb{R}^d} \boldsymbol{y}\, \mathrm{d}\gamma(\boldsymbol{y}), \quad \forall \boldsymbol{y}, \boldsymbol{y}' \in \mathbb{R}^d : (\boldsymbol{K}\boldsymbol{y}, \boldsymbol{y}') = \int_{\mathbb{R}^d} (\boldsymbol{y}, \boldsymbol{x} - \boldsymbol{\mu})(\boldsymbol{y}', \boldsymbol{x} - \boldsymbol{\mu})\, \mathrm{d}\gamma(\boldsymbol{x}).$$

The symmetric linear operator $C \in \mathcal{L}(\mathbb{R}^d \times \mathbb{R}^d, \mathbb{R})$ defined by the later relation and represented by the symmetric positive definite matrix $\boldsymbol{K}$ is the covariance operator *associated to the GM γ on $\mathbb{R}^d$.*

When we do not need to distinguish between the covariance operator C and the covariance matrix $\boldsymbol{K}$, we simply speak of "the covariance" of a GM γ. If a joint probability distribution of RVs $y_1, \dots, y_d$ is a GM on $\mathbb{R}^d$ with mean vector $\boldsymbol{\mu}$ and covariance matrix $\boldsymbol{K}$ we write $(y_1, \dots, y_d) \sim \mathcal{N}(\boldsymbol{\mu}, \boldsymbol{K})$.

In what follows, we use γ_d to denote the standard GM on $\mathbb{R}^d$. Denote by $L^2(\mathbb{R}^d; \gamma_d)$ the Hilbert space of all γ_d-measurable, real-valued functions f on $\mathbb{R}^d$ such that the norm

$$\|f\|_{L^2(\mathbb{R}^d;\gamma_d)} := \left(\int_{\mathbb{R}^d} |f(\boldsymbol{y})|^2 \,\mathrm{d}\gamma_d(\boldsymbol{y}) \right)^{1/2}$$

is finite. The corresponding inner product is denoted by $(\cdot, \cdot)_{L^2(\mathbb{R}^d;\gamma_d)}$.

2.1.3 Hermite Polynomials

A key role in the ensuing sparsity analysis of parametric solution families is taken by Wiener-Hermite PC expansions. We consider GRF inputs and, accordingly, will employ polynomial systems on $\mathbb{R}$ which are orthogonal with respect to the GM γ_1 on $\mathbb{R}$, the so-called Hermite polynomials, as pioneered for the analysis of GRFs by N. Wiener in [109]. To this end, we recapitulate basic definitions and properties, in particular the various normalizations which are met in the literature. Particular attention will be paid to estimates for Hermite coefficients of functions which are holomorphic in a strip, going back to Einar Hille in [69].

Definition 2.2 For $k \in \mathbb{N}_0$, the normalized probabilistic Hermite polynomial H_k of degree k on $\mathbb{R}$ is defined by

$$H_k(x) := \frac{(-1)^k}{\sqrt{k!}} \exp\left(\frac{x^2}{2}\right) \frac{\mathrm{d}^k}{\mathrm{d}x^k} \exp\left(-\frac{x^2}{2}\right). \tag{2.1}$$

For every multi-degree $\boldsymbol{\nu} \in \mathbb{N}_0^m$, the m-variate Hermite polynomial $H_{\boldsymbol{\nu}}$ is defined by

$$H_{\boldsymbol{\nu}}(x_1, \dots, x_m) := \prod_{j=1}^{m} H_{\nu_j}(x_j), \quad x_j \in \mathbb{R}, \ j = 1, \dots, m.$$

Remark 2.3 (Normalizations of Hermite Polynomials and Hermite Functions)

(i) Definition (2.1) provides for every $k \in \mathbb{N}_0$ a polynomial of degree k. The scaling factor in (2.1) has been chosen to ensure normalization with respect to GM γ_1, see also Lemma 2.4, item (i).

(ii) Other normalizations with at times the same notation are used. The "classical" normalization of H_k we denote by $\tilde{H}_k(x)$. It is defined by (see, e.g., [1, Page

787], and compare (2.1) with [105, Equation (5.5.3)])

$$\tilde{H}_k(x/\sqrt{2}) := 2^{k/2}\sqrt{k!}H_k(x).$$

(iii) In [7], so-called "*normalized Hermite polynomials*" are introduced as

$$\tilde{\tilde{H}}_k(x) := [\sqrt{\pi}2^k k!]^{1/2}(-1)^k \exp(x^2)\frac{\mathrm{d}^k}{\mathrm{d}x^k}\exp(-x^2).$$

The system $(\tilde{\tilde{H}}_k)_{k\in\mathbb{N}_0}$ is an orthonormal basis (ONB for short) for the space $L^2(\mathbb{R}, \tilde{\tilde{\gamma}})$ with the weight $\tilde{\tilde{\gamma}} = \exp(-x^2)$, i.e., (compare, e.g., [105, Eqn. (5.5.1)])

$$\int_{\mathbb{R}} \tilde{\tilde{H}}_n(x)\tilde{\tilde{H}}_{n'}(x)\exp(-x^2)\,\mathrm{d}x = \delta_{nn'}, \quad n, n' \in \mathbb{N}_0.$$

(iv) With the Hermite polynomials $\tilde{\tilde{H}}_k$, in [69] *Hermite functions* are introduced for $k \in \mathbb{N}_0$ as

$$h_k(x) := \exp(-x^2/2)\tilde{\tilde{H}}_k(x), \quad x \in \mathbb{R}.$$

(v) It has been shown in [69, Theorem 1] that in order for functions $f : \mathbb{C} \to \mathbb{C}$ defined in the strip $S(\rho) := \{z \in \mathbb{C} : z = x + \mathrm{i}y,\ x \in \mathbb{R},\ |y| < \rho\}$ to admit a Fourier-Hermite expansion

$$\sum_{n=0}^{\infty} f_n h_n(z), \qquad f_n := \int_{\mathbb{R}} f(x)h_n(x)\,\mathrm{d}x = \int_{\mathbb{R}} f(x)\tilde{\tilde{H}}_n(x)\exp(-x^2/2)\,\mathrm{d}x$$

which converges to $f(z)$ for $z \in S(\rho)$ a necessary and sufficient condition is that (a) f is holomorphic in $S(\rho) \subset \mathbb{C}$ and (b) for every $0 < \rho' < \rho$ there exist a finite bound $B(\rho')$ and β such that

$$|f(x + \mathrm{i}y)| \le B(\rho')\exp[-|x|(\beta^2 - y^2)^{1/2}], \quad x \in \mathbb{R}, |y| \le \rho'.$$

There is a constant $C(f) > 0$ such that for the Fourier-Hermite coefficients f_n, holds

$$|f_n| \le C\exp(-\rho\sqrt{2n+1}) \quad \forall n \in \mathbb{N}_0.$$

We state some basic properties of the Hermite polynomials H_k defined in (2.1).

Lemma 2.4 *The collection $(H_k)_{k\in\mathbb{N}_0}$ of Hermite polynomials* (2.1) *in $\mathbb{R}$ has the following properties.*

(i) $(H_k)_{k\in\mathbb{N}_0}$ *is an ONB of the space $L^2(\mathbb{R};\gamma_1)$.*
(ii) *For every $k\in\mathbb{N}$ holds: $H_k'(x)=\sqrt{k}H_{k-1}(x)=H_k(x)-\sqrt{k+1}H_{k+1}(x)$.*
(iii) *For all $x_1,\ldots,x_m\in\mathbb{R}$ holds*

$$\prod_{i=1}^{m}\sqrt{k_i!}H_{k_i}(x_i)=\frac{\partial^{k_1+\ldots+k_m}}{\partial t_1^{k_1}\ldots\partial t_m^{k_m}}\exp\left(\sum_{i=1}^{m}t_ix_i-\frac{1}{2}\sum_{i=1}^{m}t_i^2\right)|_{t_1=\ldots=t_m=0}\,.$$

(iv) *For every $f\in C^\infty(\mathbb{R})$ such that $f^{(k)}\in L^2(\mathbb{R};\gamma_1)$ for all $k\in\mathbb{N}_0$ holds*

$$\int_{\mathbb{R}} f(x)H_k(x)\,\mathrm{d}\gamma_1(x)=\frac{(-1)^k}{\sqrt{k!}\int_{\mathbb{R}} f^{(k)}(x)\,\mathrm{d}\gamma_1(x)},$$

and, hence, in $L^2(\mathbb{R};\gamma_1)$,

$$f=\sum_{k\in\mathbb{N}_0}\frac{(-1)^k}{\sqrt{k!}}(f^{(k)},1)_{L^2(\mathbb{R};\gamma_1)}H_k.$$

It follows from item (i) of this lemma in particular that

$$\left\{H_{\boldsymbol{\nu}}:\boldsymbol{\nu}\in\mathbb{N}_0^m\right\}\text{ is an ONB of }L^2(\mathbb{R}^m;\gamma_m).$$

Denote for $k\in\mathbb{N}_0$ and $m\in\mathbb{N}$ by $\mathcal{H}_k$ the space of d-variate Hermite polynomials which are homogeneous of degree k, i.e.,

$$\mathcal{H}_k:=\operatorname{span}\left\{H_{\boldsymbol{\nu}}:\boldsymbol{\nu}\in\mathbb{N}_0^m,\,|\boldsymbol{\nu}|=k\right\}.$$

Then $\mathcal{H}_k$ ("homogeneous polynomial chaos of degree k" [109]) is a closed, linear subspace of $L^2(\mathbb{R}^m;\gamma_m)$ and

$$L^2(\mathbb{R}^m;\gamma_m)=\bigoplus_{k\in\mathbb{N}_0}\mathcal{H}_k\ \text{ in }\ L^2(\mathbb{R}^m;\gamma_m).$$

2.2 Gaussian Measures on Separable Locally Convex Spaces

An important mathematical ingredient in a number of applications, in particular in UQ, Bayesian PDE inversion, risk analysis, but also in statistical learning theory applied to input-output maps for PDEs, is the construction of measures on function spaces. A particular interest is in GMs on separable Hilbert or Banach or, more generally, on locally convex spaces of uncertain input data for PDEs. Accordingly, we review constructions of such measures, in terms of suitable *bases of the input spaces*. This implies, in particular, *separability* of the spaces of admissible PDE inputs or, at least, the uncertain input data being *a separably-valued* random element of otherwise nonseparable spaces (such as, e.g., $L^\infty(D)$) of valid inputs for the PDE of interest.

Let $(\Omega, \mathcal{A}, \mu)$ be a measure space and $1 \le p \le \infty$. Recall that the normed space $L^p(\Omega, \mu)$ is defined as the space of all μ-measurable functions u from Ω to $\mathbb{R}$ such that the norm

$$\|u\|_{L^p(\Omega,\mu)} := \left(\int_\Omega |u(x)|^p \, \mathrm{d}\mu(x) \right)^{1/p} < \infty.$$

When $p = \infty$ the norm of $u \in L^\infty(\Omega, \mu)$ is given by

$$\|u\|_{L^\infty(\Omega,\mu)} := \operatorname*{ess\,sup}_{x\in\Omega} |u(x)|.$$

If $\Omega \subset \mathbb{R}^m$ and μ is the Lebesgue measure, we simply denote these spaces by $L^p(\Omega)$.

Throughout this section, X will denote a real separable and locally convex space with Borel σ-field $\mathcal{B}(X)$ and with dual space X^*.

Example 2.5 Let $\mathbb{R}^\infty$ be the linear space of all sequences $\boldsymbol{y} = (y_j)_{j\in\mathbb{N}}$ with $y_j \in \mathbb{R}$. This linear space becomes a locally convex space (still denoted by $\mathbb{R}^\infty$) equipped with the topology generated by the countable family of semi-norms

$$p_j(\boldsymbol{y}) := |y_j|, \quad j \in \mathbb{N}.$$

The locally convex space $\mathbb{R}^\infty$ is separable and complete and, therefore, a Fréchet space. However, it is not normable, and hence not a Banach space.

Example 2.6 Let $D \subset \mathbb{R}^d$ be an open bounded Lipschitz domain.

(i) The Banach spaces $C(\overline{D})$ and $L^1(D)$ are separable.
(ii) For $0 < s < 1$ we denote by $C^s(D)$ the space of s-Hölder continuous functions in D equipped with the norm and seminorm

$$\|a\|_{C^s} := \|a\|_{L^\infty} + |a|_{C^s}, \quad |a|_{C^s} := \sup_{\boldsymbol{x},\boldsymbol{x}'\in D, \boldsymbol{x}\neq\boldsymbol{x}'} \frac{|a(\boldsymbol{x}) - a(\boldsymbol{x}')|}{|\boldsymbol{x} - \boldsymbol{x}'|^s}.$$

Then the Banach space $C^s(D)$ is not separable. A separable subspace is

$$C^s_\circ(D) := \left\{ a \in C^s(D) : \forall \boldsymbol{x} \in D \lim_{D\ni\boldsymbol{x}'\to\boldsymbol{x}} \frac{|a(\boldsymbol{x}) - a(\boldsymbol{x}')|}{|\boldsymbol{x} - \boldsymbol{x}'|^s} = 0 \right\}.$$

We review and present constructions of GMs γ on X.

2.2.1 Cylindrical Sets

Cylindrical sets are subsets of X of the form

$$C = \{x \in X : (l_1(x), \ldots, l_n(x)) \in C_0 : C_0 \in \mathcal{B}(\mathbb{R}^n), l_i \in X^*\}, \text{ for some } n \in \mathbb{N}.$$

Here, the Borel set $C_0 \in \mathcal{B}(\mathbb{R}^n)$ is sometimes referred to as *basis of the cylinder* C. We denote by $\mathcal{E}(X)$ the σ-field generated by all cylindrical subsets of X. It is the smallest σ-field for which all continuous linear functionals are measurable. Evidently then $\mathcal{E}(X) \subset \mathcal{B}(X)$, with in general strict inclusion (see, e.g., [21, A.3.8]). If, however, X is separable, then $\mathcal{E}(X) = \mathcal{B}(X)$ ([21, Theorem A.3.7]).

Sets of the form

$$\{y \in \mathbb{R}^\infty : (y_1, \ldots, y_n) \in B, B \in \mathcal{B}(\mathbb{R}^n), n \in \mathbb{N}\}$$

generate $\mathcal{B}(\mathbb{R}^\infty)$ [21, Lemma 2.1.1], and a set C belongs to $\mathcal{B}(X)$ iff it is of the form

$$C = \{x \in X : (l_1(x), \ldots, l_n(x), \ldots) \in B, \text{ for } l_i \in X^*, B \in \mathcal{B}(\mathbb{R}^\infty)\},$$

(see, e.g., [21, Lemma 2.1.2]).

2.2.2 Definition and Basic Properties of Gaussian Measures

Definition 2.7 ([21, Definition 2.2.1]) A probability measure γ defined on the σ-field $\mathcal{E}(X)$ generated by X^* is called Gaussian if, for any $f \in X^*$ the induced measure $\gamma \circ f^{-1}$ on $\mathbb{R}$ is Gaussian. The measure γ is centered or symmetric if all measures $\gamma \circ f^{-1}$, $f \in X^*$ are centered.

Let $(\Omega, \mathcal{A}, \mathbb{P})$ be a probability space. A random field u taking values in X (recall that throughout, X is a separable locally convex space) is a map $u : \Omega \to X$ such that

$$\forall B \in \mathcal{B}(X) : \quad u^{-1}(B) \in \mathcal{A}.$$

The *law of the random field* u is the probability measure $\mathfrak{m}_u$ on $(X, \mathcal{B}(X))$ which is defined as

$$\mathfrak{m}_u(B) := \mathbb{P}(u^{-1}(B)), \quad B \in \mathcal{B}(X).$$

The random field u is said to be *Gaussian* if its law is a GM on $(X, \mathcal{B}(X))$.

Images of GMs under continuous affine transformations on X are Gaussian.

Lemma 2.8 ([21, Lemma 2.2.2]) *Let γ be a GM on X and let $T : X \to Y$ be a linear map to another locally convex space Y such that $l \circ T \in X^*$ for all $l \in Y^*$. Then $\gamma \circ T^{-1}$ is a GM on Y.*

This remains true for the affine map $x \mapsto Tx + \mu$ for some $\mu \in Y$.

The *Fourier transform* of a measure $\mathfrak{m}$ over $(X, \mathcal{B}(X))$ is given by

$$\hat{\mathfrak{m}} : X^* \to \mathbb{C} : f \mapsto \hat{\mathfrak{m}}(f) := \int_X \exp(\mathrm{i} f(x))\, \mathrm{d}\mathfrak{m}(x).$$

Theorem 2.9 ([21, Theorem 2.2.4]) *A measure γ on X is Gaussian iff its Fourier transform $\hat{\gamma}$ can be expressed with some linear functional $L(\cdot)$ on X^* and a symmetric bilinear form $B(.,.)$ on $X^* \times X^*$ such that $f \mapsto B(f, f)$ is nonnegative as*

$$\forall f \in X^* : \quad \hat{\gamma}(f) = \exp\left(\mathrm{i}L(f) - \frac{1}{2}B(f, f)\right). \tag{2.2}$$

A GM γ on X is therefore characterized by L and B. It also follows from (2.2) that a GM γ on X is centered iff $\gamma(A) = \gamma(-A)$ for all $A \in \mathcal{B}(X)$, i.e., iff $L = 0$ in (2.2).

Definition 2.10 Let $\mathfrak{m}$ be a measure on $\mathcal{B}(X)$ such that $X^* \subset L^2(X, \mathfrak{m})$. Then the element $a_{\mathfrak{m}} \in (X^*)'$ in the algebraic dual $(X^*)'$ defined by

$$a_{\mathfrak{m}}(f) := \int_X f(x)\, \mathrm{d}\mathfrak{m}(x), \quad f \in X^*,$$

is called *mean of* $\mathfrak{m}$.

The operator $R_{\mathfrak{m}} : X^* \to (X^*)'$ defined by

$$R_{\mathfrak{m}}(f)(g) := \int_X [f(x) - a_{\mathfrak{m}}(f)][g(x) - a_{\mathfrak{m}}(g)]\, \mathrm{d}\mathfrak{m}(x)$$

is called *covariance operator* of $\mathfrak{m}$. The quadratic form on X^* is called *covariance of* $\mathfrak{m}$.

When X is a real separable Hilbert space, one can say more.

Definition 2.11 (Nuclear Operators) Let H_1, H_2 be real separable Hilbert spaces with the norms $\| \circ \|_{H_1}$ and $\| \circ \|_{H_2}$, respectively, and with corresponding inner products $(\cdot, \cdot)_{H_i}$, $i = 1, 2$.

A linear operator $K \in \mathcal{L}(H_1, H_2)$ is called *nuclear* or *trace class* if it can be represented as

$$\forall u \in H_1 : \quad Ku = \sum_{k \in \mathbb{N}} (u, x_{1k})_{H_1} x_{2k} \text{ in } H_2.$$

Here, $(x_{ik})_{k\in\mathbb{N}} \subset H_i$, $i = 1, 2$ are such that $\sum_{k\in\mathbb{N}} \|x_{1k}\|_{H_1} \|x_{2k}\|_{H_2} < \infty$.

We denote by $\mathcal{L}_1(H_1, H_2) \subset \mathcal{L}(H_1, H_2)$ the space of all nuclear operators. This is a separable Banach space when it is endowed with *nuclear norm*

$$\|K\|_1 := \inf\left\{\sum_{k\in\mathbb{N}} \|x_{1k}\|_{H_1}\|x_{2k}\|_{H_2} : Ku = \sum_{k\in\mathbb{N}} (u, x_{1k})_{H_1} x_{2k}\right\}.$$

When $X = H_1 = H_2$, we also write $\mathcal{L}_1(X)$.

Proposition 2.12 ([21, Theorem 2.3.1]) *Let γ be a GM on a separable Hilbert space X with innerproduct $(\cdot,\cdot)_X$, and let X^* denote its dual, identified with X via the Riesz isometry.*

Then there exist $\mu \in X$ and a symmetric, nonnegative nuclear operator $K \in \mathcal{L}_1(X)$ such that the Fourier transform $\hat{\gamma}$ of γ is

$$\hat{\gamma} : X \to \mathbb{C} : x \mapsto \exp\left(\mathrm{i}(\mu, x)_X - \frac{1}{2}(Kx, x)_X\right). \tag{2.3}$$

Remark 2.13 Consider that X is a real, separable Hilbert space with innerproduct $(\cdot,\cdot)_X$ and assume given a GM γ on X.

(i) In (2.3), $K \in \mathcal{L}(X)$ and $\mu \in X$ are determined by

$$\forall u, v \in X : (\mu, v)_X = \int_X (x, v)_X \, \mathrm{d}\gamma(x),$$
$$(Ku, v)_X = \int_X (u, x-\mu)_X (v, x-\mu)_X \, \mathrm{d}\gamma(x).$$

The closure of $X = X^*$ in $L^2(X;\gamma)$ then equals the completion of X with respect to the norm $x \mapsto \|K^{1/2}x\|_X = \sqrt{(Kx, x)_X}$. Let $(e_n)_{n\in\mathbb{N}}$ denote the ONB of X formed by eigenvectors of K, with corresponding real, non-negative eigenvalues $k_n \in \mathbb{N}_0$, i.e., $Ke_n = k_n e_n$ for $n = 1, 2, \ldots$. Then the completion can be identified with the weighted sequence (Hilbert) space

$$\left\{(x_n)_{n\in\mathbb{N}} : \sum_{n\in\mathbb{N}} k_n x_n^2 < \infty\right\}.$$

The nuclear operator K is the *covariance of the GM γ on the Hilbert space X.*

(ii) In coordinates $\boldsymbol{y} = (y_j)_{j\in\mathbb{N}} \in \ell^2(\mathbb{N})$ associated to the ONB $(e_n)_{n\in\mathbb{N}}$ of X, (2.3) takes the form

$$\hat{\gamma} : \ell^2(\mathbb{N}) \to \mathbb{C} : \boldsymbol{y} \mapsto \exp\left(\mathrm{i}\sum_{n\in\mathbb{N}} a_n y_n - \frac{1}{2}\sum_{n\in\mathbb{N}} k_n y_n^2\right).$$

(iii) Consider $a = 0 \in X$ and, for finite $n \in \mathbb{N}$, a cylindrical set $C = P_n^{-1}(B)$ with P_n denoting the orthogonal projection onto $X_n := \text{span}\{e_j : j = 1, \dots, n\} \subset X$, and with $B \in \mathcal{B}(X_n)$. Then

$$\gamma(C) = \int_B \prod_{j=1}^{n} (2\pi k_j)^{-1/2} \exp\left(-\frac{1}{2k_j} y_j^2\right) \mathrm{d}y_1 \dots \mathrm{d}y_n.$$

For $f \in X^*$ and $x \in X$, one frequently writes the $X^* \times X$ duality pairing as

$$f(x) = \langle f, x \rangle.$$

With the notation from Definition 2.10, the covariance operator $C_g = R_{\gamma_g}$ in Definition 2.10 of a *centered*, Gaussian random vector $g : (\Omega, \mathcal{A}; \gamma_g) \to X$ with Gaussian law γ_g on a separable, real Banach space X admits the representations

$$R_{\gamma_g} = C_g : X^* \to X : C_g \varphi := \mathbb{E}\langle \varphi, g \rangle g,$$
$$C_g : X^* \times X^* \to \mathbb{R} : (\psi, \varphi) \mapsto \langle \psi, C_g \varphi \rangle.$$

2.3 Cameron-Martin Space

Let X be a real separable locally convex space and γ a GM on $\mathcal{E}(X)$ such that $X^* \subset L^2(X; \gamma)$. Then, for every $\varphi \in X^*$, the image measure $\varphi(\gamma)$ is a GM on $\mathbb{R}$. By Bogachev [21, Theorem 3.2.3], there exists a unique $a_\gamma \in X$, the mean of γ, such that

$$\forall \varphi \in X^* : \quad \varphi(a_\gamma) = \int_X \varphi(h) \, \mathrm{d}\gamma(h).$$

Denote by X_γ^* the closure of the set $\{\varphi - \varphi(a_\gamma), \varphi \in X^*)\}$ embedded into the normed space $L^2(X; \gamma)$ w.r.t. its norm.

The covariance operator, R_γ, of γ is formally given by

$$\forall \varphi, \psi \in X^* : \quad \langle R_\gamma \varphi, \psi \rangle = \int_X \varphi(h - a_\gamma) \psi(h - a_\gamma) \, \mathrm{d}\gamma(h). \tag{2.4}$$

As X is a separable locally convex space, [21, Theorem 3.2.3] implies that there is a unique linear operator $R_\gamma : X^* \to X$ such that (2.4) holds. We define

$$\forall \varphi \in X^* : \quad \sigma(\varphi) := \sqrt{\langle R_\gamma \varphi, \varphi \rangle}.$$

If $h = R_\gamma \varphi$ for some $\varphi \in X^*$, the map $h \mapsto \|h\| := \sigma(\varphi)$ defines a norm on range$(R_\gamma) \subset X$. There holds [21, Lemma 2.4.1] $\|h\| = \|h\|_{H(\gamma)} = \|\varphi\|_{L^2(X;\gamma)}$.

The *Cameron-Martin space* of the GM γ on X is the completion of the range of R_γ in X with respect to the norm $\| \circ \|$. The Cameron-Martin space of the GM γ on X is denoted by $H(\gamma)$. It is also called the reproducing kernel Hilbert space (RKHS for short) of γ on X.

By Bogachev [21, Theorem 3.2.7], $H(\gamma)$ is a separable Hilbert space, and $H(\gamma) \subset X$ with continuous embedding, according to [21, Proposition 2.4.6]. In case that $X \subset Y$ for another Banach space, with continuous and linear embedding, the Cameron-Martin spaces for X and Y coincide. For example, in the context of Remark 2.13, item (i), $H(\gamma) = K(X^*_\gamma)$.

Being a Hilbert space, introduce an innerproduct $(\cdot, \cdot)_{H(\gamma)}$ on $H(\gamma)$ compatible with the norm $\| \circ \|_{H(\gamma)}$ via the parallelogram law. Then there holds

$$\forall \varphi \in X^* \; \forall f \in H(\gamma) : \quad (f, R_\gamma \varphi)_{H(\gamma)} = \varphi(f).$$

Since $H(\gamma)$ is also separable, there is an ONB.

Proposition 2.14 ([21, Theorem 3.5.10, Corollary 3.5.11]) *For a centered GM on a real, separable Banach space X with norm $\| \circ \|_X$, there exists an ONB $(e_n)_{n\in\mathbb{N}}$ of the Cameron-Martin space $H(\gamma) \subset X$ such that*

$$\sum_{n\in\mathbb{N}} \|e_n\|_X^2 < \infty, \qquad \forall \varphi \in X^* : \; R_\gamma \varphi = \sum_{n\in\mathbb{N}} \varphi(e_n) e_n.$$

We remark that Proposition 2.14 is not true for arbitrary ONB $(e_n)_{n\in\mathbb{N}}$ of $H(\gamma)$.

2.4 Gaussian Product Measures

We recall a notion of product measures which gives an efficient method to construct Gaussian measures on a countable Cartesian product of locally convex spaces.

Definition 2.15 (Product Measure, [21, p. 372]) Let μ_n be probability measures defined on σ-fields $\mathcal{B}_n$ in locally convex spaces X_n. Put

$$X := \prod_{n\in\mathbb{N}} X_n.$$

Let

$$\mathcal{B} := \bigotimes_{n\in\mathbb{N}} \mathcal{B}_n$$

be the σ-field generated by all the sets of the form

$$B = B_1 \times B_2 \times \ldots \times B_n \times X_{n+1} \times X_{n+2} \times \ldots, \quad B_i \in \mathcal{B}_i. \tag{2.5}$$

The product measure

$$\mu := \bigotimes_{n\in\mathbb{N}} \mu_n$$

is the probability measure on $\mathcal{B}$ defined by $\mu(B) := \prod_{i=1}^{n} \mu_i(B_i)$ for the sets B of the form (2.5).

Example 2.16 ([21, Example 2.3.8]) Let $(\mu_n)_{n\in\mathbb{N}}$ be a sequence of GMs. Then the product measure $\mu := \otimes_{n\in\mathbb{N}}\mu_n$ is a GM on $X := \prod_{n\in\mathbb{N}} X_n$. The Cameron-Martin space $H(\mu)$ of μ is the Hilbert direct sum of spaces $H(\mu_n)$, i.e.,

$$H(\mu) := \left\{ h = (h_j)_{j\in\mathbb{N}} \in X : \, h_j \in H(\mu_j), \, \|h\|^2_{H(\mu)} = \sum_{j\in\mathbb{N}} \|h_j\|^2_{H(\mu_j)} \right\}.$$

The space X^*_μ is the set of all functions of the form

$$\varphi \mapsto \sum_{j\in\mathbb{N}} f_j(\varphi_j), \quad f_j \in X^*_{\mu_j}, \quad \sum_{j\in\mathbb{N}} \sigma(f_j)^2 < \infty,$$

and

$$a_\mu(f) = \sum_{j\in\mathbb{N}} a_{\mu_j}(f_j), \quad \forall f = (f_j)_{j\in\mathbb{N}} \in X^*.$$

Example 2.17 ([21, Example 2.3.5]) Denote by $(\gamma_{1,n})_{n\in\mathbb{N}}$ a sequence of standard GMs on $(\mathbb{R}, \mathcal{B}(\mathbb{R}))$. Then the product measure

$$\gamma = \bigotimes_{n\in\mathbb{N}} \gamma_{1,n}$$

is a centered GM on $\mathbb{R}^\infty$. Furthermore, $H(\gamma) = \ell^2(\mathbb{N})$ and $X^*_\gamma \simeq \ell^2(\mathbb{N})$. If μ is a GM on $\mathbb{R}^\infty$, then by a result of Fernique, the measures γ and μ are either mutually singular or equivalent [21, Theorem 2.12.9]. The locally convex space $\mathbb{R}^\infty$ with the product measure γ of standard GMs is the main parametric domain in the stochastic setting of UQ problems for PDEs with GRF inputs considered in this text.

2.5 Gaussian Series

A key role in the numerical analysis of PDEs with GRF inputs from separable Banach spaces E is played by representing these GRFs in terms of series with respect to suitable *representation systems* $(\psi_j)_{j\in\mathbb{N}} \in E^\infty$ of E with random coefficients. There arises the question of admissibility of $(\psi_j)_{j\in\mathbb{N}} \in E^\infty$ so as to allow (a) to transfer randomness of function space-valued inputs to a parametric, deterministic representation (as is customary, for example, in the transition from nonparametric to parametric models in statistics) and (b) to ensure suitability for numerical approximation.

Items (a) and (b) are closely related to the selection of stable bases for E, with item (b) mandating additional requirements, such as efficient accessibility for float point computations, quadrature, etc.

We first present an abstract result, Theorem 2.21 and then, in Sects. 2.5.2 and 2.5.3, we review several concrete constructions of such series. We discuss in Sects. 2.5.2 and 2.5.3 several examples, in particular the classical Karhunen-Loève Expansion [77, 101] of GRFs taking values in a separable Hilbert space. All examples will be admissible in parametrizing GRF input data for PDEs and of Gaussian priors in the ensuing sparsity and approximation rate analysis in Chap. 3 and the following sections.

2.5.1 Some Abstract Results

We place ourselves in the setting of a real separable locally convex space X, with a GM γ on X, and with associated Cameron-Martin Hilbert space $H(\gamma) \subset X$ as introduced in Sect. 2.3.

We first consider expansions of Gaussian random vectors with respect to orthonormal bases $(e_j)_{j\in\mathbb{N}}$ of the Cameron-Martin space $H(\gamma)$. As linear transformations of GM are Gaussian (see Lemma 2.8), we admit a linear transformation A.

Theorem 2.18 ([21, Theorems. 3.5.1, 3.5.7, (3.5.4)]) *Let γ be a centered GM on a real separable locally convex space X with Cameron-Martin space $H(\gamma)$ and with* some *ONB $(e_j)_{j\in\mathbb{N}}$ of $H(\gamma)$. Let further denote $(y_j)_{j\in\mathbb{N}}$ any sequence of independent standard Gaussian RVs on a probability space $(\Omega, \mathcal{A}, \mathbb{P})$ and let $A \in \mathcal{L}(H(\gamma))$ be arbitrary.*

Then the Gaussian series

$$\sum_{j\in\mathbb{N}} y_j(\omega) A e_j$$

converges $\mathbb{P}$-a.s. in X. The law of its limit is a centered GM λ with covariance R_λ given by

$$R_\lambda(f)(g) = \left(A^* R_\gamma(f), A^* R_\gamma(g)\right)_{H(\gamma)}.$$

Furthermore, there holds of independent standard Gaussian RVs on a probability space $(\Omega, \mathcal{A}, \mathbb{P})$.

$$\int_X f(x)\gamma(\mathrm{d}x) = \int_\Omega f\Big(\sum_{j\in\mathbb{N}} y_j(\omega)e_j\Big)\,\mathrm{d}\mathbb{P}(\omega).$$

If X is a real separable Banach space X with norm $\|\circ\|_X$, for all sufficiently small constants $c > 0$ holds

$$\lim_{n\to\infty}\int_\Omega \exp\left(c\left\|\sum_{j=n}^{\infty} y_j(\omega)Ae_j\right\|_X^2\right)\mathrm{d}\mathbb{P}(\omega) = 1.$$

In particular, for every $p \in [1, \infty)$ we have $\|\sum_{j=n}^{\infty} y_j Ae_j\|_X^p \to 0$ in $L^1(\Omega, \mathbb{P})$ as $n \to \infty$.

Often, in numerical applications, ensuring orthonormality of the basis elements could be computationally costly. It is therefore of some interest to consider Gaussian series with respect to more general representation systems $(\psi_j)_{j\in\mathbb{N}}$. An important notion is *admissibility* of such systems.

Definition 2.19 Let X be a real, separable locally convex space, and let $g : (\Omega, \mathcal{A}, \mathbb{P}) \to X$ be a centered Gaussian random vector with law $\gamma_g = \mathbb{P}_X$. Let further $(y_j)_{j\in\mathbb{N}}$ be a sequence of i.i.d. standard real Gaussian RVs $y_j \sim \mathcal{N}(0, 1)$.

A sequence $(\psi_j)_{j\in\mathbb{N}} \in X^\infty$ is called *admissible for g* if

$$\sum_{j\in\mathbb{N}} y_j\psi_j \text{ converges } \mathbb{P}\text{-a.s. in } X \text{ and } g = \sum_{j\in\mathbb{N}} y_j\psi_j.$$

To state the next theorem, we recall the notion of *frames in separable Hilbert space* (see, e.g., [65] and the references there for background and theory of frames. In the terminology of frame theory, Parseval frames correspond to tight frames with frame bounds equal to 1).

Definition 2.20 A sequence $(\psi_j)_{j\in\mathbb{N}} \subset H$ in a real separable Hilbert space H with inner product $(\cdot,\cdot)_H$ is a *Parseval frame of H* if

$$\forall f \in H: \quad f = \sum_{j\in\mathbb{N}} (\psi_j, f)_H\, \psi_j \text{ in } H.$$

The following result, from [88], characterizes admissible affine representation systems for GRFs u taking values in real, separable Banach spaces X.

Theorem 2.21 ([88, Theorem 1]) *We have the following.*

(i) In a real, separable Banach space X with a centered GM γ on X, a representation system $\boldsymbol{\Psi} = (\psi_j)_{j\in\mathbb{N}} \in X^\infty$ is admissible *for γ iff $\boldsymbol{\Psi}$ is a Parseval frame for the Cameron-Martin space $H(\gamma) \subset X$, i.e.,*

$$\forall f \in H(\gamma): \quad \|f\|^2_{H(\gamma)} = \sum_{j\in\mathbb{N}} |\langle f, \psi_j\rangle|^2.$$

(ii) Let u denote a GRF taking values in X with law γ and with RKHS $H(\gamma)$. For a countable collection $\boldsymbol{\Psi} = (\psi_j)_{j\in\mathbb{N}} \in X^\infty$ the following are equivalent:

(i) $\boldsymbol{\Psi}$ is a Parseval frame of $H(\gamma)$ and

(ii) there is a sequence $\boldsymbol{y} = (y_j)_{j\in\mathbb{N}}$ of i.i.d standard Gaussian RVs y_j such that there holds $\gamma - a.s.$ the representation

$$u = \sum_{j\in\mathbb{N}} y_j \psi_j \quad in \quad H(\gamma).$$

(iii) Consider a GRF u taking values in X with law γ and covariance $R_\gamma \in \mathcal{L}(X', X)$. If $R_\gamma = SS'$ with $S \in \mathcal{L}(K, X)$ for some separable Hilbert space K, for any *Parseval frame $\boldsymbol{\Phi} = (\varphi_j)_{j\in\mathbb{N}}$ of K, the countable collection $\boldsymbol{\Psi} = S\boldsymbol{\Phi} = (S\varphi_j)_{j\in\mathbb{N}}$ is a Parseval frame of the RKHS $H(\gamma)$ of u.*

The last assertion in the preceding result is [88, Proposition 1]. It generalizes the observation that for a symmetric positive definite matrix $\boldsymbol{M}$ in $\mathbb{R}^d$, any factorization $\boldsymbol{M} = \boldsymbol{L}\boldsymbol{L}^\top$ implies that for $z \sim \mathcal{N}(0, \boldsymbol{I})$ it holds $\boldsymbol{L}z \sim \mathcal{N}(0, \boldsymbol{M})$. The result is useful in building customized representation systems $\boldsymbol{\Psi}$ which are frames of a GRF u with computationally convenient properties in particular applications.

We review several widely used constructions of Parseval frames. These comprise expansions in eigenfunctions of the covariance operator K (referred to also as principal component analysis, or as "Karhunen-Loève expansions"), but also "eigenvalue-free" multiresolution constructions (generalizing the classical Lévy-Cieselski construction of the Brownian bridge) for various geometric settings, in particular bounded subdomains of euclidean space, compact manifolds without boundary etc. Any of these constructions will be admissible choices as representation system of the GRF input of PDEs to render these PDEs parametric-deterministic where, in turn, our parametric regularity results will apply.

Example 2.22 (Brownian Bridge) On the bounded time interval $[0, T]$, consider the *Brownian bridge* $(B_t)_{t\geq 0}$. It is defined in terms of a Wiener process $(W_t)_{t\geq 0}$ by conditioning as

$$(B_t)_{0\leq t\leq T} := \{(W_t)_{0\leq t\leq T} | W_T = 0\}. \tag{2.6}$$

It is a simple example of *kriging* applied to the GRF W_t.

The covariance function of the GRF B_t is easily calculated as

$$k_B(s,t) = \mathbb{E}[B_s B_t] = s(T-t)/T \text{ if } s < t.$$

Various other representations of B_t are

$$B_t = W_t - \frac{t}{T} W_T = \frac{T-t}{\sqrt{T}} W_{t/(T-t)}.$$

The RKHS $H(\gamma)$ corresponding to the GRF B_t is the Sobolev space $H_0^1(0, T)$.

2.5.2 *Karhunen-Loève Expansion*

A widely used representation system in the analysis and computation of GRFs is the so-called Karhunen-Loève expansion (KL expansion for short) of GRFs, going back to [77]. We present main ideas and definitions, in a generic setting of [79], see also [3, Chap. 3.3].

Let $\mathcal{M}$ be a compact space with metric $\rho : \mathcal{M} \times \mathcal{M} \to \mathbb{R}$ and with Borel sigma-algebra $\mathcal{B} = \mathcal{B}(\mathcal{M})$. Assume given a Borel measure μ on $(\mathcal{M}, \mathcal{B})$. Let further $(\Omega, \mathcal{A}, \mathbb{P})$ be a probability space. Examples are $\mathcal{M} = D$ a bounded domain in Euclidean space $\mathbb{R}^d$, with ρ denoting the Euclidean distance between pairs (x, x') of points in D, and $\mathcal{M}$ being a smooth, closed 2-surface in $\mathbb{R}^3$, where ρ is the geodesic distance between pairs of points in $\mathcal{M}$.

Consider a measurable map

$$Z : (\mathcal{M}, \mathcal{B}) \otimes (\Omega, \mathcal{A}) \to \mathbb{R} : (x, \omega) \mapsto Z_x(\omega) \in \mathbb{R}$$

such that for each $x \in \mathcal{M}$, Z_x is a centered, Gaussian RV. We call the collection $(Z_x)_{x\in\mathcal{M}}$ a *GRF indexed by* $\mathcal{M}$.

Assume furthermore for all $n \in \mathbb{N}$, for all $x_1, \ldots, x_n \in \mathcal{M}$ and for every $\xi_1, \ldots, \xi_n \in \mathbb{R}$

$$\sum_{i=1}^{n} \xi_i Z_{x_i} \text{ is a centered Gaussian RV.}$$

Then the *covariance function*

$$K : \mathcal{M} \times \mathcal{M} \to \mathbb{R} : (x, x') \mapsto K(x, x')$$

associated with the centered GRF $(Z_x)_{x\in\mathcal{M}}$ is defined pointwise by

$$K(x, x') := \mathbb{E}[Z_x Z_{x'}] \quad x, x' \in \mathcal{M}.$$

Evidently, the covariance function $K : \mathcal{M} \times \mathcal{M} \to \mathbb{R}$ corresponding to a Gaussian RV indexed by $\mathcal{M}$ is a real-valued, symmetric, and positive definite function, i.e., there holds

$$\forall n \in \mathbb{N}\ \forall (x_j)_{1\le j\le n} \in \mathcal{M}^n, \forall (\xi_j)_{1\le j\le n} \in \mathbb{R}^n : \sum_{1\le i,j\le n} \xi_i \xi_j K(x_i, x_j) \ge 0.$$

The operator $K \in \mathcal{L}(L^2(\mathcal{M}, \mu), L^2(\mathcal{M}, \mu))$ defined by

$$\forall f \in L^2(\mathcal{M}, \mu) : \ (Kf)(x) := \int_{\mathcal{M}} K(x, x') f(x') \,\mathrm{d}\mu(x'), \quad x \in \mathcal{M},$$

is a self-adjoint, compact positive operator on $L^2(\mathcal{M}, \mu)$. Furthermore, K is trace-class and $K(L^2(\mathcal{M}, \mu)) \subset C(\mathcal{M}, \mathbb{R})$.

The spectral theorem for compact, self-adjoint operators on the separable Hilbert space $L^2(\mathcal{M}, \mu)$ ensures the existence of a sequence $\lambda_1 \ge \lambda_2 \ge \ldots \ge 0$ of real eigenvalues of K (counted according to multiplicity and accumulating only at zero) with associated eigenfunctions $\psi_k \in L^2(\mathcal{M}, \mu)$ normalized in $L^2(\mathcal{M}, \mu)$, i.e., for all $k \in \mathbb{N}$ holds

$$K\psi_k = \lambda_k \psi_k \ \text{ in } \ L^2(\mathcal{M}, \mu), \qquad \int_{\mathcal{M}} \psi_k(x)\psi_\ell(x) \,\mathrm{d}\mu(x) = \delta_{k\ell}, \ \ k, \ell \in \mathbb{N}.$$

Then, there holds $\psi_k \in C(\mathcal{M}; \mathbb{R})$ and the sequence $(\psi_k)_{k\in\mathbb{N}}$ is an ONB of $L^2(\mathcal{M}, \mu)$. From Mercer's theorem (see, e.g., [101]), there holds the *Mercer expansion*

$$\forall x, x' \in \mathcal{M} : \quad K(x, x') = \sum_{k\in\mathbb{N}} \lambda_k \psi_k(x) \psi_k(x')$$

with absolute and uniform convergence on $\mathcal{M} \times \mathcal{M}$. This result implies that

$$\lim_{m\to\infty} \int_{\mathcal{M}\times\mathcal{M}} \left| K(x, x') - \sum_{j=1}^{m} \lambda_j \psi_j(x) \psi_j(x') \right|^2 \mathrm{d}\mu(x) \,\mathrm{d}\mu(x') = 0.$$

We denote by $H \subset L^2(\Omega, \mathbb{P})$ the $L^2(\mathcal{M}, \mu)$ closure of finite linear combinations of $(Z_x)_{x\in\mathcal{M}}$. This so-called *Gaussian* space (e.g. [75]) is a Hilbert space when equipped with the $L^2(\mathcal{M}, \mu)$ innerproduct. Then, the sequence $(B_k)_{k\in\mathbb{N}} \subset \mathbb{R}$ defined by

$$\forall k \in \mathbb{N}: \quad B_k(\omega) := \frac{1}{\sqrt{\lambda_k}} \int_{\mathcal{M}} Z_x(\omega)\psi_k(x)\,\mathrm{d}\mu(x) \in H$$

is a sequence of i.i.d, $N(0, 1)$ RVs. The expression

$$\tilde{Z}_x(\omega) := \sum_{k\in\mathbb{N}} \sqrt{\lambda_k}\psi_k(x)B_k(\omega) \tag{2.7}$$

is a modification of $Z_x(\omega)$, i.e., for every $x \in \mathcal{M}$ holds that $\mathbb{P}(\{Z_x = \tilde{Z}_x\}) = 1$, which is referred to as *Karhunen-Loève expansion of the GRF* $\{Z_x : x \in \mathcal{M}\}$.

Example 2.23 (KL Expansion of the Brownian Bridge (2.6)**)** On the compact interval $\mathcal{M} = [0, T] \subset \mathbb{R}$, the KL expansion of the Brownian bridge is

$$B_t = \sum_{k\in\mathbb{N}} Z_k \frac{\sqrt{2T}}{k\pi} \sin(k\pi t/T), \quad t \in [0, T].$$

Then

$$H(\gamma) = H_0^1(0, T) = \operatorname{span}\{\sin(k\pi t/T) : k \in \mathbb{N}\}.$$

In view of GRFs appearing as diffusion coefficients in elliptic and parabolic PDEs, criteria on their path regularity are of some interest. Many such conditions are known and we present some of these, from [3, Chapters 3.2, 3.3].

Proposition 2.24 *For any compact set* $\mathcal{M} \subset \mathbb{R}^d$*, if for* $\alpha > 0$*,* $\eta > \alpha$ *and some constant* $C > 0$ *holds*

$$\mathbb{E}[|Z_{\boldsymbol{x}+\boldsymbol{h}} - Z_{\boldsymbol{x}}|^\alpha] \le C \frac{|\boldsymbol{h}|^{2d}}{|\log|\boldsymbol{h}||^{1+\eta}}, \tag{2.8}$$

then

$$\boldsymbol{x} \to Z_{\boldsymbol{x}}(\omega) \in C^0(\mathcal{M}) \ \ \mathbb{P} - a.s.$$

Choosing $\alpha = 2$ in (2.8), we obtain for $\mathcal{M}$ such that $\mathcal{M} = \overline{D}$, where $D \subset \mathbb{R}^d$ is a bounded Lipschitz domain, the sufficient criterion that there exist $C > 0$, $\eta > 2$ with

$$\forall \boldsymbol{x} \in D: \quad K(\boldsymbol{x}+\boldsymbol{h}, \boldsymbol{x}+\boldsymbol{h}) - K(\boldsymbol{x}+\boldsymbol{h}, \boldsymbol{x}) - K(\boldsymbol{x}, \boldsymbol{x}+\boldsymbol{h}) + K(\boldsymbol{x}, \boldsymbol{x}) \le C \frac{|\boldsymbol{h}|^{2d}}{|\log|\boldsymbol{h}||^{1+\eta}}.$$

This is to hold for some $\eta > 2$ with the covariance kernel K of the GRF Z, in order to ensure that $[\boldsymbol{x} \mapsto Z_{\boldsymbol{x}}] \in C^1(\overline{D}) \subset W^1_\infty(D)$ $\mathbb{P}$-a.s., see [3, Theorem 3.2.5, page 49 bottom].

Further examples of explicit Karhunen-Loève expansions of GRFs can be found in [35, 79, 84] and a statement for $\mathbb{P}$-a.s Hölder continuity of GRFs Z on smooth manifolds $\mathcal{M}$ is proved in [4].

2.5.3 Multiresolution Representations of GRFs

Karhunen-Loève expansions (2.7) provide an important source of concrete examples of Gaussian series representations of GRFs u in Theorem 2.18. Since KL expansions involve the eigenfunctions of the covariance operators of the GRF u, all terms in these expansions are, in general, globally supported in the physical domain $\mathcal{M}$ indexing the GRF u. Often, it is desirable to have Gaussian series representations of u in Theorem 2.18 where the elements $(e_n)_{n\in\mathbb{N}}$ of the representation system are locally supported in the indexing domain $\mathcal{M}$.

Example 2.25 (Lévy-Cieselsky Representation of Brownian Bridge, [34]) Consider the Brownian bridge $(B_t)_{0\le t\le T}$ from Examples 2.22, 2.23. For $T = 1$, it may also be represented as Gaussian series (e.g. [34])

$$B_t = \sum_{j\in\mathbb{N}} \sum_{k=0}^{2^j-1} Z_{jk} 2^{-j/2} h(2^j t - k) = \sum_{j\in\mathbb{N}} \sum_{k=0}^{2^j-1} Z_{jk}\psi_{jk}(t), \quad t \in \mathcal{M} = [0, 1],$$

where

$$\psi_{jk}(t) := 2^{-j/2} h(2^j t - k),$$

with $h(s) := \max\{1 - 2|s - 1/2|, 0\}$ denoting the standard, continuous piecewise affine "hat" function on $(0, 1)$. Here, μ is the Lebesgue measure in $\mathcal{M} = [0, 1]$, and $Z_{jk} \sim \mathcal{N}(0, 1)$ are i.i.d standard normal RVs.

By suitable reordering of the index pairs (j, k), e.g., via the bijection $(j, k) \mapsto j := 2^j + k$, the representation (2.25) is readily seen to be a special case of Theorem 2.21, item ii). The corresponding system

$$\boldsymbol{\Psi} = \{\psi_{jk} : j \in \mathbb{N}_0, 0 \le k \le 2^j - 1\}$$

is, in fact, a basis for $C_0([0, 1]) := \{v \in C([0, 1]) : v(0) = v(1) = 0\}$, the so-called *Schauder basis*.

There holds

$$\sum_{j\in\mathbb{N}} \sum_{k=0}^{2^j-1} 2^{js} |\psi_{jk}(t)| < \infty, \quad t \in [0, 1],$$

for any $0 \leq s < 1/2$. The functions ψ_{jk} are localized in the sense that $|\text{supp}(\psi_{jk})| = 2^{-j}$ for $k = 0, 1, \ldots, 2^j - 1$.

Further constructions of such *multiresolution representations* of GRFs with either Riesz basis or frame properties are available on polytopal domains $M \subset \mathbb{R}^d$, (e.g. [12], for a needlet multiresolution analysis on the 2-sphere $\mathcal{M} = \mathbb{S}^2$ embedded in $\mathbb{R}^3$, where μ in Sect. 2.5.2 can be chosen as the surface measure see, also, for representation systems by so-called spherical needlets [92], [13]).

We also mention [5] for optimal approximation rates of truncated wavelet series approximations of fractional Brownian random fields, and to [79] for corresponding spectral representations.

Multiresolution constructions are also available on data-graphs M (see, e.g., [41] and the references there).

2.5.4 Periodic Continuation of a Stationary GRF

Let $(Z_x)_{x \in D}$ be a GRF indexed by $D \subset \mathbb{R}^d$, where D is a bounded domain. We aim for representations of the general form

$$Z_x = \sum_{j \in \mathbb{N}} \phi_j(x) y_j, \tag{2.9}$$

where the y_j are i.i.d. $\mathcal{N}(0, 1)$ RVs and the $(\phi_j)_{j \in \mathbb{N}}$ are a given sequence of functions defined on D. One natural choice of ϕ_j is $\phi_j = \sqrt{\lambda_j}\psi_j$, where ψ_j and are the eigen-functions and λ_j eigenvalues of the covariance operator. However, Karhunen-Loève eigenfunctions on D are typically not explicitly known and globally supported in the physical domain D. One of the strategies for deriving better representations over D is to view it as the restriction to D of a periodic Gaussian process Z_x^{ext} defined on a suitable larger torus $\mathbb{T}^d$.

Since D is bounded, without loss of generality, we may the physical domain D to be contained in the box $[-\frac{1}{2}, \frac{1}{2}]^d$. We wish to construct a periodic process Z_x^{ext} on the torus $\mathbb{T}^d$ where $\mathbb{T} = [-\ell, \ell]$ whose restriction of Z_x^{ext} on D is such that $Z_x^{\text{ext}}|_D = Z_x$. As a consequence, any representation

$$Z_x^{\text{ext}} = \sum_{j \in \mathbb{N}} y_j \tilde{\phi}_j$$

yields a representation (2.9) where $\phi_j = \tilde{\phi}_j|_D$.

Assume that $(Z_x)_{x \in D}$ is a restriction of a real-valued, stationary and centered GRF $(Z_x)_{x \in \mathbb{R}^d}$ on $\mathbb{R}^d$ whose covariance is given in the form

$$\mathbb{E}[Z_x Z_{x'}] = \rho(x - x'), \quad x, x' \in \mathbb{R}^d, \tag{2.10}$$

where ρ is a real-valued, even function and its Fourier transform is a non-negative function. The extension is feasible provided that we can find an even and $\mathbb{T}^d$-periodic function ρ^{ext} which agree with ρ over $[-1,1]^d$ such that the Fourier coefficients

$$c_{\boldsymbol{n}}(\rho^{\text{ext}}) = \int_{\mathbb{T}^d} \rho^{\text{ext}}(\boldsymbol{\xi}) \exp\Big(-\mathrm{i}\frac{\pi}{\ell}(\boldsymbol{n},\boldsymbol{\xi})\Big)\, \mathrm{d}\boldsymbol{\xi}, \qquad \boldsymbol{n} \in \mathbb{Z}^d$$

are non-negative.

A natural way of constructing the function ρ^{ext} is by truncation and periodization. First one chooses a sufficiently smooth and even cutoff function φ_κ such that $\varphi_\kappa|_{[-1,1]^d} = 1$ and $\varphi_\kappa(\boldsymbol{x}) = 0$ for $\boldsymbol{x} \notin [-\kappa,\kappa]^d$ where $\kappa = 2\ell - 1$. Then ρ^{ext} is defined as the periodization of the truncation $\rho\varphi_\kappa$, i.e.,

$$\rho^{\text{ext}}(\boldsymbol{\xi}) = \sum_{\boldsymbol{n}\in\mathbb{Z}^d} (\rho\varphi_\kappa)(\boldsymbol{\xi} + 2\ell\boldsymbol{n}).$$

It is easily seen that ρ^{ext} agrees with ρ over $[-1,1]^d$ and

$$c_{\boldsymbol{n}}(\rho^{\text{ext}}) = \widehat{\rho\varphi_\kappa}\Big(\frac{\pi}{\ell}\boldsymbol{n}\Big).$$

Therefore $c_{\boldsymbol{n}}(\rho^{\text{ext}})$ is non-negative if we can prove that $\widehat{\rho\varphi_\kappa}(\boldsymbol{\xi}) \geq 0$ for $\boldsymbol{\xi} \in \mathbb{R}^d$. The following result is shown in [12].

Theorem 2.26 *Let ρ be an even function on $\mathbb{R}^d$ such that*

$$c(1+|\boldsymbol{\xi}|^2)^{-s} \leq \hat{\rho}(\boldsymbol{\xi}) \leq C(1+|\boldsymbol{\xi}|^2)^{-r}, \qquad \boldsymbol{\xi} \in \mathbb{R}^d \tag{2.11}$$

for some $s \geq r \geq d/2$ and $0 < c \leq C$ and

$$\lim_{R\to+\infty} \int_{|\boldsymbol{x}|>R} |\partial^{\alpha}\rho(\boldsymbol{x})|\, \mathrm{d}\boldsymbol{x} = 0, \qquad |\boldsymbol{\alpha}| \leq 2\lceil s\rceil.$$

Then for κ sufficiently large, there exists φ_κ satisfying $\varphi_\kappa|_{[-1,1]^d} = 1$ and $\varphi_\kappa(\boldsymbol{x}) = 0$ for $\boldsymbol{x} \notin [-\kappa,\kappa]^d$ such that

$$0 < \widehat{\rho\varphi_\kappa}(\boldsymbol{\xi}) \leq C(1+|\boldsymbol{\xi}|^2)^{-r}, \qquad \boldsymbol{\xi} \in \mathbb{R}^d.$$

The assertion in Theorem 2.11 implies that

$$0 < c_{\boldsymbol{n}}(\rho^{\text{ext}}) \leq C(1+|\boldsymbol{n}|^2)^{-r}, \qquad \boldsymbol{n} \in \mathbb{Z}^d.$$

In the following we present an explicit construction of the function φ_κ for GRFs with Matérn covariance

$$\rho_{\lambda,\nu}(\boldsymbol{x}) := \frac{2^{1-\nu}}{\Gamma(\nu)}\left(\frac{\sqrt{2\nu}|\boldsymbol{x}|}{\lambda}\right)^{\nu} K_\nu\left(\frac{\sqrt{2\nu}|\boldsymbol{x}|}{\lambda}\right),$$

where $\lambda > 0$, $\nu > 0$ and K_ν is the modified Bessel functions of the second kind. Note that the Matérn covariances satisfy the assumption (2.11) with $s = r = \nu + d/2$.

Let $P := 2\lceil \nu + \frac{d}{2} \rceil + 1$ and N_P be the cardinal B-spline function with nodes $\{-P, \ldots, -1, 0\}$. For $\kappa > 0$ we define the even function $\varphi \in C^{P-1}(\mathbb{R})$ by

$$\varphi(t) = \begin{cases} 1 & \text{if } |t| \le \kappa/2 \\ \dfrac{2P}{\kappa} \displaystyle\int_{-\infty}^{t+\kappa/2} N_P\left(\frac{2P}{\kappa}\xi\right) \mathrm{d}\xi & \text{if } t \le -\kappa/2. \end{cases}$$

It is easy to see that $\varphi(t) = 0$ if $|t| \ge \kappa$. We now define

$$\varphi_\kappa(\boldsymbol{x}) := \varphi(|\boldsymbol{x}|).$$

With this choice of φ_κ, we have $\rho^{\text{ext}} = \rho_{\lambda,\nu}$ on $[-1,1]^d$ provided that $\ell \ge \frac{\kappa+\sqrt{d}}{2}$. The required size of κ is given in the following theorem, see [14, Theorem 10].

Theorem 2.27 *For φ_κ as defined above, there exist constants $C_1, C_2 > 0$ such that for any $0 < \lambda, \nu < \infty$, we have $\widehat{\rho_{\lambda,\nu}\varphi_\kappa} > 0$ provided that $\kappa > 1$ and*

$$\frac{\kappa}{\lambda} \ge C_1 + C_2 \max\left\{\nu^{\frac{1}{2}}(1 + |\ln \nu|), \nu^{-\frac{1}{2}}\right\}.$$

Remark 2.28 The periodic random field $Z_{\boldsymbol{x}}^{\text{ext}}$ on $\mathbb{T}^d$ provides a tool for deriving series expansions of the original random field. In contrast to the Karhunen-Loève eigenfunctions on D, which are typically not explicitly known, the corresponding eigenfunctions ψ_j^{ext} of the periodic covariance are explicitly known trigonometric functions and one has the following Karhunen-Loève expansion for the periodized random field:

$$Z_{\boldsymbol{x}}^{\text{ext}} = \sum_{j\in\mathbb{N}} y_j \sqrt{\lambda_j^{\text{ext}}}\, \psi_j^{\text{ext}}, \quad y_j \sim \mathcal{N}(0,1) \text{ i.i.d.},$$

with λ_j^{ext} denoting the eigenvalues of the periodized covariance and the ψ_j^{ext} are normalized in $L^2(\mathbb{T}^d)$. Restricting this expansion back to D, one obtains an exact expansion of the original random field on D

$$Z_{\boldsymbol{x}} = \sum_{j\in\mathbb{N}} y_j \sqrt{\lambda_j^{\text{ext}}}\, \psi_j^{\text{ext}}|_D, \quad y_j \sim \mathcal{N}(0,1) \text{ i.i.d.} \tag{2.12}$$

This provides an alternative to the standard KL expansion of $Z_{\boldsymbol{x}}$ in terms of eigenvalues λ_j and eigenfunctions ψ_j normalized in $L^2(D)$. The main difference is that the functions $\psi_j^{\text{ext}}|_D$ in (2.12) are not $L^2(D)$-orthogonal. However, these functions are given explicitly, and thus no approximate computation of eigenfunctions is required.

The KL expansion of $Z_{\boldsymbol{x}}^{\text{ext}}$ also enables the construction of alternative expansions of $Z_{\boldsymbol{x}}$ of the basic form (2.12), but with the spatial functions having additional properties. In [12], wavelet-type representations

$$Z_{\boldsymbol{x}}^{\text{ext}} = \sum_{\ell,k} y_{\ell,k}\psi_{\ell,k}, \quad y_{\ell,k} \sim \mathcal{N}(0,1) \text{ i.i.d.},$$

are constructed where the functions $\psi_{\ell,k}$ have the same multilevel-type localisation as the Meyer wavelets. This feature yields improved convergence estimates for tensor Hermite polynomial approximations of solutions of random diffusion equations with log-Gaussian coefficients .

2.5.5 Sampling Stationary GRFs

The simulation of GRFs with specified covariance is a fundamental task in computational statistics with a wide range of applications. In this section we present an efficient methods for sampling such fields. Consider a GRF $(Z_{\boldsymbol{x}})_{\boldsymbol{x}\in D}$ where D is contained in $[-1/2, 1/2]^d$. Assume that $(Z_{\boldsymbol{x}})_{\boldsymbol{x}\in D}$ is a restriction of a real-valued, stationary and centered GRF $(Z_{\boldsymbol{x}})_{\boldsymbol{x}\in\mathbb{R}^d}$ on $\mathbb{R}^d$ with covariance given in (2.10). Let $m \in \mathbb{N}$ and $\boldsymbol{x}_1, \dots, \boldsymbol{x}_M$ be $M = (m+1)^d$ uniform grid points on $[-1/2, 1/2]^d$ with grid spacing $h = 1/m$. We wish to obtain samples of the Gaussian RV

$$\boldsymbol{Z} = (Z_{\boldsymbol{x}_1}, \dots, Z_{\boldsymbol{x}_M})$$

with covariance matrix

$$\boldsymbol{\Sigma} = [\Sigma_{i,j}]_{i,j=1}^M, \qquad \Sigma_{i,j} = \rho(\boldsymbol{x}_i - \boldsymbol{x}_j), \quad i, j = 1, \dots, M. \tag{2.13}$$

Since $\boldsymbol{\Sigma}$ is symmetric positive semidefinite, this can in principle be done by performing the Cholesky factorisation $\boldsymbol{\Sigma} = \boldsymbol{F}\boldsymbol{F}^\top$ with $\boldsymbol{F} = \boldsymbol{\Sigma}^{1/2}$, from which the desired samples are provided by the product $\boldsymbol{F}\boldsymbol{Y}$ where $\boldsymbol{Y} \sim \mathcal{N}(0, \boldsymbol{I})$. However, since Σ is large and dense when m is large, this factorisation is prohibitively expensive. Since the covariance matrix $\boldsymbol{\Sigma}$ is a nested block Toeplitz matrix under appropriate ordering, an efficient approach is to extend $\boldsymbol{\Sigma}$ to a appropriate larger nested block circulant matrix whose spectral decomposition can be rapidly computed using FFT.

For any $\ell \geq 1$ we construct a 2ℓ-periodic extension of ρ as follows

$$\rho^{\text{ext}}(\boldsymbol{x}) = \sum_{\boldsymbol{n}\in\mathbb{Z}^d} \left(\rho\chi_{(-\ell,\ell]^d}\right)(\boldsymbol{x} + 2\ell\boldsymbol{n}), \quad \boldsymbol{x} \in \mathbb{R}^d.$$

Clearly, ρ^{ext} is 2ℓ-periodic and $\rho^{\text{ext}} = \rho$ on $[-1, 1]^d$. Denote $\boldsymbol{\xi}_1, \ldots, \boldsymbol{\xi}_s$, $s = (2\ell/h)^d$, the uniform grid points on $[-\ell, \ell]^d$ with grid space h. Let $\boldsymbol{Z}^{\text{ext}} = (Z_{\boldsymbol{\xi}_1}, \ldots, Z_{\boldsymbol{\xi}_s})$ be the extended GRV with covariance matrix $\boldsymbol{\Sigma}^{\text{ext}}$ whose entries is given by formula (2.13), with ρ replaced by ρ^{ext} and $\boldsymbol{x}_i$ by $\boldsymbol{\xi}_i$. Hence $\boldsymbol{\Sigma}$ is embedded into the nested circulant matrix $\boldsymbol{\Sigma}^{\text{ext}}$ which can be diagonalized using FFT (with log-linear complexity) to provide the spectral decomposition

$$\boldsymbol{\Sigma}^{\text{ext}} = \boldsymbol{Q}^{\text{ext}}\boldsymbol{\Lambda}^{\text{ext}}(\boldsymbol{Q}^{\text{ext}})^\top,$$

with Λ^{ext} diagonal and containing the eigenvalues λ_j^{ext} of $\boldsymbol{\Sigma}^{\text{ext}}$ and $\boldsymbol{Q}^{\text{ext}}$ being a Fourier matrix. Provided that these eigenvalues are non-negative, the samples of the grid values of Z can be drawn as follows. First we draw a random vector $(y_j)_{j=1,\ldots,s}$ with $y_j \sim \mathcal{N}(0, 1)$ i.i.d., then compute

$$\boldsymbol{Z}^{\text{ext}} = \sum_{j=1}^{s} y_j\sqrt{\lambda_j^{\text{ext}}}\,\boldsymbol{q}_j$$

using the FFT, with $\boldsymbol{q}_j$ the columns of $\boldsymbol{Q}^{\text{ext}}$. Finally, a sample of $\boldsymbol{Z}$ is obtained by extracting from $\boldsymbol{Z}^{\text{ext}}$ the entries corresponding to the original grid points.

The above mentioned process is feasible, provided that $\boldsymbol{\Sigma}$ is positive semidefinite. The following theorem characterizes the condition on ℓ for GRF with Matérn covariance such that $\boldsymbol{\Sigma}^{\text{ext}}$ is positive semidefinite, see [60].

Theorem 2.29 *Let* $1/2 \leq \nu < \infty$, $\lambda \leq 1$, *and* $h/\lambda \leq e^{-1}$. *Then there exist* $C_1, C_2 > 0$ *which may depend on* d *but are independent of* ℓ, h, λ, ν, *such that* Σ^{ext} *is positive definite if*

$$\frac{\ell}{\lambda} \geq C_1 + C_2\,\nu^{\frac{1}{2}}\log\left(\max\left\{\lambda/h,\ \nu^{\frac{1}{2}}\right\}\right).$$

Remark 2.30 For GRF with Matérn covariances, it is well-known (see, e.g. [59, Corollary 5], [10, eq.(64)]) that the exact KL eigenvalues λ_j of $Z_{\boldsymbol{x}}$ in $L^2(D)$ decay with the rate $\lambda_j \leq Cj^{-(1+2\nu/d)}$. It has been proved recently in [14] that the eigenvalue λ_j^{ext} maintain this rate of decay up to a factor of order $\mathcal{O}(|\log h|^\nu)$.

2.6 Finite Element Discretization

The approximation results and algorithms to be developed in the present text involve, besides the Wiener-Hermite PC expansions with respect to Gaussian coordinates $\boldsymbol{y} \in \mathbb{R}^\infty$, also certain numerical approximations in the physical domain D. Due to their wide use in the numerical solution of elliptic and parabolic PDEs, we opt for considering standard, primal Lagrangian finite element discretizations. We confine the presentation and analysis to Lipschitz polytopal domains $D \subset \mathbb{R}^d$ with principal interest in $d = 2$ (D is a polygon with straight sides) and $d = 3$ (D is a polyhedron with plane faces). We confine the presentation to so-called primal FE discretizations in D but hasten to add that with minor extra mathematical effort, similar results could be developed also for so-called mixed, or dual FE discretizations (see,e.g., [20] and the references there).

In presenting (known) results on finite element method (FEM for short) convergence rates, we consider separately FEM in polytopal domains $D \subset \mathbb{R}^d, d = 1, 2, 3$, and FEM on smooth d-surfaces $\Gamma \subset \mathbb{R}^{d+1}, d = 1, 2$. See [22, 49].

2.6.1 Function Spaces

For a bounded domain $D \subset \mathbb{R}^d$, the usual Sobolev function spaces of integer order $s \in \mathbb{N}_0$ and integrability $q \in [1, \infty]$ are denoted by $W^s_q(D)$ with the understanding that $L^q(D) = W^0_q(D)$. The norm of $v \in W^s_q(D)$ is defined by

$$\|v\|_{W^s_q} := \sum_{\boldsymbol{\alpha} \in \mathbb{Z}^d_+ : |\boldsymbol{\alpha}| \le s} \|D^{\boldsymbol{\alpha}} v\|_{L^q}.$$

Here $D^{\boldsymbol{\alpha}}$ denotes the partial weak derivative of order $\boldsymbol{\alpha}$. We refer to any standard text such as [2] for basic properties of these spaces. Hilbertian Sobolev spaces are given for $s \in \mathbb{N}_0$ by $H^s(D) = W^s_2(D)$, with the usual understanding that $L^2(D) = H^0(D)$.

For $s \in \mathbb{N}$, we call a C^s-domain $D \subset \mathbb{R}^d$ a bounded domain whose boundary ∂D is locally parameterized in a finite number of co-ordinate systems as a graph of a C^s function. In a similar way, we shall call $D \subset \mathbb{R}^d$ a Lipschitz domain, when ∂D is, locally, the graph of a Lipschitz function. We refer to [2, 55] and the references there or to [61].

We call *polygonal domain* a domain $D \subset \mathbb{R}^2$ that is a polygon with Lipschitz boundary ∂D (which precludes cusps and slits) and with a finite number of straight sides.

Let $D \subset \mathbb{R}^2$ denote an open bounded polygonal domain. We introduce in D a nonnegative function $r_D : D \to \mathbb{R}_+$ which is smooth in D, and which coincides for $\boldsymbol{x}$ in a vicinity of each corner $\boldsymbol{c} \in \partial D$ with the Euclidean distance $|\boldsymbol{x} - \boldsymbol{c}|$.

To state elliptic regularity shifts in D, we require certain corner-weighted Sobolev spaces. We require these only for integrability $q = 2$ and for $q = \infty$.

For $s \in \mathbb{N}_0$ and $\varkappa \in \mathbb{R}$ we define

$$\mathcal{K}^s_\varkappa(D) := \left\{u : D \to \mathbb{C} : r_D^{|\alpha|-\varkappa} D^\alpha u \in L^2(D), |\alpha| \le s\right\}$$

and

$$\mathcal{W}^s_\infty(D) := \left\{u : D \to \mathbb{C} : r_D^{|\alpha|} D^\alpha u \in L^\infty(D), \ |\alpha| \le s\right\}.$$

Here, for $\boldsymbol{\alpha} \in \mathbb{N}_0^2$ and as before $D^{\boldsymbol{\alpha}}$ denotes the partial weak derivative of order $\boldsymbol{\alpha}$.

The *corner-weighted norms in these spaces* are given by

$$\|u\|_{\mathcal{K}^s_\varkappa} := \sum_{|\alpha| \le s} \|r_D^{|\alpha|-\varkappa} D^\alpha u\|_{L^2} \qquad \text{and} \qquad \|u\|_{\mathcal{W}^s_\infty} := \sum_{|\alpha| \le s} \|r_D^{|\alpha|} D^\alpha u\|_{L^\infty}.$$

The function spaces $\mathcal{K}^s_\varkappa(D)$ and $\mathcal{W}^s_\infty(D)$ endowed with these norms are Banach spaces, and $\mathcal{K}^s_\varkappa(D)$ are separable Hilbert spaces. These corner-weighted Sobolev spaces are called Kondrat'ev spaces.

An embedding of these spaces is $H_0^1(D) \hookrightarrow \mathcal{K}_0^1(D)$. This follows from the existence of a constant $c(D) > 0$ such that for every $\boldsymbol{x} \in D$ holds $r_D(\boldsymbol{x}) \ge c(D)\mathrm{dist}(\boldsymbol{x}, \partial D)$.

2.6.2 Finite Element Interpolation

In this section, we review some results on FE approximations in polygonal domains D on locally refined triangulations $\mathcal{T}$ in D. These results are in principle known for the standard Sobolev spaces $H^s(D)$ and available in the standard texts [23, 33]. For spaces with corner weights in polygonal domains $D \subset \mathbb{R}^2$, such as $\mathcal{K}^s_\varkappa$ and $\mathcal{W}^s_\infty$, however, which arise in the regularity of the Wiener-Hermite PC expansion coefficient functions for elliptic PDEs in corner domains in Sect. 3.8 ahead, we provide references to corresponding FE approximation rate bounds.

The corresponding FE spaces involve suitable mesh refinement to compensate for the reduced regularity caused by corner and edge singularities which occur in solutions to elliptic and parabolic boundary value problems in these domains.

We define the FE spaces in a polygonal domain $D \subset \mathbb{R}^2$ (see [23, 33] for details). Let $\mathcal{T}$ denote a regular triangulation of $\overline{D}$, i.e., a partition of $\overline{D}$ into a finite number $N(\mathcal{T})$ of closed, nondegenerate triangles $T \in \mathcal{T}$ (i.e., $|T| > 0$) such that for any two $T, T' \in \mathcal{T}$, the intersection $T \cap T'$ is either empty, a vertex or an entire edge. We denote the *meshwidth* of $\mathcal{T}$ as

$$h(\mathcal{T}) := \max\{h(T) : T \in \mathcal{T}\}, \ \text{ where } \ h(T) := \mathrm{diam}(T).$$

For $T \in \mathcal{T}$, denote $\rho(T)$ the diameter of the largest circle that can be inscribed into T. We say $\mathcal{T}$ is κ *shape-regular*, if

$$\forall T \in \mathcal{T}: \quad \frac{h(T)}{\rho(T)} \leq \kappa.$$

A sequence $\mathfrak{T} := (\mathcal{T}_n)_{n\in\mathbb{N}}$ is κ shape-regular if each $\mathcal{T} \in \mathfrak{T}$ is κ shape-regular, with one common constant $\kappa > 1$ for all $\mathcal{T} \in \mathfrak{T}$.

In a polygon D, with a regular, simplicial triangulation $\mathcal{T}$, and for a polynomial degree $m \in \mathbb{N}$, the Lagrangian FE space $S^m(D, \mathcal{T})$ of continuous, piecewise polynomial functions of degree m on $\mathcal{T}$ is defined as

$$S^m(D, \mathcal{T}) = \{v \in H^1(D) : \forall T \in \mathcal{T} : v|_T \in \mathbb{P}_m\}.$$

Here, $\mathbb{P}_m := \operatorname{span}\{\boldsymbol{x}^{\boldsymbol{\alpha}} : |\boldsymbol{\alpha}| \leq m\}$ denotes the space of polynomials of $\boldsymbol{x} \in \mathbb{R}^2$ of total degree at most m. We also define $S_0^m(D, \mathcal{T}) := S^m(D, \mathcal{T}) \cap H_0^1(D)$.

The main result on FE approximation rates in a polygon $D \subset \mathbb{R}^2$ in corner-weighted spaces $\mathcal{K}_\kappa^s(D)$ reads as follows.

Proposition 2.31 *Consider a bounded polygonal domain $D \subset \mathbb{R}^2$. Then, for every polynomial degree $m \in \mathbb{N}$, there exists a sequence $(\mathcal{T}_n)_{n\in\mathbb{N}}$ of κ shape-regular, simplicial triangulations of D such that for every $u \in (H_0^1 \cap \mathcal{K}_\lambda^{m+1})(D)$ for some $\lambda > 0$, the FE interpolation error converges at rate m. More precisely, there exists a constant $C(D, \kappa, \lambda, m) > 0$ such that for all $\mathcal{T} \in (\mathcal{T}_n)_{n\in\mathbb{N}}$ and for all $u \in (H_0^1 \cap \mathcal{K}_\lambda^{m+1})(D)$ holds*

$$\|u - I_{\mathcal{T}}^m u\|_{H^1} \leq C h(\mathcal{T})^m \|u\|_{\mathcal{K}_\lambda^{m+1}}.$$

Equivalently, in terms of the number $n := \#(\mathcal{T})$ of triangles, there holds

$$\|u - I_{\mathcal{T}}^m u\|_{H^1} \leq C n^{-m/2} \|u\|_{\mathcal{K}_\lambda^{m+1}}. \tag{2.14}$$

Here, $I_{\mathcal{T}}^m : C^0(\overline{D}) \to S^m(D, \mathcal{T})$ denotes the nodal, Lagrangian interpolant. The constant $C > 0$ depends on m, D and the shape regularity of $\mathcal{T}$, but is independent of u.

For a proof of this proposition, we refer, for example, to [25, Theorems 4.2, 4.4].

We remark that due to $\mathcal{K}_\lambda^2(D) \subset C^0(\overline{D})$, the nodal interpolant $I_{\mathcal{T}}^m$ in (2.14) is well-defined. We also remark that the triangulations $\mathcal{T}_n$ need not necessarily be nested (the constructions in [6, 25] do not provide nestedness; for a bisection tree construction of $(\mathcal{T}_n)_{n\in\mathbb{N}}$ which are nested, such as typically produced by adaptive FE algorithms, with the error bounds (2.14), we refer to [54].

For similar results in polyhedral domains in space dimension $d = 3$, we refer to [26, 27, 86] and to the references there.

Chapter 3
Elliptic Divergence-Form PDEs with Log-Gaussian Coefficient

We present a model second order linear divergence-form PDE with log-Gaussian input data. We review known results on its well-posedness, and Lipschitz continuous dependence on the input data. Particular attention is placed on regularity results in polygonal domains $D \subset \mathbb{R}^2$. Here, solutions belong to Kondrat'ev spaces. We discuss regularity results for parametric coefficients, and establish in particular parametric holomorphy results for the coefficient-to-solution maps.

The outline of this section is as follows. In Sect. 3.1, we present the strong and variational forms of the PDE, its well-posedness and the continuity of the data-to-solution map in appropriate spaces. Importantly, we do not aim at the most general setting, but to ease notation and for simplicity of presentation we address a rather simple, particular case: in a bounded domain D in Euclidean space $\mathbb{R}^d$. All the ensuing derivations will directly generalize to linear second order elliptic systems. A stronger Lipschitz continuous dependence on data result is stated in Sect. 3.2. Higher regularity and fractional regularity of the solution provided correspondingly by higher regularity of data are discussed in Sect. 3.3.

Sections 3.4 and 3.5 describe uncertainty modelling by placing GMs on sets of admissible, countably parametric input data, i.e., formalizing mathematically aleatoric uncertainty in input data. Here, the Gaussian series introduced in Sect. 2.5 will be seen to take a key role in converting operator equations with GRF inputs to infinitely-parametric, deterministic operator equations. The Lipschitz continuous dependence of the solutions on input data from function spaces will imply strong measurability of corresponding random solutions, and render well-defined the *uncertainty propagation*, i.e., the push-forward of the GM on the input data.

In Sects. 3.6–3.8, we connect quantified holomorphy of the parametric, deterministic solution manifold $\{u(\boldsymbol{y}) : \boldsymbol{y} \in \mathbb{R}^\infty\}$ with sparsity of the coefficients $(\|u_\boldsymbol{\nu}\|_H)_{\boldsymbol{\nu}\in\mathcal{F}}$ of Wiener-Hermite PC expansion as elements of certain Sobolev spaces: We start with the case $H = H_0^1(D)$ in Sect. 3.6 and subsequently discuss higher regularity $H = H^s(D)$, $s \in \mathbb{N}$, in Sect. 3.7 and finally H being a Kondrat'ev space on a bounded polygonal domain $D \subset \mathbb{R}^2$ in Sect. 3.8.

D. Dũng et al., *Analyticity and Sparsity in Uncertainty Quantification for PDEs with Gaussian Random Field Inputs*, Lecture Notes in Mathematics 2334,
https://doi.org/10.1007/978-3-031-38384-7_3

3.1 Statement of the Problem and Well-Posedness

In a bounded Lipschitz domain $D \subset \mathbb{R}^d$ ($d = 1, 2$ or 3), consider the linear second order elliptic PDE in divergence-form

$$P_a u := \begin{Bmatrix} -\operatorname{div}(a(\boldsymbol{x})\nabla u(\boldsymbol{x})) \\ \tau_0(u) \end{Bmatrix} = \begin{cases} f(\boldsymbol{x}) \text{ in } D, \\ 0 \text{ on } \partial D. \end{cases} \tag{3.1}$$

Here, $\tau_0 : H^1(D) \to H^{1/2}(\partial D)$ denotes the trace map. With the notation $V := H_0^1(D)$ and $V^* = H^{-1}(D)$, for any $f \in V^*$, by the Lax-Milgram lemma the weak formulation given by

$$u \in V : \quad \int_D a\nabla u \cdot \nabla v \, \mathrm{d}\boldsymbol{x} = \langle f, v \rangle_{V^*,V}, \qquad v \in V, \tag{3.2}$$

admits a unique solution $u \in V$ whenever the coefficient a satisfies the ellipticity assumption

$$0 < a_{\min} := \operatorname*{ess\,inf}_{\boldsymbol{x} \in D} a(\boldsymbol{x}) \le a_{\max} = \|a\|_{L^\infty} < \infty. \tag{3.3}$$

With $\|v\|_V := \|\nabla v\|_{L^2}$ denoting the norm of $v \in V$, there holds the a-priori estimate

$$\|u\|_V \le \frac{\|f\|_{V^*}}{a_{\min}}. \tag{3.4}$$

In particular, with

$$L_+^\infty(D) := \{a \in L^\infty(D) : a_{\min} > 0\},$$

the *data-to-solution operator*

$$\mathcal{S} : L_+^\infty(D) \times V^* \to V : (a, f) \mapsto u \tag{3.5}$$

is continuous.

3.2 Lipschitz Continuous Dependence

The continuity (3.5) of the data-to-solution map $\mathcal{S}$ allows to infer already strong measurability of solutions of (3.1) with respect to random coefficients a. For purposes of stable numerical approximation, we will be interested in quantitative bounds of the effect of perturbations of the coefficient a in (3.2) and of the source term data f on the solution $u = \mathcal{S}(a, f)$. Mere continuity of $\mathcal{S}$ as a map from

$L^\infty_+(D) \times V^*$ to $V = H^1_0(D)$ will not be sufficient to this end. To quantify the impact of uncertainty in the coefficient a on the solution $u \in V$, local Hölder or, preferably, Lipschitz continuity of the map $\mathcal{S}$ is required, at least locally, close to nominal values of the data (a, f).

To this end, consider given $a_1, a_2 \in L^\infty_+(D)$, $f_1, f_2 \in L^2(D) \subset V^*$ with corresponding unique solutions $u_i = \mathcal{S}(a_i, f_i) \in V$, $i = 1, 2$.

Proposition 3.1 *In a bounded Lipschitz domain $D \subset \mathbb{R}^d$, for given data bounds $r_a, r_f \in (0, \infty)$, there exist constants c_a and c_f such that for every $a_i \in L^\infty_+(D)$ with $\|\log(a_i)\|_{L^\infty} \le r_a$, and for every $f_i \in L^2(D)$ with $\|f_i\|_{L^2} \le r_f$, $i = 1, 2$, it holds*

$$\|u_1 - u_2\|_V \le \frac{c_P}{a_{1,\min} \wedge a_{2,\min}} \|f_1 - f_2\|_{L^2} + \frac{\|f_1\|_{V^*} \vee \|f_2\|_{V^*}}{a_{1,\min} a_{2,\min}} \|a_1 - a_2\|_{L^\infty}. \tag{3.6}$$

Therefore

$$\|\mathcal{S}(a_1, f_1) - \mathcal{S}(a_2, f_2)\|_V \le c_a \|a_1 - a_2\|_{L^\infty} + c_f \|f_1 - f_2\|_{L^2}, \tag{3.7}$$

and

$$\|\mathcal{S}(a_1, f_1) - \mathcal{S}(a_2, f_2)\|_V \le \tilde{c}_a \|\log(a_1) - \log(a_2)\|_{L^\infty} + c_f \|f_1 - f_2\|_{L^2}. \tag{3.8}$$

Here, we may take $c_f = c_P \exp(r_a)$, $c_a = c_P r_f \exp(2r_a)$ and $\tilde{c}_a = c_P r_f \exp(3r_a)$. The constant $c_P = c(D) > 0$ denotes the $V - L^2(D)$ Poincaré constant of D.

The bounds (3.7) and (3.8) follow from the continuous dependence estimates in [15] by elementary manipulations. For a proof (in a slightly more general setting), we also refer to Sect. 4.3.1 ahead.

3.3 Regularity of the Solution

It is well known that weak solutions $u \in V$ of the linear elliptic boundary value problem (BVP for short) (3.1) admit higher regularity for more regular data (i.e., coefficient $a(\boldsymbol{x})$, source term $f(\boldsymbol{x})$ and domain D). Standard references for corresponding results are [55, 61]. The proofs in these references cover general, linear elliptic PDEs, with possibly matrix-valued coefficients, and aim at sharp results on the Sobolev and Hölder regularity of solutions, in terms of corresponding regularity of coefficients, source term and boundar ∂D. In order to handle the dependence of solutions on random field and parametric coefficients in a quantitative manner, we develop presently self-contained, straightforward arguments for solution regularity of (3.1).

Here is a first regularity statement, which will be used in several places subsequently. To state it, we denote by W the normed space of all functions $v \in V$ such that $\Delta v \in L^2(D)$. The norm in W is defined by

$$\|v\|_W := \|\Delta v\|_{L^2}.$$

The map $v \mapsto \|v\|_W$ is indeed a norm on W due to the homogeneous Dirichlet boundary condition of $v \in V$: $\|v\|_W = 0$ implies that v is harmonic in D, and $v \in V$ implies that the trace of v on ∂D vanishes, whence $v = 0$ in D by the maximum principle.

Proposition 3.2 *Consider the boundary value problem* (3.1) *in a bounded domain D with Lipschitz boundary, and with $a \in W^1_\infty(D)$, $f \in L^2(D)$. Then the weak solution $u \in V$ of* (3.1) *belongs to the space W and there holds the a-priori estimate*

$$\|u\|_W \leq \frac{1}{a_{\min}}\left(\|f\|_{L^2} + \|f\|_{V^*}\frac{\|\nabla a\|_{L^\infty}}{a_{\min}}\right) \leq \frac{c}{a_{\min}}\left(1+\frac{\|\nabla a\|_{L^\infty}}{a_{\min}}\right)\|f\|_{L^2}, \tag{3.9}$$

where $a_{\min} = \min\{a(\boldsymbol{x}) : \boldsymbol{x} \in \overline{D}\}$.

Proof That $u \in V$ belongs to W is verified by observing that under these assumptions, there holds

$$-a\Delta u = f + \nabla a \cdot \nabla u \ \text{ in the sense of } L^2(D). \tag{3.10}$$

The first bound (3.9) follows by elementary argument using (3.4), the second bound by an application of the $L^2(D)$-V^* Poincaré inequality in D. □

Remark 3.3 The relevance of the space W stems from the relation to the corner-weighted Kondrat'ev spaces $\mathcal{K}^m_\kappa(D)$ which were introduced in Sect. 2.6.1. When the domain $D \subset \mathbb{R}^2$ is a polygon with straight sides, in the presently considered homogeneous Dirichlet boundary conditions on all of ∂D, it holds that $W \subset \mathcal{K}^2_\kappa(D)$ with continuous injection provided that $|\kappa| < \pi/\omega$ where $0 < \omega < 2\pi$ is the largest interior opening angle at the vertices of D. Membership of u in $\mathcal{K}^2_\kappa(D)$ in turn implies optimal approximation rates for standard, Lagrangian FE approximations in D with suitable, corner-refined triangulations in D, see Proposition 2.31.

Remark 3.4 If the physical domain D is convex or of type $C^{1,1}$, then $u \in W$ implies that $u \in (H^2 \cap H^1_0)(D)$ and (3.9) gives rise to an H^2 a-priori estimate (see, e.g., [61, Theorem 2.2.2.3]).

The regularity in Proposition 3.2 is adequate for diffusion coefficients $a(\boldsymbol{x})$ which are Lipschitz continuous in D, which is essentially (up to modification) $W^1_\infty(D) \simeq C^{0,1}(D)$. In view of our interest in admitting diffusion coefficients which are (realizations of) GRF (see Sect. 3.4), it is clear from Example 2.25 that relevant GRF models may exhibit mere Hölder path regularity.

The Hölder spaces $C^s(D)$ on Lipschitz domains D can be obtained as interpolation spaces, via the so-called K-method of function space interpolation which we briefly recapitulate (see, e.g., [107, Chapter 1.3], [18]). Two Banach spaces A_0, A_1 with continuous embedding $A_1 \hookrightarrow A_0$ with respective norms $\| \circ \|_{A_i}$, $i = 0, 1$, constitute an interpolation couple. For $0 < s < 1$, the *interpolation space* $[A_0, A_1]_{s,q}$ of smoothness order s with fine index $q \in [1, \infty]$ is defined via the K-functional: for $a \in A_0$, this functional is given by

$$K(a, t; A_0, A_1) := \inf_{a_1 \in A_1} \{\|a - a_1\|_{A_0} + t\|a_1\|_{A_1}\}, \quad t > 0. \tag{3.11}$$

For $0 < s < 1$ the intermediate, "interpolation" space of order s and fine index q is denoted by $[A_0, A_1]_{s,q}$. It is the set of functions $a \in A_0$ such that the quantity

$$\|a\|_{[A_0,A_1]_{s,q}} := \begin{cases} \left(\int_0^\infty (t^{-s} K(a, t, A_0, A_1))^q \frac{dt}{t}\right)^{1/q}, & 1 \le q < \infty, \\ \sup_{t>0} t^{-s} K(a, t, A_0, A_1), & q = \infty, \end{cases} \tag{3.12}$$

is finite. When the A_i are Banach spaces, the sets $[A_0, A_1]_{s,q}$ are Banach spaces with norm given by (3.12). In particular (see, e.g., [2, Lemma 7.36]), in the bounded Lipschitz domain D

$$C^s(D) = [L^\infty(D), W^1_\infty(D)]_{s,\infty}, \quad 0 < s < 1. \tag{3.13}$$

With the spaces $V := H^1_0(D)$ and $W \subset V$, we define the (non-separable, non-reflexive) Banach space

$$W^s := [V, W]_{s,\infty}, \quad 0 < s < 1. \tag{3.14}$$

Then there holds the following generalization of (3.9).

Proposition 3.5 *For a bounded Lipschitz domain $D \subset \mathbb{R}^d$, $d \ge 2$, for every $f \in L^2(D)$ and $a \in C^s(D)$ for some $0 < s < 1$ with*

$$a_{\min} = \min\{a(\boldsymbol{x}) : \boldsymbol{x} \in \overline{D}\} > 0,$$

the solution $u \in V$ of (3.1), (3.2) *belongs to W^s, and there exists a constant $c(s, D)$ such that*

$$\|u\|_{W^s} \le \frac{c}{a_{\min}} \left(1 + \|a\|_{C^s}^{1/s} a_{\min}^{-1/s}\right) \|f\|_{L^2}. \tag{3.15}$$

Proof The estimate follows from the a-priori bounds for $s = 0$ and $s = 1$, i.e., (3.4) and (3.9), by interpolation with the Lipschitz continuity (3.6) of the solution operator.

Let $a \in C^s(D)$ with $a_{\min} > 0$ be given. From (3.13), for every $\delta > 0$ exists $a_\delta \in W^{1,\infty}(D)$ with

$$\|a - a_\delta\|_{C^0} \leq C\delta^s \|a\|_{C^s}, \quad \|a_\delta\|_{W^1_\infty} \leq C\delta^{s-1}\|a\|_{C^s}.$$

From

$$\min_{x\in D} a_\delta(x) \geq \min_{x\in D} a(x) - \|a - a_\delta\|_{C^0} \geq a_{\min} - C\delta^s \|a\|_{C^s}$$

follows for $0 < \delta \leq 2^{-1/s} \|a/a_{\min}\|_{C^s}^{-1/s}$, that

$$\min_{x\in D} a_\delta(x) \geq a_{\min}/2.$$

For such δ and for $f \in L^2(D)$, (3.1) with a_δ admits a unique solution $u_\delta \in V$ and from (3.9)

$$\|u_\delta\|_W \leq \frac{2c}{a_{\min}} \left(1 + \frac{\|\nabla a_\delta\|_{L^\infty}}{a_{\min}}\right) \|f\|_{L^2}.$$

From (3.6) (with $f_1 = f_2 = f$) we find

$$\|u - u_\delta\|_V \leq \frac{2c}{a_{\min}^2}\|a - a_\delta\|_{L^\infty}\|f\|_{L^2} \leq C\frac{\delta^s}{a_{\min}^2}\|a\|_{C^s}\|f\|_{L^2}.$$

This implies in (3.11) that for some constant $C > 0$ (depending only on D and on s)

$$K(u, t, V, W) \leq \frac{C}{a_{\min}} \left(\delta^s A_s + t\left(1 + \delta^{s-1}A_s\right)\right) \|f\|_{L^2}, \quad t > 0 \tag{3.16}$$

where we have set $A_s := \left\|\frac{a}{a_{\min}}\right\|_{C^s} \in [1, \infty)$.

To complete the proof, by (3.14) we bound $\|u\|_{W^s} = \sup_{t>0} t^{-s} K(u, t, V, W)$. To this end, it suffices to bound $K(u, t, V, W)$ for $0 < t < 1$. Given such t, we choose in the bound (3.16) $\delta = t\delta_0 \in (0, \delta_0)$ with $\delta_0 := 2^{-1/s} A_s^{-1/s}$. This yields

$$\begin{aligned} \delta^s A_s + t\left(1 + \delta^{s-1}A_s\right) &= t^s \left(\delta_0^s A_s + t^{1-s} + \delta_0^{s-1}A_s\right) \\ &= t^s \left(2^{-1} + t^{1-s} + 2^{-(s-1)/s} A_s^{1-(s-1)/s}\right) \end{aligned}$$

and we obtain for $0 < t < 1$ the bound

$$t^{-s} K(u, t, V, W) \leq \frac{C}{a_{\min}} \left(2 + 2^{-(s-1)/s} A_s^{1/s}\right) \|f\|_{L^2}.$$

Adjusting the value of the constant C, we arrive at (3.15). □

3.4 Random Input Data

We are in particular interested in the input data a and f of the elliptic divergence-form PDE (3.1) being not precisely known. The Lipschitz continuous data-dependence in Proposition 3.1 of the variational solution $u \in V$ of (3.1) will ensure that small variations in the data $(a, f) \in L^\infty_+(D) \times V^*$ imply corresponding small changes in the (unique) solution $u \in V$. A natural paradigm is to model uncertain data probabilistically. To this end, we work with a base probability space $(\Omega, \mathcal{A}, \mathbb{P})$. Given a known right hand side $f \in L^2(D)$, and uncertain diffusion coefficient $a \in E \subseteq L^\infty_+(D)$, where E denotes a suitable subset of $L^\infty_+(D)$ of admissible diffusion coefficients, we model the function a or $\log a$ as RVs taking values in a subset E of $L^\infty(D)$. We will assume the random data a to be separably-valued, more precisely, the set E of admissible random data will almost surely belong to a subset of a separable subspace of $L^\infty(D)$. See [21, Chap. 2.6] for details on separable-valuedness. Separability of E is natural from the point of view of numerical approximation of (samples of) random input a and simplifies many technicalities in the mathematical description; we refer in particular to the construction of GMs on E in Sects. 2.2–2.5. One valid choice for the space of admissible input data E consists in $E = C(\overline{D}) \cap L^\infty_+(D)$. In the log-Gaussian models to be analyzed subsequently, $E \subset L^\infty_+(D)$ will be ensured by modelling $\log(a)$ as a GRF, i.e., we assume the probability measure $\mathbb{P}$ to be such that the law of $\log(a)$ is a GM on $L^\infty(D)$ which charges E, so that the random element $\log(a(\cdot, \omega)) \in L^\infty_+(D)$ $\mathbb{P}$-a.s.. This, in turn, implies with the well-posedness result in Sect. 3.1 that there exists a unique random solution $u(\omega) = \mathcal{S}(a, f) \in V$ $\mathbb{P}$-a.s.. Furthermore, the Lipschitz continuity (3.8) then implies that the corresponding map $\omega \mapsto u(\omega)$ is a composition of the measurable map $\omega \mapsto \log(a(\cdot, \omega))$ with the Lipschitz continuous deterministic data-to-solution map $\mathcal{S}$, hence strongly measurable, and thus a RV on $(\Omega, \mathcal{A}, \mathbb{P})$ taking values in V.

3.5 Parametric Deterministic Coefficient

A key step in the deterministic numerical approximation of the elliptic divergence-form PDE (3.1) with log-Gaussian random inputs (i.e., $\log(a)$ is a GRF on a suitable locally convex space E of admissible input data) is to place a GM on E and to describe the realizations of GRF b in terms of affine-parametric representations discussed in Sect. 2.5. In Sect. 3.5.1, we briefly describe this and in doing so extend a-priori estimates to this resulting deterministic parametric version of elliptic PDE (3.1). Subsequently, in Sect. 3.5.3, we show that the resulting, countably-parametric, linear elliptic problem admits an extension to certain complex parameter domains, while still remaining well-posed.

3.5.1 Deterministic Countably Parametric Elliptic PDEs

Placing a Gaussian probability measure on the random inputs $\log(a)$ to the elliptic divergence-form PDE (3.1) can be achieved via Gaussian series as discussed in Sect. 2.5. Affine-parametric representations which are admissible in the sense of Definition 2.19 of the random input $\log(a)$ of (3.1), subject to a Gaussian law on the corresponding input locally convex space E, render the elliptic divergence-form PDE (3.1) with random inputs a deterministic parametric elliptic PDE. More precisely, $b := \log(a)$ will depend on the sequence $\boldsymbol{y} = (y_j)_{j\in\mathbb{N}}$ of parameters from the parameter space $\mathbb{R}^\infty$. Accordingly, we consider parametric diffusion coefficients $a = a(\boldsymbol{y})$, where

$$\boldsymbol{y} = (y_j)_{j\in\mathbb{N}} \in U.$$

Here and throughout the rest of this book we make use of the notation

$$U := \mathbb{R}^\infty.$$

We develop the holomorphy-based analysis of parametric regularity and Wiener-Hermite PC expansion coefficient sparsity for the model parametric linear second order elliptic divergence-form PDE with so-called "log-affine coefficients"

$$-\operatorname{div}\big(\exp(b(\boldsymbol{y}))\nabla u(\boldsymbol{y})\big) = f \quad \text{in} \quad D, \quad u(\boldsymbol{y})|_{\partial D} = 0, \tag{3.17}$$

i.e.,

$$a(\boldsymbol{y}) = \exp(b(\boldsymbol{y})).$$

Here, the coefficient $b(\boldsymbol{y}) = \log(a(\boldsymbol{y}))$ is assumed to be affine-parametric

$$b(\boldsymbol{y}) = \sum_{j\in\mathbb{N}} y_j\psi_j(\boldsymbol{x}), \quad \boldsymbol{x} \in D, \quad \boldsymbol{y} \in U. \tag{3.18}$$

We assume that $\psi_j \in E \subset L^\infty(D)$ for every $j \in \mathbb{N}$. For any $\boldsymbol{y} \in U$ such that $b(\boldsymbol{y}) \in L^\infty(D)$, by (3.4) we have the estimate

$$\|u(\boldsymbol{y})\|_V \leq \|f\|_{V^*}\|a(\boldsymbol{y})^{-1}\|_{L^\infty} \leq \exp(\|b(\boldsymbol{y})\|_{L^\infty})\|f\|_{V^*}. \tag{3.19}$$

For every $\boldsymbol{y} \in U$ satisfying $b(\boldsymbol{y}) \in L^\infty(D)$, the variational form (3.2) of (3.17) gives rise to the *parametric energy norm* $\|v\|_{a(\boldsymbol{y})}$ on V which is defined by

$$\|v\|_{a(\boldsymbol{y})}^2 := \int_D a(\boldsymbol{y})|\nabla v|^2 \, \mathrm{d}\boldsymbol{x}, \quad v \in V.$$

The norms $\|\circ\|_{a(\boldsymbol{y})}$ and $\|\circ\|_V$ are equivalent on V but not uniformly w.r.t. $\boldsymbol{y}$. It holds

$$\exp(-\|b(\boldsymbol{y})\|_{L^\infty})\|v\|_V^2 \leq \|v\|_{a(\boldsymbol{y})}^2 \leq \exp(\|b(\boldsymbol{y})\|_{L^\infty})\|v\|_V^2, \quad v \in V. \tag{3.20}$$

3.5.2 Probabilistic Setting

In a probabilistic setting, the parameter sequence $\boldsymbol{y}$ is chosen as a sequence of i.i.d. standard Gaussian RVs $\mathcal{N}(0,1)$ and $(\psi_j)_{j\in\mathbb{N}}$ a given sequence of functions in the Banach space $L^\infty(D)$ to which we refer as *representation system* of the uncertain input. We then treat (3.17) as the stochastic linear second order elliptic divergence-form PDE with so-called "log-Gaussian coefficients". We refer to Sect. 2.5 for the construction of GMs based on affine representation systems $(\psi_j)_{j\in\mathbb{N}}$. Due to $L^\infty(D)$ being non-separable, we consider GRFs $b(\boldsymbol{y})$ which take values in separable subspaces $E \subset L^\infty(D)$, such as $E = C^0(\overline{D})$.

The probability space $(\Omega, \mathcal{A}, \mathbb{P})$ from Sect. 3.4 on the parametric solutions $\{u(\boldsymbol{y}) : \boldsymbol{y} \in U\}$ is chosen as $(U, \mathcal{B}(U); \gamma)$. Here and throughout the rest of this book, we make use of the notation: $\mathcal{B}(U)$ is the σ-field on the locally convex space U generated by cylinders of Borel sets on $\mathbb{R}$, and γ is the product measure of the standard GM γ_1 on $\mathbb{R}$ (see the definition in Example 2.17). We shall refer to γ as the *standard GM on* U.

It follows from the a-priori estimate (3.19) that for $f \in V^*$ the parametric elliptic diffusion problem (3.17) admits a unique solution for parameters $\boldsymbol{y}$ in the set

$$U_0 := \{\boldsymbol{y} \in U : b(\boldsymbol{y}) \in L^\infty(D)\}. \tag{3.21}$$

The measure $\gamma(U_0)$ of the set $U_0 \subset U$ depends on the structure of $\boldsymbol{y} \mapsto b(\boldsymbol{y})$. The following sufficient condition on the representation system $(\psi_j)_{j\in\mathbb{N}}$ will be assumed throughout.

Assumption 3.6 *For every $j \in \mathbb{N}$, $\psi_j \in L^\infty(D)$, and there exists a positive sequence $(\lambda_j)_{j\in\mathbb{N}}$ such that $\left(\exp(-\lambda_j^2)\right)_{j\in\mathbb{N}} \in \ell^1(\mathbb{N})$ and the series $\sum_{j\in\mathbb{N}} \lambda_j|\psi_j|$ converges in $L^\infty(D)$.*

For the statement of the next result, we recall a notion of Bochner spaces. For a measure space $(\Omega, \mathcal{A}, \mu)$ let X a Banach space and $1 \leq p < \infty$. Then the Bochner space $L^p(\Omega, X; \mu)$ is defined as the space of all strongly μ-measurable mappings u from Ω to X such that the norm

$$\|u\|_{L^p(\Omega,X;\mu)} := \left(\int_\Omega \|u(y)\|_X^p \,\mathrm{d}\mu(y)\right)^{1/p} < \infty. \tag{3.22}$$

In particular, when $(\Omega, \mathcal{A}, \mu) = (U, \mathcal{B}(U); \gamma)$, X is separable and $p = 2$, the hilbertian space $L^2(U, X; \gamma)$ is one of the most important for the problems considered in this book.

The following result was shown in [9, Theorem 2.2].

Proposition 3.7 *Under Assumption 3.6, the set U_0 has full GM, i.e., $\gamma(U_0) = 1$. For all $k \in \mathbb{N}$ there holds, with $\mathbb{E}(\cdot)$ denoting expectation with respect to γ,*

$$\mathbb{E}\left(\exp(k\|b(\cdot)\|_{L^\infty})\right) < \infty.$$

The solution family $\{u(\boldsymbol{y}) : \boldsymbol{y} \in U_0\}$ of the parametric elliptic boundary value problem (3.17) *is in $L^k(U, V; \gamma)$ for every finite $k \in \mathbb{N}$.*

3.5.3 Deterministic Complex-Parametric Elliptic PDEs

Towards the aim of establishing sparsity of Wiener-Hermite PC expansions of the parametric solutions $\{u(\boldsymbol{y}) : \boldsymbol{y} \in U_0\}$ of (3.17), we extend the deterministic parametric elliptic problem (3.17) from real-valued to complex-valued parameters.

Formally, replacing $\boldsymbol{y} = (y_j)_{j\in\mathbb{N}} \in U$ in the coefficient $a(\boldsymbol{y})$ by $\boldsymbol{z} = (z_j)_{j\in\mathbb{N}} = (y_j + \mathrm{i}\xi_j)_{j\in\mathbb{N}} \in \mathbb{C}^\infty$, the real part of $a(\boldsymbol{z})$ is

$$\Re[a(\boldsymbol{z})] = \exp\left(\sum_{j\in\mathbb{N}} y_j\psi_j(\boldsymbol{x})\right)\cos\left(\sum_{j\in\mathbb{N}} \xi_j\psi_j(\boldsymbol{x})\right). \tag{3.23}$$

We find that $\Re[a(\boldsymbol{z})] > 0$ if

$$\left\|\sum_{j\in\mathbb{N}} \xi_j\psi_j\right\|_{L^\infty} < \frac{\pi}{2}.$$

This observation and Proposition 3.7 motivate the study of the analytic continuation of the solution map $\boldsymbol{y} \mapsto u(\boldsymbol{y})$ to $\boldsymbol{z} \mapsto u(\boldsymbol{z})$ for complex parameters $\boldsymbol{z} = (z_j)_{j\in\mathbb{N}}$

by formally replacing the parameter y_j by z_j in the definition of the parametric coefficient a, where each z_j lies in the strip

$$\mathcal{S}_j(\boldsymbol{\rho}) := \{z_j \in \mathbb{C} : |\mathfrak{Im} z_j| < \rho_j\} \tag{3.24}$$

and where $\rho_j > 0$ and $\boldsymbol{\rho} = (\rho_j)_{j\in\mathbb{N}} \in (0,\infty)^\infty$ is any sequence of positive numbers such that

$$\left\| \sum_{j\in\mathbb{N}} \rho_j |\psi_j| \right\|_{L^\infty} < \frac{\pi}{2}.$$

3.6 Analyticity and Sparsity

We address the analyticity (holomorphy) of the parametric solutions $\{u(\boldsymbol{y}) : \boldsymbol{y} \in U_0\}$. We analyze the sparsity by estimating, in particular, the size of the domains of holomorphy to which the parametric solutions can be extended. We also treat the weighted ℓ^2-summability and ℓ^p-summability (sparsity) for the series of Wiener-Hermite the PC expansion coefficients $(u_{\boldsymbol{\nu}})_{\boldsymbol{\nu}\in\mathcal{F}}$ of $u(\boldsymbol{y})$.

3.6.1 Parametric Holomorphy

In this section we establish holomorphic parametric dependence u on a and on f as in [39] by verifying complex differentiability of a suitable complex-parametric extension of $\boldsymbol{y} \mapsto u(\boldsymbol{y})$. We observe that the Lax-Milgram theory can be extended to the case where the coefficient function a is complex-valued. In this case, $V := H_0^1(D,\mathbb{C})$ in (3.2) and the ellipticity assumption (3.3) is extended to the complex domain as

$$0 < \rho(a) := \operatorname*{ess\,inf}_{\boldsymbol{x}\in D} \Re(a(\boldsymbol{x})) \le |a(\boldsymbol{x})| \le \|a\|_{L^\infty} < \infty, \qquad \boldsymbol{x} \in D. \tag{3.25}$$

Under this condition, there exists a unique variational solution $u \in V$ of (3.1) and for this solution, the estimate (3.4) remains valid, i.e.,

$$\|u\|_V \le \frac{\|f\|_{V^*}}{\rho(a)}. \tag{3.26}$$

Let $\boldsymbol{\rho} = (\rho_j)_{j\in\mathbb{N}} \in [0,\infty)^\infty$ be a sequence of non-negative numbers and assume that $\mathfrak{u} \subseteq \operatorname{supp}(\boldsymbol{\rho})$ is finite. Define

$$\mathcal{S}_{\mathfrak{u}}(\boldsymbol{\rho}) := \bigtimes_{j\in\mathfrak{u}} \mathcal{S}_j(\boldsymbol{\rho}), \tag{3.27}$$

where the strip $\mathcal{S}_j(\boldsymbol{\rho})$ is given in (3.24). For $\boldsymbol{y} \in U$, put

$$\mathcal{S}_{\mathrm{u}}(\boldsymbol{y}, \boldsymbol{\rho}) := \big\{(z_j)_{j\in\mathbb{N}} : z_j \in \mathcal{S}_j(\boldsymbol{\rho}) \text{ if } j \in \mathrm{u} \text{ and } z_j = y_j \text{ if } j \notin \mathrm{u}\big\}.$$

Proposition 3.8 *Let the sequence $\boldsymbol{\rho} = (\rho_j)_{j\in\mathbb{N}} \in [0, \infty)^\infty$ satisfy*

$$\Big\| \sum_{j\in\mathbb{N}} \rho_j |\psi_j| \Big\|_{L^\infty} \le \kappa < \frac{\pi}{2}. \tag{3.28}$$

Let $\boldsymbol{y}_0 = (y_{0,1}, y_{0,2}, \ldots) \in U$ be such that $b(\boldsymbol{y}_0)$ belongs to $L^\infty(D)$, and let $\mathrm{u} \subseteq$ $\mathrm{supp}(\boldsymbol{\rho})$ be a finite set.

Then the solution u of the variational form of (3.17) *is holomorphic on $\mathcal{S}_{\mathrm{u}}(\boldsymbol{\rho})$ as a function of the parameters $\boldsymbol{z}_{\mathrm{u}} = (z_j)_{j\in\mathbb{N}} \in \mathcal{S}_{\mathrm{u}}(\boldsymbol{y}_0, \boldsymbol{\rho})$ taking values in V with $z_j = y_{0,j}$ for $j \notin \mathrm{u}$ held fixed.*

Proof Let $N \in \mathbb{N}$. We denote

$$\mathcal{S}_{\mathrm{u},N}(\boldsymbol{\rho}) := \big\{(y_j + \mathrm{i}\xi_j)_{j\in\mathrm{u}} \in \mathcal{S}_{\mathrm{u}}(\boldsymbol{\rho}) : |y_j - y_{0,j}| < N\big\}. \tag{3.29}$$

For $\boldsymbol{z}_{\mathrm{u}} = (y_j + \mathrm{i}\xi_j)_{j\in\mathbb{N}} \in \mathcal{S}_{\mathrm{u}}(\boldsymbol{y}_0, \boldsymbol{\rho})$ with $(y_j + \mathrm{i}\xi_j)_{j\in\mathrm{u}} \in \mathcal{S}_{\mathrm{u},N}(\boldsymbol{\rho})$ we have

$$\begin{aligned}\Big\| \sum_{j\in\mathbb{N}} y_j\psi_j \Big\|_{L^\infty} &\le \|b(\boldsymbol{y}_0)\|_{L^\infty} + \Big\| \sum_{j\in\mathrm{u}} |(y - y_{0,j})\psi_j| \Big\|_{L^\infty} \\ &\le \|b(\boldsymbol{y}_0)\|_{L^\infty} + N \Big\| \sum_{j\in\mathrm{u}} |\psi_j| \Big\|_{L^\infty} =: M < \infty\end{aligned}$$

and

$$\Big\| \sum_{j\in\mathrm{u}} \xi_j\psi_j \Big\|_{L^\infty} \le \Big\| \sum_{j\in\mathrm{u}} |\rho_j\psi_j| \Big\|_{L^\infty} \le \kappa.$$

Consequently, we obtain from (3.23)

$$\rho(a(\boldsymbol{z}_{\mathrm{u}})) \ge \exp\Big(- \Big\| \sum_{j\in\mathbb{N}} y_j\psi_j \Big\|_{L^\infty} \Big) \cos\Big(\Big\| \sum_{j\in\mathrm{u}} \xi_j\psi_j \Big\|_{L^\infty} \Big) \ge \exp(-M)\cos\kappa \tag{3.30}$$

for all $\boldsymbol{z}_{\mathrm{u}} \in \mathcal{S}_{\mathrm{u}}(\boldsymbol{y}_0, \boldsymbol{\rho})$ with $(y_j + \mathrm{i}\xi_j)_{j\in\mathrm{u}} \in \mathcal{S}_{\mathrm{u},N}(\boldsymbol{\rho})$. From this and the analyticity of exponential functions we conclude that the map $\boldsymbol{z}_{\mathrm{u}} \to u(\boldsymbol{z}_{\mathrm{u}})$ is holomorphic on the set $\mathcal{S}_{\mathrm{u},N}(\boldsymbol{\rho})$, see [37, Pages 22, 23]. Since N is arbitrary we deduce that the map $\boldsymbol{z}_{\mathrm{u}} \to u(\boldsymbol{z}_{\mathrm{u}})$ is holomorphic on $\mathcal{S}_{\mathrm{u}}(\boldsymbol{\rho})$. □

The analytic continuation of the parametric solutions $\{u(\boldsymbol{y}) : \boldsymbol{y} \in U\}$ to $\mathcal{S}_{\mathfrak{u}}(\boldsymbol{\rho})$ leads to a result on parametric V-regularity.

Lemma 3.9 *Let $\boldsymbol{\rho} = (\rho_j)_{j\in\mathbb{N}}$ be a non-negative sequence satisfying* (3.28). *Let $\boldsymbol{y} \in U$ with $b(\boldsymbol{y}) \in L^\infty(D)$ and $\boldsymbol{\nu} \in \mathcal{F}$ such that* $\operatorname{supp}(\boldsymbol{\nu}) \subseteq \operatorname{supp}(\boldsymbol{\rho})$. *Then we have*

$$\|\partial^{\boldsymbol{\nu}} u(\boldsymbol{y})\|_V \le C_0 \frac{\boldsymbol{\nu}!}{\boldsymbol{\rho}^{\boldsymbol{\nu}}} \exp\big(\|b(\boldsymbol{y})\|_{L^\infty}\big),$$

where $C_0 = e^\kappa (\cos\kappa)^{-1} \|f\|_{V^*}$.

Proof Let $\boldsymbol{\nu} \in \mathcal{F}$ such that $\operatorname{supp}(\boldsymbol{\nu}) \subseteq \operatorname{supp}(\boldsymbol{\rho})$. Denote $\mathfrak{u} = \operatorname{supp}(\boldsymbol{\nu})$. For fixed variable y_j with $j \notin \mathfrak{u}$, the map $\mathcal{S}_{\mathfrak{u}}(\boldsymbol{y}, \boldsymbol{\rho}) \ni \boldsymbol{z}_{\mathfrak{u}} \to u(\boldsymbol{z}_{\mathfrak{u}})$ is holomorphic on the domain $\mathcal{S}_{\mathfrak{u}}(\boldsymbol{y}, \kappa'\boldsymbol{\rho})$ where $\kappa' > 1$ is sufficiently small such that $\kappa < \kappa\kappa' < \pi/2$, see Proposition 3.8. Applying Cauchy's integral formula gives

$$\partial^{\boldsymbol{\nu}} u(\boldsymbol{y}) = \frac{\boldsymbol{\nu}!}{(2\pi i)^{|\mathfrak{u}|}} \int_{\mathcal{C}_{\boldsymbol{y},\mathfrak{u}}(\boldsymbol{\rho})} \frac{u(\boldsymbol{z}_{\mathfrak{u}})}{\prod_{j\in\mathfrak{u}} (z_j - y_j)^{\nu_j+1}} \prod_{j\in\mathfrak{u}} \mathrm{d}z_j,$$

where

$$\mathcal{C}_{\boldsymbol{y},\mathfrak{u}}(\boldsymbol{\rho}) := \bigtimes_{j\in\mathfrak{u}} \mathcal{C}_{\boldsymbol{y},j}(\boldsymbol{\rho}), \qquad \mathcal{C}_{\boldsymbol{y},j}(\boldsymbol{\rho}) := \big\{z_j \in \mathbb{C} : |z_j - y_j| = \rho_j\big\}. \tag{3.31}$$

This leads to

$$\|\partial^{\boldsymbol{\nu}} u(\boldsymbol{y})\|_V \le \frac{\boldsymbol{\nu}!}{\boldsymbol{\rho}^{\boldsymbol{\nu}}} \sup_{\boldsymbol{z}_{\mathfrak{u}} \in \mathcal{C}_{\mathfrak{u}}(\boldsymbol{y},\boldsymbol{\rho})} \|u(\boldsymbol{z}_{\mathfrak{u}})\|_V \tag{3.32}$$

with

$$\mathcal{C}_{\mathfrak{u}}(\boldsymbol{y}, \boldsymbol{\rho}) = \big\{(z_j)_{j\in\mathbb{N}} \in \mathcal{S}_{\mathfrak{u}}(\boldsymbol{y}, \boldsymbol{\rho}) : \ (z_j)_{j\in\mathfrak{u}} \in \mathcal{C}_{\boldsymbol{y},\mathfrak{u}}(\boldsymbol{\rho})\big\}. \tag{3.33}$$

Notice that for $\boldsymbol{z}_{\mathfrak{u}} = (z_j)_{j\in\mathbb{N}} \in \mathcal{C}_{\mathfrak{u}}(\boldsymbol{y}, \boldsymbol{\rho})$ we can write $z_j = y_j + \eta_j + \mathrm{i}\xi_j \in \mathcal{C}_{\boldsymbol{y},j}(\boldsymbol{\rho})$ with $|\eta_j| \le \rho_j$, $|\xi_j| \le \rho_j$ if $j \in \mathfrak{u}$ and $\eta_j = \xi_j = 0$ if $j \notin \mathfrak{u}$. By denoting $\boldsymbol{\eta} = (\eta_j)_{j\in\mathbb{N}}$ and $\boldsymbol{\xi} = (\xi_j)_{j\in\mathbb{N}}$ we see that $\|b(\boldsymbol{\eta})\|_{L^\infty} \le \kappa$ and $\|b(\boldsymbol{\xi})\|_{L^\infty} \le \kappa$. Hence we deduce from (3.26) that

$$\|u(\boldsymbol{z}_{\mathfrak{u}})\|_V \le \frac{\exp\big(\|b(\boldsymbol{y}+\boldsymbol{\eta})\|_{L^\infty}\big)}{\cos\big(\|b(\boldsymbol{\xi})\|_{L^\infty}\big)} \|f\|_{V^*} \le \frac{\exp\big(\kappa + \|b(\boldsymbol{y})\|_{L^\infty}\big)}{\cos\kappa} \|f\|_{V^*}.$$

Inserting this into (3.32) we obtain the desired estimate. □

3.6.2 Sparsity of Wiener-Hermite PC Expansion Coefficients

In this section, we will exploit the analyticity of u to prove a weighted ℓ^2-summability result for the V-norms of the coefficients in the Wiener-Hermite PC expansion of the solution map $\boldsymbol{y} \to u(\boldsymbol{y})$. Our analysis yields the same ℓ^p-summability result as in the papers [8, 9] in the case ψ_j have arbitrary supports. In this case, our result implies that the ℓ^p-summability of $(\|u_{\boldsymbol{\nu}}\|_V)_{\boldsymbol{\nu}\in\mathcal{F}}$ for $0 < p \le 1$ (the sparsity of parametric solutions) follows from the ℓ^p-summability of the sequence $(j^\alpha \|\psi_j\|_{L^\infty})_{j\in\mathbb{N}}$ for some $\alpha > 1/2$ which is an improvement over the condition $(j\|\psi_j\|_{L^\infty})_{j\in\mathbb{N}} \in \ell^p(\mathbb{N})$ in [71], see [9, Section 6.3]. In the case of disjoint or finitely overlapping supports our analysis obtains a weaker result compared to [8, 9]. As observed in [39], one advantage of establishing sparsity of Wiener-Hermite PC expansion coefficients via holomorphy rather than by successive differentiation is that it allows to derive, in a unified way, summability bounds for the coefficients of Wiener-Hermite PC expansion whose size is measured in scales of Sobolev and Besov spaces in the domain D. Using real-variable arguments as, e.g., in [8, 9], establishing sparsity of parametric solutions in Besov spaces in D of higher smoothness seems to require more involved technical and notational developments, according to [8, Comment on Page 2157].

The parametric solution $\{u(\boldsymbol{y}) : \boldsymbol{y} \in U\}$ of (3.17) belongs to the space $L^2(U, V; \gamma)$ or more generally, $L^2(U, (H^{1+s} \cap H_0^1)(D); \gamma)$ for s-order of extra differentiability provided by higher data regularity. We recall from Sect. 2.1.3 the normalized probabilistic Hermite polynomials $(H_k)_{k\in\mathbb{N}_0}$. Every $u \in L^2(U, X; \gamma)$ admits the *Wiener-Hermite PC expansion*

$$\sum_{\boldsymbol{\nu}\in\mathcal{F}} u_{\boldsymbol{\nu}} H_{\boldsymbol{\nu}}(\boldsymbol{y}), \tag{3.34}$$

where for $\boldsymbol{\nu} \in \mathcal{F}$,

$$H_{\boldsymbol{\nu}}(\boldsymbol{y}) = \prod_{j\in\mathbb{N}} H_{\nu_j}(y_j),$$

and

$$u_{\boldsymbol{\nu}} := \int_U u(\boldsymbol{y})\, H_{\boldsymbol{\nu}}(\boldsymbol{y})\, \mathrm{d}\gamma(\boldsymbol{y})$$

are called *Wiener-Hermite PC expansion coefficients*. Notice that $(H_{\boldsymbol{\nu}})_{\boldsymbol{\nu}\in\mathcal{F}}$ forms an ONB of $L^2(U; \gamma)$.

For every $u \in L^2(U, X; \gamma)$, there holds the Parseval-type identity

$$\|u\|^2_{L^2(U,X;\gamma)} = \sum_{\boldsymbol{\nu}\in\mathcal{F}} \|u_{\boldsymbol{\nu}}\|^2_X, \quad u \in L^2(U, X; \gamma). \tag{3.35}$$

The error of approximation of the parametric solution $\{u(\boldsymbol{y}) : \boldsymbol{y} \in U\}$ of (3.17) will be measured in the Bochner space $L^2(U, V; \gamma)$. A basic role in this approximation is taken by the Wiener-Hermite PC expansion (3.34) of u in the space $L^2(U, V; \gamma)$.

For a finite set $\Lambda \subset \mathcal{F}$, we denote by $u_\Lambda = \sum_{\boldsymbol{\nu}\in\Lambda} u_{\boldsymbol{\nu}} H_{\boldsymbol{\nu}}$ the corresponding partial sum of the Wiener-Hermite PC expansion (3.34). It follows from (3.35) that

$$\|u - u_\Lambda\|^2_{L^2(U,V;\gamma)} = \sum_{\boldsymbol{\nu}\in\mathcal{F}\setminus\Lambda} \|u_{\boldsymbol{\nu}}\|^2_V.$$

Therefore, summability results of the coefficients $(\|u_{\boldsymbol{\nu}}\|_V)_{\boldsymbol{\nu}\in\mathcal{F}}$ imply convergence rate estimates of finitely truncated expansions u_{Λ_n} for suitable sequences $(\Lambda_n)_{n\in\mathbb{N}}$ of sets of n indices $\boldsymbol{\nu}$ (see [9, 43, 71]). We next recapitulate some weighted summability results for Wiener-Hermite expansions.

For $r \in \mathbb{N}$ and a sequence of nonnegative numbers $\boldsymbol{\varrho} = (\varrho_j)_{j\in\mathbb{N}}$, we define the *Wiener-Hermite weights*

$$\beta_{\boldsymbol{\nu}}(r, \boldsymbol{\varrho}) := \sum_{\|\boldsymbol{\nu}'\|_{\ell^\infty}\le r} \binom{\boldsymbol{\nu}}{\boldsymbol{\nu}'} \boldsymbol{\varrho}^{2\boldsymbol{\nu}'} = \prod_{j\in\mathbb{N}} \left(\sum_{\ell=0}^{r} \binom{\nu_j}{\ell} \varrho_j^{2\ell} \right), \quad \boldsymbol{\nu} \in \mathcal{F}. \tag{3.36}$$

The following identity was proved in [9, Theorem 3.3]. For convenience to the reader, we present the proof from that paper.

Lemma 3.10 *Let Assumption 3.6 hold. Let $r \in \mathbb{N}$ and $\boldsymbol{\varrho} = (\varrho_j)_{j\in\mathbb{N}}$ be a sequence of nonnegative numbers. Then*

$$\sum_{\boldsymbol{\nu}\in\mathcal{F}} \beta_{\boldsymbol{\nu}}(r, \boldsymbol{\varrho}) \|u_{\boldsymbol{\nu}}\|^2_V = \sum_{\|\boldsymbol{\nu}\|_{\ell^\infty}\le r} \frac{\boldsymbol{\varrho}^{2\boldsymbol{\nu}}}{\boldsymbol{\nu}!} \int_U \|\partial^{\boldsymbol{\nu}} u(\boldsymbol{y})\|^2_V \, d\gamma(\boldsymbol{y}). \tag{3.37}$$

Proof Recall that $p(y) := p(y, 0, 1) = -\frac{1}{\sqrt{2\pi}} \exp(-y^2/2)$ is the density function of the standard GM on $\mathbb{R}$. Let $\mu \in \mathbb{N}$. For a sufficiently smooth, univariate function $v \in L^2(\mathbb{R}; \gamma)$, from $H_\nu(y) = \frac{(-1)^\nu}{\sqrt{\nu!}} \frac{p^{(\nu)}(y)}{p(y)}$ we have for $\nu \ge \mu$

$$v_\nu := \int_{\mathbb{R}} v(y) H_\nu(y) p(y) \, dy = \frac{(-1)^\nu}{\sqrt{\nu!}} \int_{\mathbb{R}} v(y) p^{(\nu)}(y) \, dy$$

$$= \frac{(-1)^{\nu-\mu}}{\sqrt{\nu!}} \int_{\mathbb{R}} v^{(\mu)}(y) p^{(\nu-\mu)}(y) \, dy = \sqrt{\frac{(\nu-\mu)!}{\nu!}} \int_{\mathbb{R}} v^{(\mu)}(y) H_{\nu-\mu}(y) p(y) \, dy.$$

Hence

$$\sqrt{\frac{\nu!}{\mu!(\nu-\mu)!}} v_\nu = \sqrt{\frac{1}{\mu!}} \int_{\mathbb{R}} v^{(\mu)}(y) H_{\nu-\mu}(y) \, d\gamma(y).$$

By Parseval's identity, we have

$$\frac{1}{\mu!}\int_{\mathbb{R}} |v^{(\mu)}(y)|^2\,\mathrm{d}\gamma(y) = \sum_{\nu\geq\mu} \frac{\nu!}{\mu!(\nu-\mu)!}|v_\nu|^2 = \sum_{\nu\in\mathbb{N}_0}\binom{\nu}{\mu}|v_\nu|^2,$$

where we use the convention $\binom{\nu}{\mu} = 0$ if $\mu > \nu$.

For multi-indices and for $u \in L^2(U, V; \gamma)$, if $\boldsymbol{\mu} \leq \boldsymbol{\nu}$, applying the above argument in coordinate-wise for the coefficients

$$u_{\boldsymbol{\nu}} = \sqrt{\frac{(\boldsymbol{\nu}-\boldsymbol{\mu})!}{\boldsymbol{\nu}!}}\int_U \partial^{\boldsymbol{\mu}} u(\boldsymbol{y}) H_{\boldsymbol{\nu}-\boldsymbol{\mu}}(\boldsymbol{y})\,\mathrm{d}\gamma(\boldsymbol{y})$$

we get

$$\frac{1}{\boldsymbol{\mu}!}\int_U \|\partial^{\boldsymbol{\mu}} u(\boldsymbol{y})\|_V^2\,\mathrm{d}\gamma(\boldsymbol{y}) = \sum_{\boldsymbol{\nu}\geq\boldsymbol{\mu}} \frac{\boldsymbol{\nu}!}{(\boldsymbol{\mu}!(\boldsymbol{\nu}-\boldsymbol{\mu})!)}\|u_{\boldsymbol{\nu}}\|_V^2 = \sum_{\boldsymbol{\nu}\in\mathcal{F}}\binom{\boldsymbol{\nu}}{\boldsymbol{\mu}}\|u_{\boldsymbol{\nu}}\|_V^2.$$

Multiplying both sides by $\boldsymbol{\varrho}^{2\boldsymbol{\mu}}$ and summing over $\boldsymbol{\mu}$ with $\|\boldsymbol{\mu}\|_{\ell^\infty} \leq r$, we obtain

$$\begin{aligned}\sum_{\|\boldsymbol{\mu}\|_{\ell^\infty}\leq r} \frac{\boldsymbol{\varrho}^{2\boldsymbol{\mu}}}{\boldsymbol{\mu}!}\int_U \|\partial^{\boldsymbol{\mu}} u(\boldsymbol{y})\|_V^2\,\mathrm{d}\gamma(\boldsymbol{y}) &= \sum_{\|\boldsymbol{\mu}\|_{\ell^\infty}\leq r}\sum_{\boldsymbol{\nu}\in\mathcal{F}}\binom{\boldsymbol{\nu}}{\boldsymbol{\mu}}\boldsymbol{\varrho}^{2\boldsymbol{\mu}}\|u_{\boldsymbol{\nu}}\|_V^2 \\ &= \sum_{\boldsymbol{\nu}\in\mathcal{F}} \beta_{\boldsymbol{\nu}}(r,\boldsymbol{\varrho})\|u_{\boldsymbol{\nu}}\|_V^2.\end{aligned}$$

□

We recall a summability property of the sequence $(\beta_{\boldsymbol{\nu}}(r,\boldsymbol{\varrho})^{-1})_{j\in\mathbb{N}}$ and its proof, given in [9, Lemma 5.1].

Lemma 3.11 *Let* $0 < p < \infty$ *and* $q := \frac{2p}{2-p}$. *Let* $\boldsymbol{\varrho} = (\varrho_j)_{j\in\mathbb{N}} \in [0,\infty)^\infty$ *be a sequence of positive numbers such that*

$$(\varrho_j^{-1})_{j\in\mathbb{N}} \in \ell^q(\mathbb{N}).$$

Then for any $r \in \mathbb{N}$ *such that* $\frac{2}{r+1} < p$, *the family* $(\beta_{\boldsymbol{\nu}}(r,\boldsymbol{\varrho}))_{\boldsymbol{\nu}\in\mathcal{F}}$ *defined in* (3.36) *for this* r *satisfies*

$$\sum_{\boldsymbol{\nu}\in\mathcal{F}} \beta_{\boldsymbol{\nu}}(r,\boldsymbol{\varrho})^{-q/2} < \infty. \tag{3.38}$$

Proof First we have the decomposition

$$\sum_{\boldsymbol{\nu}\in\mathcal{F}} b_{\boldsymbol{\nu}}(r,\boldsymbol{\varrho})^{-q/2} = \sum_{\boldsymbol{\nu}\in\mathcal{F}} \prod_{j\in\mathbb{N}} \left(\sum_{\ell=0}^{r} \binom{\nu_j}{\ell} \varrho_j^{2\ell}\right)^{-q/2} = \prod_{j\in\mathbb{N}} \sum_{n\in\mathbb{N}_0} \left(\sum_{\ell=0}^{r} \binom{n}{\ell} \varrho_j^{2\ell}\right)^{-q/2}.$$

For each $j \in \mathbb{N}$ we have

$$\begin{aligned}\sum_{n\in\mathbb{N}_0} \left(\sum_{\ell=0}^{r} \binom{n}{\ell} \varrho_j^{2\ell}\right)^{-q/2} &\le \sum_{n\in\mathbb{N}_0} \left[\binom{n}{\min\{n,r\}} \varrho_j^{2\min\{n,r\}}\right]^{-q/2} \\ &= \sum_{n=0}^{r-1} \varrho_j^{-nq} + C_{r,q}\varrho_j^{-rq},\end{aligned} \tag{3.39}$$

where

$$C_{r,q} := \sum_{n=r}^{+\infty} \binom{n}{r}^{-q/2} = (r!)^{q/2} \sum_{n\in\mathbb{N}_0} \left[(n+1)\dots(n+r)\right]^{-q/2}.$$

Since $\lim_{n\to+\infty} \frac{(n+1)\dots(n+r)}{n^r} = 1$, we find that $C_{r,q}$ is finite if and only if $q > 2/r$. This is equivalent to $\frac{2}{r+1} < p$. From the assumption $(\varrho_j^{-1})_{j\in\mathbb{N}} \in \ell^q(\mathbb{N})$ we find some $J > 1$ such that $\varrho_j > 1$ for all $j > J$. This implies $\varrho_j^{-nq} \le \varrho_j^{-q}$ for $n = 1, \dots, r$ and $j > J$. Therefore, one can bound the right side of (3.39) by $1+(C_{r,q}+r-1)\varrho_j^{-q}$. Hence we obtain

$$\begin{aligned}\sum_{\boldsymbol{\nu}\in\mathcal{F}} b_{\boldsymbol{\nu}}(r,\boldsymbol{\varrho})^{-q/2} &\le C \prod_{j>J} \left[1 + (C_{r,q} + r - 1)\varrho_j^{-q}\right] \\ &\le C \prod_{j>J} \exp\left((C_{r,q} + r - 1)\varrho_j^{-q}\right) \\ &\le C \exp\left((C_{r,q} + r - 1)\|(\varrho_j^{-1})_{j\in\mathbb{N}}\|_{\ell^q}^q\right)\end{aligned}$$

which is finite since $(\varrho_j^{-1})_{j\in\mathbb{N}} \in \ell^q(\mathbb{N})$. □

In what follows, we denote by $(\boldsymbol{e}_j)_{j\in\mathbb{N}}$ the standard basis of $\ell^2(\mathbb{N})$, i.e., $\boldsymbol{e}_j = (e_{j,i})_{i\in\mathbb{N}}$ with $e_{j,i} = 1$ for $i = j$ and $e_{j,i} = 0$ for $i \ne j$. The following lemma was obtained in [38, Lemma 7.1, Theorem 7.2] and [37, Lemma 3.17].

Lemma 3.12 *Let $\boldsymbol{\alpha} = (\alpha_j)_{j\in\mathbb{N}}$ be a sequence of nonnegative numbers. Then we have the following.*

(i) *For $0 < p < \infty$, the family $(\boldsymbol{\alpha}^{\boldsymbol{\nu}})_{\boldsymbol{\nu}\in\mathcal{F}}$ belongs to $\ell^p(\mathcal{F})$ if and only if $\|\boldsymbol{\alpha}\|_{\ell^p} < \infty$ and $\|\boldsymbol{\alpha}\|_{\ell^\infty} < 1$.*

(ii) *For* $0 < p \leq 1$, *the family* $(\boldsymbol{\alpha}^{\boldsymbol{\nu}}|\boldsymbol{\nu}|!/\boldsymbol{\nu}!)_{\boldsymbol{\nu}\in\mathcal{F}}$ *belongs to* $\ell^p(\mathcal{F})$ *if and only if* $\|\boldsymbol{\alpha}\|_{\ell^p} < \infty$ *and* $\|\boldsymbol{\alpha}\|_{\ell^1} < 1$.

Proof

Step 1. We prove the first statement. Assume that $\|\boldsymbol{\alpha}\|_{\ell^\infty} < 1$. Then we have

$$\sum_{\boldsymbol{\nu}\in\mathcal{F}} \boldsymbol{\alpha}^{\boldsymbol{\nu}p} = \prod_{j\in\mathbb{N}}\sum_{n\in\mathbb{N}_0} \alpha_j^{pn} = \prod_{j\in\mathbb{N}} \frac{1}{1-\alpha_j^p}$$
$$= \prod_{j\in\mathbb{N}}\left(1+\frac{\alpha_j^p}{1-\alpha_j^p}\right) \leq \prod_{j\in\mathbb{N}} \exp\left(\frac{\alpha_j^p}{1-\alpha_j^p}\right)$$
$$\leq \prod_{j\in\mathbb{N}} \exp\left(\frac{1}{1-\|\boldsymbol{\alpha}\|_{\ell^\infty}^p}\alpha_j^p\right) = \exp\left(\frac{1}{1-\|\boldsymbol{\alpha}\|_{\ell^\infty}^p}\|\boldsymbol{\alpha}\|_{\ell^p}^p\right),$$

where in the last equality we have used $(\boldsymbol{\alpha}^{\boldsymbol{\nu}})_{\boldsymbol{\nu}\in\mathcal{F}} \in \ell^p(\mathcal{F})$.
Since the sequence $(\alpha_j)_{j\in\mathbb{N}} = (\boldsymbol{\alpha}^{\boldsymbol{e}_j})_{j\in\mathbb{N}}$ is a subsequence of $(\alpha^{\boldsymbol{\nu}})_{\boldsymbol{\nu}\in\mathcal{F}}$, $(\boldsymbol{\alpha}^{\boldsymbol{\nu}})_{\boldsymbol{\nu}\in\mathcal{F}} \in \ell^p(\mathcal{F})$ implies that $\boldsymbol{\alpha}$ belongs to $\ell^p(\mathbb{N})$. Moreover we have for any $j \geq 1$

$$\sum_{n\in\mathbb{N}_0} \alpha_j^{np} = \sum_{n\in\mathbb{N}_0} \boldsymbol{\alpha}^{n\boldsymbol{e}_j p} \leq \sum_{\boldsymbol{\nu}\in\mathcal{F}} \boldsymbol{\alpha}^{\boldsymbol{\nu}p} < \infty.$$

From this we have $\alpha_j^p < 1$ which implies $\alpha_j < 1$ for all $j \in \mathbb{N}$. Since $\boldsymbol{\alpha} \in \ell^p(\mathbb{N})$ it is easily seen that $\|\boldsymbol{\alpha}\|_{\ell^\infty} < 1$.
Step 2. We prove the second statement. We observe that

$$\sum_{\boldsymbol{\nu}\in\mathcal{F}} \frac{|\boldsymbol{\nu}|!}{\boldsymbol{\nu}!}\boldsymbol{\alpha}^{\boldsymbol{\nu}} = \sum_{k\in\mathbb{N}_0}\sum_{|\boldsymbol{\nu}|=k} \frac{|\boldsymbol{\nu}|!}{\boldsymbol{\nu}!}\boldsymbol{\alpha}^{\boldsymbol{\nu}} = \sum_{k\in\mathbb{N}_0}\left(\sum_{j\in\mathbb{N}}\alpha_j\right)^k.$$

From this we deduce that $(\boldsymbol{\alpha}^{\boldsymbol{\nu}}|\boldsymbol{\nu}|!/\boldsymbol{\nu}!)_{\boldsymbol{\nu}\in\mathcal{F}}$ belongs to $\ell^1(\mathcal{F})$ if and only if $\boldsymbol{\alpha} \in \ell^1(\mathbb{N})$ and $\|\boldsymbol{\alpha}\|_{\ell^1} < 1$.
Suppose that

$$(\boldsymbol{\alpha}^{\boldsymbol{\nu}}|\boldsymbol{\nu}|!/\boldsymbol{\nu}!)_{\boldsymbol{\nu}\in\mathcal{F}} \in \ell^p(\mathcal{F})$$

for some $p \in (0,1)$. As in Step 1, the sequence $(\alpha_j)_{j\in\mathbb{N}} = (\boldsymbol{\alpha}^{\boldsymbol{e}_j})_{j\in\mathbb{N}}$ and $(\alpha_j^n)_{n\in\mathbb{N}_0} = (\boldsymbol{\alpha}^{n\boldsymbol{e}_j})_{n\in\mathbb{N}_0}$ are subsequences of $(\boldsymbol{\alpha}^{\boldsymbol{\nu}})_{\boldsymbol{\nu}\in\mathcal{F}}$. Therefore $(\boldsymbol{\alpha}^{\boldsymbol{\nu}}|\boldsymbol{\nu}|!/\boldsymbol{\nu}!)_{\boldsymbol{\nu}\in\mathcal{F}}$ belongs to $\ell^p(\mathcal{F})$ implies that $\|\boldsymbol{\alpha}\|_{\ell^p} < \infty$ and $\|\boldsymbol{\alpha}\|_{\ell^1} < 1$.
Conversely, assume that $\|\boldsymbol{\alpha}\|_{\ell^p} < \infty$ and $\|\boldsymbol{\alpha}\|_{\ell^1} < 1$. We put $\delta := 1 - \|\boldsymbol{\alpha}\|_{\ell^1} > 0$ and $\eta := \frac{\delta}{3}$. Take J large enough such that $\sum_{j>J} \alpha_j^p \leq \eta$. We define the sequence $\boldsymbol{c}$ and $\boldsymbol{d}$ by

$$c_j = (1+\eta)\alpha_j, \qquad d_j = \frac{1}{1+\eta}$$

if $j \leq J$ and

$$c_j = \alpha_j^p, \qquad d_j = \alpha_j^{1-p}$$

if $j > J$. By this construction we have $\alpha_j = c_j d_j$ for all $j \in \mathbb{N}$. For the sequence $\boldsymbol{c}$ we have

$$\|\boldsymbol{c}\|_{\ell^1} \leq (1+\eta)\|\boldsymbol{\alpha}\|_{\ell^1} + \sum_{j>J} \alpha_j^p \leq (1+\eta)(1-\delta) + \eta < 1-\eta.$$

Next we show that $\|\boldsymbol{d}\|_{\ell^\infty} < 1$. Indeed, for $1 \leq j \leq J$ we have $d_j = \frac{1}{1+\eta} < 1$ and for $j > J$ we have

$$d_j = (\alpha_j^p)^{(1-p)/p} \leq \eta^{(1-p)/p} < 1.$$

Moreover since $d_j^{(p/(1-p))} = \alpha_j^p$ for $j > J$ we have $\boldsymbol{d} \in \ell^{p/(1-p)}(\mathbb{N})$. Now we get from Hölder's inequality

$$\sum_{\boldsymbol{\nu}\in\mathcal{F}} \left(\frac{|\boldsymbol{\nu}|!}{\boldsymbol{\nu}!}\boldsymbol{\alpha}^{\boldsymbol{\nu}}\right)^p = \sum_{\boldsymbol{\nu}\in\mathcal{F}} \left(\frac{|\boldsymbol{\nu}|!}{\boldsymbol{\nu}!}\boldsymbol{c}^{\boldsymbol{\nu}}\right)^p \boldsymbol{d}^{p\boldsymbol{\nu}} \leq \left(\sum_{\boldsymbol{\nu}\in\mathcal{F}} \frac{|\boldsymbol{\nu}|!}{\boldsymbol{\nu}!}\boldsymbol{c}^{\boldsymbol{\nu}}\right)^p \left(\sum_{\boldsymbol{\nu}\in\mathcal{F}} \boldsymbol{d}^{\boldsymbol{\nu}p/(1-p)}\right)^{1-p}.$$

Observe that the first factor on the right side is finite since $\boldsymbol{c} \in \ell^1(\mathbb{N})$ and $\|\boldsymbol{c}\|_{\ell^1} < 1$. Applying the first statement, the second factor on the right side is finite, whence $(\boldsymbol{\alpha}^{\boldsymbol{\nu}}|\boldsymbol{\nu}|!/\boldsymbol{\nu}!)_{\boldsymbol{\nu}\in\mathcal{F}} \in \ell^p(\mathcal{F})$. □

With these sequence summability results at hand, we are now in position to formulate Wiener-Hermite summation results for parametric solution families of PDEs with log-Gaussian random field data.

Theorem 3.13 (General Case) *Let Assumption 3.6 hold and assume that* $\boldsymbol{\varrho} = (\varrho_j)_{j\in\mathbb{N}} \in [0,\infty)^\infty$ *is a sequence satisfying* $(\varrho_j^{-1})_{j\in\mathbb{N}} \in \ell^q(\mathbb{N})$ *for some* $0 < q < \infty$. *Assume that, for each* $\boldsymbol{\nu} \in \mathcal{F}$, *there exists a sequence* $\boldsymbol{\rho}_{\boldsymbol{\nu}} = (\rho_{\boldsymbol{\nu},j})_{j\in\mathbb{N}} \in [0,\infty)^\infty$ *such that* $\operatorname{supp}(\boldsymbol{\nu}) \subseteq \operatorname{supp}(\boldsymbol{\rho}_{\boldsymbol{\nu}})$,

$$\sup_{\boldsymbol{\nu}\in\mathcal{F}} \left\| \sum_{j\in\mathbb{N}} \rho_{\boldsymbol{\nu},j}|\psi_j| \right\|_{L^\infty} \leq \kappa < \frac{\pi}{2}, \qquad \text{and} \qquad \sum_{\|\boldsymbol{\nu}\|_{\ell^\infty}\leq r} \frac{\boldsymbol{\nu}!\boldsymbol{\varrho}^{2\boldsymbol{\nu}}}{\boldsymbol{\rho}_{\boldsymbol{\nu}}^{2\boldsymbol{\nu}}} < \infty \tag{3.40}$$

with $r \in \mathbb{N}$, $r > 2/q$. *Then*

$$\sum_{\boldsymbol{\nu}\in\mathcal{F}} \beta_{\boldsymbol{\nu}}(r,\boldsymbol{\varrho})\|u_{\boldsymbol{\nu}}\|_V^2 < \infty \quad \text{with} \quad \left(\beta_{\boldsymbol{\nu}}(r,\boldsymbol{\varrho})^{-1/2}\right)_{\boldsymbol{\nu}\in\mathcal{F}} \in \ell^q(\mathcal{F}). \tag{3.41}$$

Furthermore,

$$(\|u_\nu\|_V)_{\nu\in\mathcal{F}} \in \ell^p(\mathcal{F}) \quad \textit{with} \quad \frac{1}{p} = \frac{1}{q} + \frac{1}{2}.$$

Proof By Proposition 3.7 Assumption 3.6 implies that $b(\boldsymbol{y})$ belongs to $L^\infty(D)$ for γ-a.e. $\boldsymbol{y} \in U$ and $\mathbb{E}(\exp(k\|b(\boldsymbol{y})\|_{L^\infty}))$ is finite for all $k \in [0,\infty)$.

For $\boldsymbol{y} \in U$ such that $b(\boldsymbol{y}) \in L^\infty(D)$ and $\boldsymbol{\nu} \in \mathcal{F}$ with $\mathfrak{u} = \operatorname{supp}(\boldsymbol{\nu})$, the solution u of (3.17) is holomorphic in $\mathcal{S}_{\mathfrak{u}}(\boldsymbol{\rho}_\nu)$, see Proposition 3.8. This, (3.40) and Lemmata 3.9 and 3.10 yield that

$$\begin{aligned}\sum_{\nu\in\mathcal{F}} \beta_\nu(r,\boldsymbol{\varrho})\|u_\nu\|_V^2 &= \sum_{\|\nu\|_{\ell^\infty}\le r} \frac{\boldsymbol{\varrho}^{2\nu}}{\boldsymbol{\nu}!}\int_U \|\partial^{\nu} u(\boldsymbol{y})\|_V^2 \,\mathrm{d}\gamma(\boldsymbol{y}) \\ &\le C_0^2 \sum_{\|\nu\|_{\ell^\infty}\le r} \frac{\boldsymbol{\nu}!\boldsymbol{\varrho}^{2\nu}}{\boldsymbol{\rho}_\nu^{2\nu}} \mathbb{E}\left(\exp\left(2\|b(\boldsymbol{y})\|_{L^\infty}\right)\right) < \infty.\end{aligned}$$

Since $r > \frac{2}{q}$ and $(\varrho_j^{-1})_{j\in\mathbb{N}} \in \ell^q(\mathbb{N})$, by Lemma 3.11 the family $\left(\beta_\nu(r,\boldsymbol{\varrho})^{-1/2}\right)_{\nu\in\mathcal{F}}$ belongs to $\ell^q(\mathcal{F})$. The relation (3.41) is proven.

From (3.41), by Hölder's inequality we get that

$$\sum_{\nu\in\mathcal{F}} \|u_\nu\|_V^p \le \left(\sum_{\nu\in\mathcal{F}} \beta_\nu(r,\boldsymbol{\varrho})\|u_\nu\|_V^2\right)^{p/2} \left(\sum_{\nu\in\mathcal{F}} \beta_\nu(r,\boldsymbol{\varrho})^{-q/2}\right)^{1-p/2} < \infty.$$

□

Corollary 3.14 (The Case of Global Supports) *Assume that there exists a sequence of positive numbers* $\boldsymbol{\lambda} = (\lambda_j)_{j\in\mathbb{N}}$ *such that*

$$\left(\lambda_j\|\psi_j\|_{L^\infty}\right)_{j\in\mathbb{N}} \in \ell^1(\mathbb{N}) \quad \textit{and} \quad (\lambda_j^{-1})_{j\in\mathbb{N}} \in \ell^q(\mathbb{N}),$$

for some $0 < q < \infty$. *Then we have* $(\|u_\nu\|_V)_{\nu\in\mathcal{F}} \in \ell^p(\mathcal{F})$ *with* $\frac{1}{p} = \frac{1}{q} + \frac{1}{2}$.

Proof Let $\boldsymbol{\nu} \in \mathcal{F}$. We define the sequence $\boldsymbol{\rho}_\nu = (\rho_{\nu,j})_{j\in\mathbb{N}}$ by $\rho_{\nu,j} := \frac{\nu_j}{|\nu|\|\psi_j\|_{L^\infty}}$ for $j \in \operatorname{supp}(\boldsymbol{\nu})$ and $\rho_{\nu,j} = 0$ if $j \notin \operatorname{supp}(\boldsymbol{\nu})$ and choose $\boldsymbol{\varrho} = \tau\boldsymbol{\lambda}$, τ is an appropriate positive constant. It is obvious that

$$\sup_{\nu\in\mathcal{F}} \left\| \sum_{j\in\mathbb{N}} \rho_{\nu,j}|\psi_j| \right\|_{L^\infty} \le 1.$$

We first show that Assumption 3.6 is satisfied for the sequence $\boldsymbol{\lambda}' = (\lambda'_j)_{j\in\mathbb{N}}$ with $\lambda'_j := \lambda_j^{1/2}$ by a similar argument as in [9, Remark 2.5]. From the assumption

$(\lambda_j^{-1})_{j\in\mathbb{N}} \in \ell^q(\mathbb{N})$ we derive that up to a nondecreasing rearrangement, $\lambda'_j \geq Cj^{1/(2q)}$ for some $C > 0$. Therefore, $\big(\exp(-\lambda'^2_j)\big)_{j\in\mathbb{N}} \in \ell^1(\mathbb{N})$. The convergence in $L^\infty(D)$ of $\sum_{j\in\mathbb{N}} \lambda'_j |\psi_j|$ can be proved as follows:

$$\Big\| \sum_{j\in\mathbb{N}} \lambda'_j |\psi_j| \Big\|_{L^\infty} \leq \sup_{j\in\mathbb{N}} \lambda_j^{-1/2} \sum_{j\in\mathbb{N}} \lambda_j \|\psi_j\|_{L^\infty} < \infty.$$

With $r > 2/q$ we have

$$\begin{aligned}
\sum_{\|\boldsymbol{\nu}\|_{\ell^\infty}\leq r} \frac{\boldsymbol{\nu}!\boldsymbol{\varrho}^{2\boldsymbol{\nu}}}{\boldsymbol{\rho}_{\boldsymbol{\nu}}^{2\boldsymbol{\nu}}} &\leq \sum_{\|\boldsymbol{\nu}\|_{\ell^\infty}\leq r} \frac{|\boldsymbol{\nu}|^{2|\boldsymbol{\nu}|}}{\boldsymbol{\nu}^{2\boldsymbol{\nu}}} \prod_{j\in\operatorname{supp}(\boldsymbol{\nu})} \big(\tau\sqrt{r!}\lambda_j \|\psi_j\|_{L^\infty}\big)^{2\nu_j} \\
&\leq \Bigg(\sum_{\|\boldsymbol{\nu}\|_{\ell^\infty}\leq r} \frac{|\boldsymbol{\nu}|^{|\boldsymbol{\nu}|}}{\boldsymbol{\nu}^{\boldsymbol{\nu}}} \prod_{j\in\operatorname{supp}(\boldsymbol{\nu})} \big(\tau\sqrt{r!}\lambda_j \|\psi_j\|_{L^\infty}\big)^{\nu_j} \Bigg)^2 \qquad (3.42) \\
&\leq \Bigg(\sum_{\|\boldsymbol{\nu}\|_{\ell^\infty}\leq r} \frac{|\boldsymbol{\nu}|!}{\boldsymbol{\nu}!} \prod_{j\in\operatorname{supp}(\boldsymbol{\nu})} \big(e\tau\sqrt{r!}\lambda_j \|\psi_j\|_{L^\infty}\big)^{\nu_j} \Bigg)^2 .
\end{aligned}$$

In the last step we used the inequality

$$\frac{|\boldsymbol{\nu}|^{|\boldsymbol{\nu}|}}{\boldsymbol{\nu}^{\boldsymbol{\nu}}} \leq \frac{e^{|\boldsymbol{\nu}|}|\boldsymbol{\nu}|!}{\boldsymbol{\nu}!},$$

which is immediately derived from the inequalities $m! \leq m^m \leq e^m m!$. Since $\big(\tau\sqrt{r!}\lambda_j \|\psi_j\|_{L^\infty}\big)_{j\in\mathbb{N}} \in \ell^1(\mathbb{N})$, we can choose a positive number τ so that

$$\big\| \big(e\tau\sqrt{r!}\lambda_j \|\psi_j\|_{L^\infty}\big)_{j\in\mathbb{N}} \big\|_{\ell^1} < 1.$$

This implies by Lemma 3.12(ii) that the last sum in (3.42) is finite. Applying Theorem 3.13 the desired result follows. □

Corollary 3.15 (The Case of Disjoint Supports) *Assuming $\psi_j \in L^\infty(D)$ for all $j \in \mathbb{N}$ with disjoint supports and, furthermore, that there exists a sequence of positive numbers $\boldsymbol{\lambda} = (\lambda_j)_{j\in\mathbb{N}}$ such that*

$$\big(\lambda_j \|\psi_j\|_{L^\infty}\big)_{j\in\mathbb{N}} \in \ell^2(\mathbb{N}) \text{ and } (\lambda_j^{-1})_{j\in\mathbb{N}} \in \ell^q(\mathbb{N}),$$

for some $0 < q < \infty$. Then $(\|u_{\boldsymbol{\nu}}\|_V)_{\boldsymbol{\nu}\in\mathcal{F}} \in \ell^p(\mathcal{F})$ with $\frac{1}{p} = \frac{1}{q} + \frac{1}{2}$.

Proof Fix $\boldsymbol{\nu} \in \mathcal{F}$, arbitrary. For this $\boldsymbol{\nu}$ we define the sequence $\boldsymbol{\rho}_{\boldsymbol{\nu}} = (\rho_j)_{j\in\mathbb{N}}$ by $\rho_j := \frac{1}{\|\psi_j\|_{L^\infty}}$ for $j \in \mathbb{N}$ and $\boldsymbol{\varrho} = \tau\boldsymbol{\lambda}$, where a positive number τ will be chosen later on. It is clear that

$$\left\| \sum_{j\in\mathbb{N}} \rho_j |\psi_j| \right\|_{L^\infty} \leq 1.$$

Since $\left(\lambda_j \rho_j^{-1}\right)_{j\in\mathbb{N}} \in \ell^2(\mathbb{N})$ and $(\lambda_j^{-1})_{j\in\mathbb{N}} \in \ell^q(\mathbb{N})$, by Hölder's inequality we get $(\rho_j^{-1})_{j\in\mathbb{N}} \in \ell^{q_0}(\mathbb{N})$ with $\frac{1}{q_0} = \frac{1}{2} + \frac{1}{q}$. Hence, similarly to the proof of Corollary 3.14, we can show that Assumption 3.6 holds for the sequence $\boldsymbol{\lambda}' = (\lambda'_j)_{j\in\mathbb{N}}$ with $\lambda'_j := \lambda_j^{1/2}$. In addition, with $r > 2/q$ we have by Lemma 3.12(i)

$$\sum_{\|\boldsymbol{\nu}\|_{\ell^\infty}\leq r} \frac{\boldsymbol{\nu}!\boldsymbol{\varrho}^{2\boldsymbol{\nu}}}{\boldsymbol{\rho}_{\boldsymbol{\nu}}^{2\boldsymbol{\nu}}} \leq \sum_{\|\boldsymbol{\nu}\|_{\ell^\infty}\leq r} \left(\prod_{j\in\operatorname{supp}(\boldsymbol{\nu})} \left(\tau\sqrt{r!}\lambda_j \|\psi_j\|_{L^\infty}\right)^{2\nu_j} \right) < \infty,$$

since by the condition $\left(\tau\sqrt{r!}\lambda_j\|\psi_j\|_{L^\infty}\right)_{j\in\mathbb{N}} \in \ell^2(\mathbb{N})$ a positive number τ can be chosen so that $\sup_{j\in\mathbb{N}}(\tau\sqrt{r!}\lambda_j\|\psi_j\|_{L^\infty}) < 1$. Finally, we apply Theorem 3.13 to obtain the desired results. □

Remark 3.16 We comment on the situation when there exists $\boldsymbol{\rho} = (\rho_j)_{j\in\mathbb{N}} \in (0,\infty)^\infty$ such that

$$\left\| \sum_{j\in\mathbb{N}} \rho_j |\psi_j| \right\|_{L^\infty} = \kappa < \frac{\pi}{2}$$

and $(\rho_j^{-1})_{j\in\mathbb{N}} \in \ell^{q_0}(\mathbb{N})$ for some $0 < q_0 < \infty$ as given in [9, Theorem 1.2]. We choose $\boldsymbol{\varrho} = (\varrho_j)_{j\in\mathbb{N}}$ by

$$\varrho_j = \rho_j^{1-q_0/2} \frac{1}{\sqrt{r!}\|(\rho_j^{-1})_{j\in\mathbb{N}}\|_{\ell^{q_0}}^{q_0/2}}$$

and $\boldsymbol{\rho}_{\boldsymbol{\nu}} = (\rho_j)_{j\in\mathbb{N}}$. Then we obtain $(\varrho_j^{-1})_{j\in\mathbb{N}} \in \ell^{q_0/(1-q_0/2)}(\mathbb{N})$ and

$$\sum_{\|\boldsymbol{\nu}\|_{\ell^\infty}\leq r} \frac{\boldsymbol{\nu}!\boldsymbol{\varrho}^{2\boldsymbol{\nu}}}{\boldsymbol{\rho}_{\boldsymbol{\nu}}^{2\boldsymbol{\nu}}} = \sum_{\|\boldsymbol{\nu}\|_{\ell^\infty}\leq r} \boldsymbol{\nu}! \prod_{j\in\operatorname{supp}\boldsymbol{\nu}} \left(\frac{\rho_j^{-q_0}}{r!\|(\rho_j^{-1})_{j\in\mathbb{N}}\|_{\ell^{q_0}}^{q_0}} \right)^{\nu_j} < \infty.$$

This implies $(\|u_{\boldsymbol{\nu}}\|_V)_{\boldsymbol{\nu}\in\mathcal{F}} \in \ell^p(\mathcal{F})$ with $p = q_0$.

Remark 3.17 The ℓ^p-summability $(\|u_{\boldsymbol{\nu}}\|_V)_{\boldsymbol{\nu}\in\mathcal{F}} \in \ell^p(\mathcal{F})$ proven in Theorem 3.13, has been used in establishing the convergence rate of the best n-term approximation of the solution u to the parametric elliptic PDE (3.17) [9]. However, such a property cannot be used for estimating convergence rates of high-dimensional deterministic numerical approximation *constructive* schemes such as single-level and multi-level versions of anisotropic sparse-grid Hermite-Smolyak interpolation and quadrature in Chap. 7. In the latter situation, the weighted ℓ^2-summability presented in Theorem 3.13

$$\sum_{\boldsymbol{\nu}\in\mathcal{F}} \beta_{\boldsymbol{\nu}}(r,\boldsymbol{\varrho})\|u_{\boldsymbol{\nu}}\|_V^2 < \infty \quad \text{with} \quad \left(\beta_{\boldsymbol{\nu}}(r,\boldsymbol{\varrho})^{-1/2}\right)_{\boldsymbol{\nu}\in\mathcal{F}} \in \ell^q(\mathcal{F})$$

and its generalization

$$\sum_{\boldsymbol{\nu}\in\mathcal{F}} (\sigma_{\boldsymbol{\nu}}\|u_{\boldsymbol{\nu}}\|_X)^2 < \infty \qquad \text{with} \quad \left(p_{\boldsymbol{\nu}}(\tau,\lambda)\sigma_{\boldsymbol{\nu}}^{-1}\right)_{\boldsymbol{\nu}\in\mathcal{F}} \in \ell^q(\mathcal{F}) \tag{3.43}$$

for a Hilbert space X are efficiently applied, where $0 < q < \infty$, $\lambda, \tau \geq 0$, $\left(\sigma_{\boldsymbol{\nu}}\right)_{\boldsymbol{\nu}\in\mathcal{F}}$ is a family of positive numbers and

$$p_{\boldsymbol{\nu}}(\tau,\lambda) := \prod_{j\in\mathbb{N}} (1+\lambda\nu_j)^\tau, \quad \boldsymbol{\nu}\in\mathcal{F}.$$

Weighted summability properties such as (3.43) have been employed in [31, 43, 45, 52] for particular questions in quadrature and interpolation with respect to GMs.

In Chaps. 6 and 7, we will see that weighted ℓ^2-summabilities of the form (3.43) play a basic role in constructing approximation algorithms of sparse-grid interpolation and quadrature and in establishing their convergence rates.

3.7 Parametric $H^s(D)$-Analyticity and Sparsity

Whereas the previous results were, in principle, already known from the real-variable analyses in [8, 9, 16], in this and the subsequent sections, we prove via analytic continuation the sparsity of the Wiener-Hermite PC expansion coefficients of the parametric solutions of (3.17) with log-Gaussian coefficient $a(\boldsymbol{y}) = \exp(b(\boldsymbol{y}))$ when the Wiener-Hermite PC expansion coefficients of the parametric solution family $\{u(\boldsymbol{y}) : \boldsymbol{y} \in U\}$ are measured in higher Sobolev norms. In Sect. 3.8 we shall establish corresponding results when the physical domain D is a plane Lipschitz polygon whose sides are analytic arcs.

3.7.1 $H^s(D)$-Analyticity

As $H^s(D)$ regularity in D is relevant in particular in conjunction with Galerkin discretization in D by continuous, piecewise polynomial, Lagrangian FEM, we review the elementary regularity results from Sect. 3.3. To state them, we recall the

Sobolev spaces $H^s(D)$, and $W^s_\infty(D)$ of functions v on D for $s \in \mathbb{N}_0$, equipped with the respective norms

$$\|v\|_{H^s} := \sum_{\boldsymbol{k}\in\mathbb{Z}^d_+:|\boldsymbol{k}|\le s} \|D^{\boldsymbol{k}} v\|_{L^2}, \qquad \|v\|_{W^s_\infty} := \sum_{\boldsymbol{k}\in\mathbb{Z}^d_+:|\boldsymbol{k}|\le s} \|D^{\boldsymbol{k}} v\|_{L^\infty}.$$

With these definitions $H^0(D) = L^2(D)$ and $W^0_\infty(D) = L^\infty(D)$. We recall from Sect. 3.3 that we identify $L^2(D)$ with its own dual, so that the space $H^{-1}(D)$ is defined as the dual of $H^1_0(D)$ with respect to the pivot space $L^2(D)$.

Lemma 3.18 *Let $s \in \mathbb{N}$ and D be a bounded domain in $\mathbb{R}^d$ with either C^∞-boundary or with convex C^{s-1}-boundary. Assume that there holds the ellipticity condition* (3.25)*, $a \in W^{s-1}_\infty(D)$ and $f \in H^{s-2}(D)$. Then the solution u of* (3.1) *belongs to $H^s(D)$ and there holds*

$$\|u\|_{H^s} \le \begin{cases} \frac{\|f\|_{H^{-1}}}{\rho(a)} & s = 1, \\ \frac{C_{d,s}}{\rho(a)}\left(\|f\|_{H^{s-2}} + \|a\|_{W^{s-1}_\infty}\|u\|_{H^{s-1}}\right) & s > 1, \end{cases} \tag{3.44}$$

with $C_{d,s}$ depending on d, s, and $\rho(a)$ given as in (3.25).

Proof Defining, for $s \in \mathbb{N}$, $H^s_0(D) := (H^s \cap H^1_0)(D)$, since D is a bounded domain in $\mathbb{R}^d$ with either C^∞-boundary or convex C^{s-1}-boundary, we have the following norm equivalence

$$\|v\|_{H^s} \asymp \begin{cases} \|v\|_{H^1_0}, & s = 1, \\ \|\Delta v\|_{H^{s-2}}, & s > 1, \end{cases} \qquad \forall v \in H^s_0, \tag{3.45}$$

see [61, Theorem 2.5.1.1]. The lemma for the case $s = 1$ and $s = 2$ is given in (3.4) and (3.9). We prove the case $s > 2$ by induction on s. Suppose that the assertion holds true for all $s' < s$. We will prove it for s. Let a $\boldsymbol{k} \in \mathbb{Z}^d_+$ with $|\boldsymbol{k}| = s - 2$ be given. Differentiating both sides of (3.10) and applying the Leibniz rule of multivariate differentiation we obtain

$$-\sum_{0\le\boldsymbol{k}'\le\boldsymbol{k}} \binom{\boldsymbol{k}}{\boldsymbol{k}'} D^{\boldsymbol{k}'} a D^{\boldsymbol{k}-\boldsymbol{k}'}(\Delta u) = D^{\boldsymbol{k}} f + \sum_{0\le\boldsymbol{k}'\le\boldsymbol{k}} \binom{\boldsymbol{k}}{\boldsymbol{k}'} (\nabla D^{\boldsymbol{k}'} a, \nabla D^{\boldsymbol{k}-\boldsymbol{k}'} u),$$

see also [8, Lemma 4.3]. Hence,

$$
\begin{aligned}
-a\,D^k \Delta u \;=\; & D^k f + \sum_{0\le k'\le k} \binom{k}{k'} \left(\nabla D^{k'} a, \nabla D^{k-k'} u\right) \\
& + \sum_{0\le k'\le k,\, k'\neq 0} \binom{k}{k'} D^{k'} a D^{k-k'} \Delta u .
\end{aligned}
$$

Taking the L^2-norm of both sides, by the ellipticity condition (3.3) we derive the inequality

$$
\rho(a)\, \|\Delta u\|_{H^{s-2}} \;\le\; C'_{d,s} \left(\|f\|_{H^{s-2}} + \|a\|_{W^{s-1}_\infty} \|u\|_{H^{s-1}} + \|a\|_{W^{s-2}_\infty} \|\Delta u\|_{H^{s-3}} \right)
$$

which yields (3.44) due to (3.45) and the inequality

$$
\|a\|_{W^{s-2}_\infty} \|\Delta u\|_{H^{s-3}} \le \|a\|_{W^{s-1}_\infty} \|u\|_{H^{s-1}},
$$

where $C'_{d,s}$ is a constant depending on d, s only. By induction, this proves that u belongs to H^s. □

Corollary 3.19 *Let $s \in \mathbb{N}$ and D be a bounded domain in $\mathbb{R}^d$ with either C^∞-boundary or convex C^{s-1}-boundary. Assume that there holds the ellipticity condition* (3.3)*, $a \in W^{s-1}_\infty(D)$ and $f \in H^{s-2}(D)$. Then the solution u of* (3.1) *belongs to $H^s(D)$ and there holds the estimate*

$$
\|u\|_{H^s} \;\le\; \frac{\|f\|_{H^{s-2}}}{\rho(a)} \begin{cases} 1, & s = 1, \\ C_{d,s}\left(1 + \frac{\|a\|_{W^{s-1}_\infty}}{\rho(a)}\right)^{s-1}, & s > 1, \end{cases}
$$

where $C_{d,s}$ is a constant depending on d, s only, and $\rho(a)$ is given as in (3.25).

We need the following lemma.

Lemma 3.20 *Let $s \in \mathbb{N}$ and assume that $b(\boldsymbol{y})$ belongs to $W^s_\infty(D)$. Then we have*

$$
\|a(\boldsymbol{y})\|_{W^s_\infty} \le C \|a(\boldsymbol{y})\|_{L^\infty} \left(1 + \|b(\boldsymbol{y})\|_{W^s_\infty}\right)^s,
$$

where the constant C depends on s and m but is independent of $\boldsymbol{y}$.

Proof For $\boldsymbol{\alpha} = (\alpha_1, \ldots, \alpha_d) \in \mathbb{N}_0^d$ with $1 \le |\boldsymbol{\alpha}| \le s$, we observe that for $\alpha_j > 0$ the product rule implies

$$D^{\boldsymbol{\alpha}} a(\boldsymbol{y}) = D^{\boldsymbol{\alpha}-\boldsymbol{e}_j}\big[a(\boldsymbol{y}) D^{\boldsymbol{e}_j} b(\boldsymbol{y})\big] = \sum_{0 \le \boldsymbol{\gamma} \le \boldsymbol{\alpha}-\boldsymbol{e}_j} \binom{\boldsymbol{\alpha}-\boldsymbol{e}_j}{\boldsymbol{\gamma}} D^{\boldsymbol{\alpha}-\boldsymbol{\gamma}} b(\boldsymbol{y}) D^{\boldsymbol{\gamma}} a(\boldsymbol{y}). \tag{3.46}$$

Here, we recall that $(\boldsymbol{e}_j)_{j=1}^d$ is the standard basis of $\mathbb{R}^d$. Taking norms on both sides, we can estimate

$$\begin{aligned}
\|D^{\boldsymbol{\alpha}} a(\boldsymbol{y})\|_{L^\infty} &= \big\|D^{\boldsymbol{\alpha}-\boldsymbol{e}_j}\big[a(\boldsymbol{y}) D^{\boldsymbol{e}_j} b(\boldsymbol{y})\big]\big\|_{L^\infty} \\
&\le \sum_{0 \le \boldsymbol{\gamma} \le \boldsymbol{\alpha}-\boldsymbol{e}_j} \binom{\boldsymbol{\alpha}-\boldsymbol{e}_j}{\boldsymbol{\gamma}} \|D^{\boldsymbol{\alpha}-\boldsymbol{\gamma}} b(\boldsymbol{y})\|_{L^\infty} \|D^{\boldsymbol{\gamma}} a(\boldsymbol{y})\|_{L^\infty} \\
&\le C\Bigg(\sum_{0 \le \boldsymbol{\gamma} \le \boldsymbol{\alpha}-\boldsymbol{e}_j} \|D^{\boldsymbol{\gamma}} a(\boldsymbol{y})\|_{L^\infty}\Bigg)\Bigg(\sum_{|\boldsymbol{k}| \le s} \|D^{\boldsymbol{k}} b(\boldsymbol{y})\|_{L^\infty}\Bigg).
\end{aligned}$$

Similarly, each term $\|D^{\boldsymbol{\gamma}} a(\boldsymbol{y})\|_{L^\infty}$ with $|\boldsymbol{\gamma}| > 0$ can be estimated

$$\|D^{\boldsymbol{\gamma}} a(\boldsymbol{y})\|_{L^\infty} \le C\Bigg(\sum_{0 \le \boldsymbol{\gamma}' \le \boldsymbol{\gamma}-\boldsymbol{e}_j} \|D^{\boldsymbol{\gamma}'} a(\boldsymbol{y})\|_{L^\infty}\Bigg)\Bigg(\sum_{|\boldsymbol{k}| \le s} \|D^{\boldsymbol{k}} b(\boldsymbol{y})\|_{L^\infty}\Bigg)$$

if $\gamma_j > 0$. This implies

$$\|D^{\boldsymbol{\alpha}} a(\boldsymbol{y})\|_{L^\infty} \le C \|a(\boldsymbol{y})\|_{L^\infty} \Bigg(1 + \sum_{|\boldsymbol{k}| \le s} \|D^{\boldsymbol{k}} b(\boldsymbol{y})\|_{L^\infty}\Bigg)^{|\boldsymbol{\alpha}|},$$

for $1 \le |\boldsymbol{\alpha}| \le s$. Summing up these terms with $\|a(\boldsymbol{y})\|_{L^\infty}$ we obtain the desired estimate. □

Proposition 3.21 *Let $s \in \mathbb{N}$ and D be a bounded domain in $\mathbb{R}^d$ with either C^∞-boundary or convex C^{s-1}-boundary. Assume that (3.28) holds and all the functions ψ_j belong to $W_\infty^{s-1}(D)$. Let $\mathfrak{u} \subseteq \operatorname{supp}(\boldsymbol{\rho})$ be a finite set and let $\boldsymbol{y}_0 = (y_{0,1}, y_{0,2}, \ldots) \in U$ be such that $b(\boldsymbol{y}_0)$ belongs to $W_\infty^{s-1}(D)$. Then the solution u of (3.17) is holomorphic in $\mathcal{S}_{\mathfrak{u}}(\boldsymbol{\rho})$ as a function in variables $\boldsymbol{z}_{\mathfrak{u}} = (z_j)_{j \in \mathbb{N}} \in \mathcal{S}_{\mathfrak{u}}(\boldsymbol{y}_0, \boldsymbol{\rho})$ taking values in $H^s(D)$ where $z_j = y_{0,j}$ for $j \notin \mathfrak{u}$ held fixed.*

Proof Let $\mathcal{S}_{\mathfrak{u},N}(\boldsymbol{\rho})$ be given in (3.29) and $\boldsymbol{z}_{\mathfrak{u}} = (y_j + \mathrm{i}\xi_j)_{j \in \mathbb{N}} \in \mathcal{S}_{\mathfrak{u}}(\boldsymbol{y}_0, \boldsymbol{\rho})$ with $(y_j + \mathrm{i}\xi_j)_{j \in \mathfrak{u}} \in \mathcal{S}_{\mathfrak{u},N}(\boldsymbol{\rho})$. Then we have from Corollary 3.19

$$\|u(\boldsymbol{z}_{\mathfrak{u}})\|_{H^s} \le C\rho(a(\boldsymbol{z}_{\mathfrak{u}}))\Big(1 + \rho(a(\boldsymbol{z}_{\mathfrak{u}}))\|a(\boldsymbol{z}_{\mathfrak{u}})\|_{W_\infty^{s-1}}\Big)^{s-1}.$$

Using Lemma 3.20 we find

$$\begin{aligned}
\|a(z_{\mathfrak{u}})\|_{W_\infty^{s-1}} &\le C\|a(z_{\mathfrak{u}})\|_{L^\infty}\left(1+\|b(z_{\mathfrak{u}})\|_{W_\infty^{s-1}}\right)^{s-1} \\
&\le C\|a(z_{\mathfrak{u}})\|_{L^\infty}\Bigg(1+\|b(y_0)\|_{W_\infty^{s-1}} \\
&\quad + \Big\|\sum_{j\in\mathfrak{u}}(y_j-y_{0,j}+\mathrm{i}\xi_j)\psi_j\Big\|_{W_\infty^{s-1}}\Bigg)^{s-1} \\
&\le C\|a(z_{\mathfrak{u}})\|_{L^\infty}\Bigg(1+\|b(y_0)\|_{W_\infty^{s-1}}+\sum_{j\in\mathfrak{u}}(N+\rho_j)\|\psi_j\|_{W_\infty^{s-1}}\Bigg)^{s-1}
\end{aligned}$$

and

$$\|a(z_{\mathfrak{u}})\|_{L^\infty} \le \exp\left(\|b(y_0)\|_{L^\infty}+\Big\|\sum_{j\in\mathfrak{u}}(y_j-y_{0,j}+\mathrm{i}\xi_j)\psi_j\Big\|_{L^\infty}\right) < \infty. \tag{3.47}$$

From this and (3.30) we obtain

$$\|u(z_{\mathfrak{u}})\|_{H^s} \le C < \infty$$

which implies the map $z_{\mathfrak{u}} \to u(z_{\mathfrak{u}})$ is holomorphic on the set $\mathcal{S}_{\mathfrak{u},N}(\boldsymbol{\rho})$ as a consequence of [24, Lemma 2.2]. For more details we refer the reader to [111, Examples 1.2.38 and 1.2.39]. Since N is arbitrary we conclude that the map $z_{\mathfrak{u}} \to u(z_{\mathfrak{u}})$ is holomorphic on $\mathcal{S}_{\mathfrak{u}}(\boldsymbol{\rho})$. □

3.7.2 *Sparsity of Wiener-Hermite PC Expansion Coefficients*

For sparsity of H^s-norms of Wiener-Hermite PC expansion coefficients we need the following assumption.

Assumption 3.22 *Let $s \in \mathbb{N}$. For every $j \in \mathbb{N}$, $\psi_j \in W_\infty^{s-1}(D)$ and there exists a positive sequence $(\lambda_j)_{j\in\mathbb{N}}$ such that $\left(\exp(-\lambda_j^2)\right)_{j\in\mathbb{N}} \in \ell^1(\mathbb{N})$ and the series*

$$\sum_{j\in\mathbb{N}} \lambda_j |D^{\boldsymbol{\alpha}}\psi_j|$$

converges in $L^\infty(D)$ for all $\boldsymbol{\alpha} \in \mathbb{N}_0^d$ with $|\boldsymbol{\alpha}| \le s-1$.

As a consequence of [9, Theorem 2.2] we have the following

Lemma 3.23 *Let Assumption 3.22 hold. Then the set* $U_{s-1} := \{\boldsymbol{y} \in U : b(\boldsymbol{y}) \in W^{s-1}_\infty(D)\}$ *has full measure, i.e.,* $\gamma(U_{s-1}) = 1$. *Furthermore,* $\mathbb{E}(\exp(k\|b(\cdot)\|_{W^{s-1}_\infty}))$ *is finite for all* $k \in [0, \infty)$.

The H^s-analytic continuation of the parametric solutions $\{u(\boldsymbol{y}) : \boldsymbol{y} \in U\}$ to $\mathcal{S}_{\mathfrak{u}}(\boldsymbol{\rho})$ leads to the following result on parametric H^s-regularity.

Lemma 3.24 *Let* $D \subset \mathbb{R}^d$ *be a bounded domain with either* C^∞*-boundary or convex* C^{s-1}*-boundary. Assume that for each* $\boldsymbol{\nu} \in \mathcal{F}$*, there exists a sequence* $\boldsymbol{\rho}_{\boldsymbol{\nu}} = (\rho_{\boldsymbol{\nu},j})_{j\in\mathbb{N}} \in [0,\infty)^\infty$ *such that* $\operatorname{supp}(\boldsymbol{\nu}) \subseteq \operatorname{supp}(\boldsymbol{\rho}_{\boldsymbol{\nu}})$*, and such that*

$$\sup_{\boldsymbol{\nu}\in\mathcal{F}} \sum_{|\boldsymbol{\alpha}|\le s-1} \left\| \sum_{j\in\mathbb{N}} \rho_{\boldsymbol{\nu},j} |D^{\boldsymbol{\alpha}} \psi_j| \right\|_{L^\infty} \le \kappa < \frac{\pi}{2}.$$

Then we have

$$\|\partial^{\boldsymbol{\nu}} u(\boldsymbol{y})\|_{H^s} \le C \frac{\boldsymbol{\nu}!}{\boldsymbol{\rho}_{\boldsymbol{\nu}}^{\boldsymbol{\nu}}} \exp\left(\|b(\boldsymbol{y})\|_{L^\infty}\right) \left\{ 1 + \exp(2\|b(\boldsymbol{y})\|_{L^\infty}) \left(1 + \|b(\boldsymbol{y})\|_{W^{s-1}_\infty}\right)^{s-1} \right\}^{s-1}, \tag{3.48}$$

where C is a constant depending on κ, d, s *only.*

Proof Let $\boldsymbol{\nu} \in \mathcal{F}$ with $\mathfrak{u} = \operatorname{supp}(\boldsymbol{\nu})$ and $\boldsymbol{y} \in U$ such that $b(\boldsymbol{y}) \in W^{s-1}_\infty(D)$. Let furthermore $\mathcal{C}_{\boldsymbol{y},\mathfrak{u}}(\boldsymbol{\rho}_{\boldsymbol{\nu}})$ and $\mathcal{C}_{\mathfrak{u}}(\boldsymbol{y}, \boldsymbol{\rho}_{\boldsymbol{\nu}})$ be given as in (3.31) and (3.33). Using Cauchy's formula as in the proof of Lemma 3.9 we obtain

$$\|\partial^{\boldsymbol{\nu}} u(\boldsymbol{y})\|_{H^s} \le \frac{\boldsymbol{\nu}!}{\boldsymbol{\rho}_{\boldsymbol{\nu}}^{\boldsymbol{\nu}}} \sup_{\boldsymbol{z}_{\mathfrak{u}} \in \mathcal{C}_{\mathfrak{u}}(\boldsymbol{y},\boldsymbol{\rho}_{\boldsymbol{\nu}})} \|u(\boldsymbol{z}_{\mathfrak{u}})\|_{H^s}. \tag{3.49}$$

For $\boldsymbol{z}_{\mathfrak{u}} = (z_j)_{j\in\mathbb{N}} \in \mathcal{C}_{\mathfrak{u}}(\boldsymbol{y}, \boldsymbol{\rho}_{\boldsymbol{\nu}})$ we can write $z_j = y_j + \eta_j + \mathrm{i}\xi_j \in \mathcal{C}_{\boldsymbol{y},j}(\boldsymbol{\rho}_{\boldsymbol{\nu}})$ with $|\eta_j| \le \rho_{\boldsymbol{\nu},j}$ and $|\xi_j| \le \rho_{\boldsymbol{\nu},j}$ for $j \in \mathfrak{u}$ and hence we get

$$\begin{aligned} \|D^{\boldsymbol{\alpha}} b(\boldsymbol{z}_{\mathfrak{u}})\|_{L^\infty} &= \left\| D^{\boldsymbol{\alpha}} \Big(b(\boldsymbol{y}) + \sum_{j\in\mathfrak{u}} (\eta_j + \mathrm{i}\xi_j)\psi_j \Big) \right\|_{L^\infty} \\ &\le \|D^{\boldsymbol{\alpha}} b(\boldsymbol{y})\|_{L^\infty} + \sqrt{2} \left\| \sum_{j\in\mathfrak{u}} \rho_{\boldsymbol{\nu},j} |D^{\boldsymbol{\alpha}} \psi_j| \right\|_{L^\infty} \\ &\le \|D^{\boldsymbol{\alpha}} b(\boldsymbol{y})\|_{L^\infty} + \kappa\sqrt{2}. \end{aligned}$$

In addition we have

$$\frac{1}{\rho(a(z_{\mathfrak{u}}))} \leq \frac{\exp(\|b(y+\sum_{j\in\mathfrak{u}}\eta_j\psi_j\|_{L^\infty})}{\cos(\|\sum_{j\in\mathfrak{u}}\xi_j\psi_j\|_{L^\infty})} \leq \frac{\exp\left(\kappa+\|b(y)\|_{L^\infty}\right)}{\cos\kappa} \tag{3.50}$$

and

$$\|a(z_{\mathfrak{u}})\|_{L^\infty} = \left\|\exp\left(b(y)+\sum_{j\in\mathfrak{u}}(\eta_j+\mathrm{i}\xi_j)\psi_j\right)\right\|_{L^\infty} \leq e^{\kappa\sqrt{2}}\exp(\|b(y)\|_{L^\infty}). \tag{3.51}$$

Consequently, we can bound

$$\begin{aligned}\|a(z_{\mathfrak{u}})\|_{W^{s-1}_\infty} &\leq C\|a(z_{\mathfrak{u}})\|_{L^\infty}\left(1+\|b(z_{\mathfrak{u}})\|_{W^{s-1}_\infty}\right)^{s-1}\\ &\leq C\exp(\|b(y)\|_{L^\infty})\left(1+\|b(y)\|_{W^{s-1}_\infty}\right)^{s-1}.\end{aligned}$$

Now Corollary 3.19 implies the inequality

$$\begin{aligned}\sup_{z_{\mathfrak{u}}\in\mathcal{C}_{\mathfrak{u}}(y,\rho_\nu)}\|u(z_{\mathfrak{u}})\|_{H^s} &\leq C\exp\left(\|b(y)\|_{L^\infty}\right)\\ &\quad\times\left\{1+\exp(2\|b(y)\|_{L^\infty})\left(1+\|b(y)\|_{W^{s-1}_\infty}\right)^{s-1}\right\}^{s-1},\end{aligned} \tag{3.52}$$

which together with (3.49) proves the lemma. □

We are now in position to formulate sparsity results for the H^s-norms of Wiener-Hermite PC expansion coefficients of the solution u.

Theorem 3.25 (General Case) *Let $s, r \in \mathbb{N}$ and $D \subset \mathbb{R}^d$ denote a bounded domain with either C^∞-boundary or convex C^{s-1}-boundary. Let further Assumption 3.22 hold, assume that $f \in H^{s-2}(D)$, and assume given a sequence $\boldsymbol{\varrho} = (\varrho_j)_{j\in\mathbb{N}} \subset (0,\infty)^\infty$ that satisfies $(\varrho_j^{-1})_{j\in\mathbb{N}} \in \ell^q(\mathbb{N})$ for some $0 < q < \infty$. Assume in addition that, for each $\boldsymbol{\nu} \in \mathcal{F}$, there exists a sequence $\boldsymbol{\rho}_\nu = (\rho_{\nu,j})_{j\in\mathbb{N}} \in [0,\infty)^\infty$ such that* $\operatorname{supp}(\boldsymbol{\nu}) \subseteq \operatorname{supp}(\boldsymbol{\rho}_\nu)$, *and such that, with $r > 2/q$,*

$$\sup_{\nu\in\mathcal{F}}\sum_{|\alpha|\leq s-1}\left\|\sum_{j\in\mathbb{N}}\rho_{\nu,j}|D^\alpha\psi_j|\right\|_{L^\infty} \leq \kappa < \frac{\pi}{2}, \quad \textit{and} \quad \sum_{\|\nu\|_{\ell^\infty}\leq r}\frac{\nu!\varrho^{2\nu}}{\rho_\nu^{2\nu}} < \infty.$$

Then there holds, with $\beta_\nu(r,\boldsymbol{\varrho})$ as in (3.36),

$$\sum_{\nu\in\mathcal{F}}\beta_\nu(r,\boldsymbol{\varrho})\|u_\nu\|^2_{H^s} < \infty \quad \textit{with} \quad \left(\beta_\nu(r,\boldsymbol{\varrho})^{-1/2}\right)_{\nu\in\mathcal{F}} \in \ell^q(\mathcal{F}). \tag{3.53}$$

Furthermore,

$$(\|u_\nu\|_{H^s})_{\nu\in\mathcal{F}} \in \ell^p(\mathcal{F}) \quad \textit{with} \quad \frac{1}{p} = \frac{1}{q} + \frac{1}{2}.$$

Proof Arguing as in the proof of [9, Theorem 3.3] we obtain that for any $r \in \mathbb{N}$ there holds following generalization of the Parseval-type identity

$$\sum_{\nu\in\mathcal{F}} \beta_\nu(r, \varrho)\|u_\nu\|_{H^s}^2 = \sum_{\|\nu\|_{\ell^\infty}\le r} \frac{\varrho^{2\nu}}{\nu!} \int_U \|\partial^\nu u(\mathbf{y})\|_{H^s}^2 \,\mathrm{d}\gamma(\mathbf{y}). \tag{3.54}$$

By (3.52), Lemma 3.23 and Hölder's inequality we derive that

$$\int_U \left(\sup_{z_{\mathfrak{u}}\in C_{\mathfrak{u}}(\mathbf{y},\boldsymbol{\rho}_\nu)} \|u(z_{\mathfrak{u}})\|_{H^s} \right)^2 \mathrm{d}\gamma(\mathbf{y}) \le C$$

and in particular, $\mathbb{E}(\|u(\mathbf{y})\|_{H^s}^k)$ is finite for all $k \in [0, \infty)$. Now (3.54), Lemma 3.24 and our assumption give

$$\begin{aligned}\sum_{\nu\in\mathcal{F}} \beta_\nu(r, \varrho)\|u_\nu\|_{H^s}^2 &= \sum_{\|\nu\|_{\ell^\infty}\le r} \frac{\varrho^{2\nu}}{\nu!} \int_U \|\partial^\nu u(\mathbf{y})\|_{H^s}^2 \,\mathrm{d}\gamma(\mathbf{y}) \\ &\le C^2 \sum_{\|\nu\|_{\ell^\infty}\le r} \frac{\nu!\varrho^{2\nu}}{\rho_\nu^{2\nu}} \int_U \mathrm{d}\gamma(\mathbf{y}) = C^2 \sum_{\|\nu\|_{\ell^\infty}\le r} \frac{\nu!\varrho^{2\nu}}{\rho_\nu^{2\nu}} < \infty,\end{aligned}$$

where C is the constant in (3.48). As in the proof of Theorem 3.13, by Lemma 3.11 the family $\left(\beta_\nu(r, \varrho)^{-1/2}\right)_{\nu\in\mathcal{F}}$ belongs to $\ell^q(\mathcal{F})$. The relation (3.53) is proven.

The assertion $(\|u_\nu\|_{H^s})_{\nu\in\mathcal{F}} \in \ell^p(\mathcal{F})$ can be proved in the same way as in the proof of Theorem 3.13. □

Similarly to Corollaries 3.14 and 3.15 from Theorem 3.25 we obtain

Corollary 3.26 (The Case of Global Supports) *Let $s \in \mathbb{N}$ and $D \subset \mathbb{R}^d$ denote a bounded domain with either C^∞-boundary or convex C^{s-1}-boundary. Assume that for all $j \in \mathbb{N}$ holds $\psi_j \in W_\infty^{s-1}(D)$, and that $f \in H^{s-2}(D)$. Assume further that there exists a sequence of positive numbers $\boldsymbol{\lambda} = (\lambda_j)_{j\in\mathbb{N}}$ such that*

$$\left(\lambda_j\|\psi_j\|_{W_\infty^{s-1}}\right)_{j\in\mathbb{N}} \in \ell^1(\mathbb{N}) \quad \textit{and} \quad (\lambda_j^{-1})_{j\in\mathbb{N}} \in \ell^q(\mathbb{N}),$$

for some $0 < q < \infty$. Then we have $(\|u_\nu\|_{H^s})_{\nu\in\mathcal{F}} \in \ell^p(\mathcal{F})$ with $\frac{1}{p} = \frac{1}{q} + \frac{1}{2}$.

Corollary 3.27 (The Case of Disjoint Supports) *Let $s \in \mathbb{N}$ and $D \subset \mathbb{R}^d$ denote a bounded domain with either C^∞-boundary or convex C^{s-1}-boundary. Assume that $f \in H^{s-2}(D)$ and for all $j \in \mathbb{N}$ holds $\psi_j \in W_\infty^{s-1}(D)$ with disjoint supports.*

Assume further that there exists a sequence of positive numbers $\boldsymbol{\lambda} = (\lambda_j)_{j\in\mathbb{N}}$ *such that*

$$\left(\lambda_j \|\psi_j\|_{W^{s-1}_\infty}\right)_{j\in\mathbb{N}} \in \ell^2(\mathbb{N}) \text{ and } (\lambda_j^{-1})_{j\in\mathbb{N}} \in \ell^q(\mathbb{N}),$$

for some $0 < q < \infty$.

Then $(\|u_{\boldsymbol{\nu}}\|_{H^s})_{\boldsymbol{\nu}\in\mathcal{F}} \in \ell^p(\mathcal{F})$ *with* $\frac{1}{p} = \frac{1}{q} + \frac{1}{2}$.

3.8 Parametric Kondrat'ev Analyticity and Sparsity

In the previous section, we investigated the weighted ℓ^2-summability and ℓ^p-summability of Wiener-Hermite PC expansion coefficients of parametric solutions measured in the standard Sobolev spaces $H^s(D)$. We assumed that $D \subset \mathbb{R}^d$ with boundary ∂D of sufficient smoothness (depending on s). In this section we consider in space dimension $d = 2$ the case when the physical domain D is a polygonal domain. In such domains, elliptic regularity shift results and shift theorems in D hold in Kondrat'ev spaces which are corner-weighted Sobolev spaces. We refer to [61, 90] and the references there for an extensive survey.

To state corresponding results for the log-Gaussian parametric elliptic problems, we first review definitions of the weighted Sobolev spaces of Kondrat'ev type and results from [24] on the holomorphy of parametric solutions in weighted Kondrat'ev spaces in polygonal domains D. Then, we establish summability results of the coefficients of Wiener-Hermite PC expansions of the parametric solutions in Kondrat'ev spaces. FE approximation results for Wiener-Hermite PC expansion coefficient functions which are in these spaces were provided in Sect. 2.6.

3.8.1 Parametric $K^s_\varkappa(D)$-Analyticity

We recall the Kondrat'ev spaces in a bounded polygonal domain D introduced in Sect. 2.6.1: for $s \in \mathbb{N}_0$ and $\varkappa \in \mathbb{R}$,

$$\mathcal{K}^s_\varkappa(D) := \left\{u : D \to \mathbb{C} : r_D^{|\boldsymbol{\alpha}|-\varkappa} D^{\boldsymbol{\alpha}} u \in L^2(D), |\boldsymbol{\alpha}| \le s\right\}$$

and

$$\mathcal{W}^s_\infty(D) := \left\{u : D \to \mathbb{C} : r_D^{|\boldsymbol{\alpha}|} D^{\boldsymbol{\alpha}} u \in L^\infty(D), \ |\boldsymbol{\alpha}| \le s\right\}.$$

The weighted Sobolev norms in these spaces are given in Sect. 2.6.1.

Lemma 3.28 *Let* $s \in \mathbb{N}_0$. *Assume that* $\boldsymbol{y} \in U$ *is such that* $b(\boldsymbol{y}) \in \mathcal{W}^s_\infty(D)$.

Then

$$\|a(y)\|_{\mathcal{W}^s_\infty} \leq C\|a(y)\|_{L^\infty}\left(1 + \|b(y)\|_{\mathcal{W}^s_\infty}\right)^s,$$

where the constant C depends on s and m.

Proof The proof proceeds along the lines of the proof of Lemma 3.20. Let $\boldsymbol{\alpha} = (\alpha_1, \ldots, \alpha_d) \in \mathbb{N}_0^d$ with $1 \leq |\boldsymbol{\alpha}| \leq s$ and recall that $(\boldsymbol{e}_j)_{j=1}^d$ is the standard basis of $\mathbb{R}^d$. Assuming that $\alpha_j > 0$ we have (3.46). We apply corner-weighted norms to both sides of (3.46). This implies

$$\begin{aligned}
\|r_D^{|\boldsymbol{\alpha}|} D^{\boldsymbol{\alpha}} a(y)\|_{L^\infty} &= \left\| D^{\boldsymbol{\alpha}-\boldsymbol{e}_j}\left[a(y) D^{\boldsymbol{e}_j} b(y)\right]\right\|_{L^\infty} \\
&\leq \sum_{0 \leq \boldsymbol{\gamma} \leq \boldsymbol{\alpha}-\boldsymbol{e}_j} \binom{\boldsymbol{\alpha}-\boldsymbol{e}_j}{\boldsymbol{\gamma}} \|r_D^{|\boldsymbol{\alpha}-\boldsymbol{\gamma}|} D^{\boldsymbol{\alpha}-\boldsymbol{\gamma}} b(y)\|_{L^\infty} \|r_D^{|\boldsymbol{\gamma}|} D^{\boldsymbol{\gamma}} a(y)\|_{L^\infty} \\
&\leq C\Bigg(\sum_{0 \leq \boldsymbol{\gamma} \leq \boldsymbol{\alpha}-\boldsymbol{e}_j} \|r_D^{|\boldsymbol{\gamma}|} D^{\boldsymbol{\gamma}} a(y)\|_{L^\infty}\Bigg)\Bigg(\sum_{|\boldsymbol{k}| \leq s} \|r_D^{|\boldsymbol{k}|} D^{\boldsymbol{k}} b(y)\|_{L^\infty}\Bigg) \\
&= C\Bigg(\sum_{0 \leq \boldsymbol{\gamma} \leq \boldsymbol{\alpha}-\boldsymbol{e}_j} \|r_D^{|\boldsymbol{\gamma}|} D^{\boldsymbol{\gamma}} a(y)\|_{L^\infty}\Bigg)\|b(y)\|_{\mathcal{W}^s_\infty}.
\end{aligned}$$

Similarly, if $\gamma_j > 0$, each term $\|r_D^{|\boldsymbol{\gamma}|} D^{\boldsymbol{\gamma}} a(y)\|_{L^\infty}$ with $|\boldsymbol{\gamma}| > 0$ can be estimated

$$\|r_D^{|\boldsymbol{\gamma}|} D^{\boldsymbol{\gamma}} a(y)\|_{L^\infty} \leq C\Bigg(\sum_{0 \leq \boldsymbol{\gamma}' \leq \boldsymbol{\gamma}-\boldsymbol{e}_j} \|r_D^{|\boldsymbol{\gamma}'|} D^{\boldsymbol{\gamma}'} a(y)\|_{L^\infty}\Bigg)\|b(y)\|_{\mathcal{W}^s_\infty}.$$

This implies

$$\|r_D^{|\boldsymbol{\alpha}|} D^{\boldsymbol{\alpha}} a(y)\|_{L^\infty} \leq C\|a(y)\|_{L^\infty}\left(1 + \|b(y)\|_{\mathcal{W}^s_\infty}\right)^{|\boldsymbol{\alpha}|},$$

for $1 \leq |\boldsymbol{\alpha}| \leq s$. This finishes the proof. □

We recall the following result from [24, Theorem 1].

Theorem 3.29 *Let $D \subset \mathbb{R}^2$ be a polygonal domain, $\eta_0 > 0$, $s \in \mathbb{N}$ and $N_s = 2^{s+1} - s - 2$. Let $a \in L^\infty(D, \mathbb{C})$.*

Then there exist τ and C_s with the following property: for any $a \in \mathcal{W}^{s-1}_\infty(D)$ and for any $\varkappa \in \mathbb{R}$ such that

$$|\varkappa| < \eta := \min\{\eta_0, \tau^{-1}\|a\|_{L^\infty}^{-1}\rho(a)\},$$

the operator P_a *defined in* (3.1) *induces an isomorphism*

$$P_a : \mathcal{K}^s_{\varkappa+1}(D) \cap \{u|_{\partial D} = 0\} \to \mathcal{K}^{s-2}_{\varkappa-1}(D)$$

such that P_a^{-1} *depends analytically on the coefficients a and has norm*

$$\|P_a^{-1}\| \le C_s\big(\rho(a) - \tau|\varkappa|\|a\|_{L^\infty}\big)^{-N_s-1}\|a\|^{N_s}_{\mathcal{W}^{s-1}_\infty}.$$

The bound of τ *and* C_s *depends only on s, D and* η_0.

Applying this result to our setting, we obtain the following parametric regularity.

Theorem 3.30 *Suppose* $\eta_0 > 0$, $\psi_j \in \mathcal{W}^{s-1}_\infty(D)$ *for all* $j \in \mathbb{N}$ *and that* (3.28) *holds. Let* $\mathfrak{u} \subseteq \operatorname{supp}(\boldsymbol{\rho})$ *be a finite set. Let further* $\boldsymbol{y}_0 = (y_{0,1}, y_{0,2}, \dots) \in U$ *be such that* $b(\boldsymbol{y}_0)$ *belongs to* $\mathcal{W}^{s-1}_\infty(D)$. *We denote*

$$\vartheta := \inf_{\boldsymbol{z}_\mathfrak{u} \in \mathcal{S}_\mathfrak{u}(\boldsymbol{y}_0,\boldsymbol{\rho})} \rho\big(a(\boldsymbol{z}_\mathfrak{u})\big)\|a(\boldsymbol{z}_\mathfrak{u})\|^{-1}_{L^\infty}.$$

Let $\tau > 0$ *be as given in Theorem 3.29.*

Then there exists a positive constant C_s *such that for* $\varkappa \in \mathbb{R}$ *with* $|\varkappa| \le \min\{\eta_0, \tau^{-1}\vartheta/2\}$, *and for* $f \in \mathcal{K}^{s-2}_{\varkappa-1}(D)$, *the solution u of* (3.17) *is holomorphic in the cylinder* $\mathcal{S}_\mathfrak{u}(\boldsymbol{\rho})$ *as a function in variables* $\boldsymbol{z}_\mathfrak{u} = (z_j)_{j\in\mathbb{N}} \in \mathcal{S}_\mathfrak{u}(\boldsymbol{y}_0, \boldsymbol{\rho})$ *taking values in* $\mathcal{K}^s_{\varkappa+1}(D) \cap V$, *where* $z_j = y_{0,j}$ *for* $j \notin \mathfrak{u}$ *held fixed. Furthermore, we have the estimate*

$$\|u(\boldsymbol{z}_\mathfrak{u})\|_{\mathcal{K}^s_{\varkappa+1}} \le C_s \frac{1}{\big(\rho(a(\boldsymbol{z}_\mathfrak{u}))\big)^{N_s+1}}\|a(\boldsymbol{z}_\mathfrak{u})\|^{N_s}_{\mathcal{W}^{s-1}_\infty}.$$

Proof Observe first that for the parametric coefficient $a(\boldsymbol{z}_\mathfrak{u})$, the conditions of Proposition 3.8 are satisfied.

Thus, the solution u is holomorphic in $\mathcal{S}_\mathfrak{u}(\boldsymbol{\rho})$ as a V-valued map in variables $\boldsymbol{z}_\mathfrak{u} = (z_j)_{j\in\mathbb{N}} \in \mathcal{S}_\mathfrak{u}(\boldsymbol{y}_0, \boldsymbol{\rho})$. We assume that $\vartheta > 0$. Let $\mathcal{S}_{\mathfrak{u},N}(\boldsymbol{\rho})$ be given in (3.29) and $\boldsymbol{z}_\mathfrak{u} = (y_j + \mathrm{i}\xi_j)_{j\in\mathbb{N}} \in \mathcal{S}_\mathfrak{u}(\boldsymbol{y}_0, \boldsymbol{\rho})$ with $(y_j + \mathrm{i}\xi_j)_{j\in\mathfrak{u}} \in \mathcal{S}_{\mathfrak{u},N}(\boldsymbol{\rho})$. From Lemma 3.28 we have

$$\|a(\boldsymbol{z}_\mathfrak{u})\|_{\mathcal{W}^{s-1}_\infty} \le C\|a(\boldsymbol{z}_\mathfrak{u})\|_{L^\infty}\Big(1 + \|b(\boldsymbol{z}_\mathfrak{u})\|_{\mathcal{W}^{s-1}_\infty}\Big)^{s-1}.$$

Furthermore

$$\|b(\boldsymbol{z}_\mathfrak{u})\|_{\mathcal{W}^{s-1}_\infty} = \sum_{|\boldsymbol{\alpha}|\le s-1}\Big\|r_D^{|\boldsymbol{\alpha}|}\sum_{j\in\mathbb{N}}(y_j + \mathrm{i}\xi_j)D^{\boldsymbol{\alpha}}\psi_j\Big\|_{L^\infty}$$
$$\le \sum_{j\in\mathfrak{u}}(|y_j - y_{0,j}| + \rho_j)\|\psi_j\|_{\mathcal{W}^{s-1}_\infty} + \|b(\boldsymbol{y}_0)\|_{\mathcal{W}^{s-1}_\infty} < \infty.$$

This together with (3.47) implies $\|a(z_{\mathrm{u}})\|_{\mathcal{W}_\infty^{s-1}} \le C$. From the condition of $\varkappa$ we infer $|\varkappa|\tau \le \vartheta/2$ which leads to

$$\tau|\varkappa|\|a(z_{\mathrm{u}})\|_{L^\infty} \le \rho(a(z_{\mathrm{u}}))/2.$$

As a consequence we obtain

$$\big(\rho(a(z_{\mathrm{u}})) - \tau|\varkappa|\|a(z)\|_{L^\infty}\big)^{-1} \le \frac{1}{\rho(a(z_{\mathrm{u}})}.$$

Since the function exp is analytic in $\mathcal{S}_{\mathrm{u},N}(\boldsymbol{\rho})$, the assertion follows for the case $\vartheta > 0$ by applying Theorem 3.29. In addition, for $z_{\mathrm{u}} = (z_j)_{j\in\mathbb{N}} \in \mathcal{S}_{\mathrm{u}}(\boldsymbol{y}_0, \boldsymbol{\rho})$ with $(z_j)_{j\in\mathrm{u}} \in \mathcal{S}_{\mathrm{u},N}(\boldsymbol{\rho})$, we have

$$\rho(a(z_{\mathrm{u}}))\|a(z_{\mathrm{u}})\|_{L^\infty}^{-1} \ge C > 0,$$

From this we conclude that u is holomorphic in the cylinder $\mathcal{S}_{\mathrm{u},N}(\boldsymbol{\rho})$ as a $\mathcal{K}_1^s(D) \cap V$-valued map, by again Theorem 3.29. This completes the proof. □

Remark 3.31 The value of ϑ depends on the system $(\psi_j)_{j\in\mathbb{N}}$. Assume that $\psi_j = j^{-\alpha}$ for some $\alpha > 1$. Then for any $\boldsymbol{y} \in U$, $\boldsymbol{\rho}$ satisfying (3.28), and finite set $\mathrm{u} \subset \mathrm{supp}(\boldsymbol{\rho})$ we have

$$\vartheta = \inf_{z_{\mathrm{u}}\in\mathcal{S}_{\mathrm{u}}(\boldsymbol{y},\boldsymbol{\rho})} \frac{\Re[\exp(\sum_{j\in\mathbb{N}}(y_j + \mathrm{i}\xi_j)j^{-\alpha})]}{\exp(\sum_{j\in\mathbb{N}} y_j j^{-\alpha})} \ge \cos\kappa.$$

We consider another case when there exists some ψ_j such that $\psi_j \ge C > 0$ in an open set Ω in D and $\|\exp(y_j\psi_j)\|_{L^\infty} \ge 1$ for all $y_j \le 0$. With $\boldsymbol{y}_0 = (\ldots, 0, y_j, 0, \ldots)$ and $v_0 \in C_0^\infty(\Omega)$ we have in this case

$$\vartheta \le \rho(\exp(y_j\psi_j)) \to 0 \quad \text{when} \quad y_j \to -\infty.$$

Hence, only for $\varkappa = 0$ is satisfied Theorem 3.30 in this situation.

Due to this observation, for Kondrat'ev regularity we consider only the case $\varkappa = 0$. In Sect. 7.6.1, we will present a stronger regularity result for a polygonal domain $D \subset \mathbb{R}^2$.

Lemma 3.32 *Let $\boldsymbol{\nu} \in \mathcal{F}$, $f \in \mathcal{K}_{-1}^{s-2}(D)$, and assume that $\psi_j \in \mathcal{W}_\infty^{s-1}(D)$ for $j \in \mathbb{N}$. Let $\boldsymbol{y} \in U$ with $b(\boldsymbol{y}) \in \mathcal{W}_\infty^{s-1}(D)$. Assume further that there exists a non-negative sequence $\boldsymbol{\rho}_{\boldsymbol{\nu}} = (\rho_{\boldsymbol{\nu},j})_{j\in\mathbb{N}}$ such that* $\mathrm{supp}(\boldsymbol{\nu}) \subset \mathrm{supp}(\boldsymbol{\rho}_{\boldsymbol{\nu}})$ *and*

$$\sum_{|\alpha|\le s-1} \Big\| \sum_{j\in\mathbb{N}} \rho_{\boldsymbol{\nu},j} |r_D^{|\alpha|} D^\alpha \psi_j| \Big\|_{L^\infty} \le \kappa < \frac{\pi}{2}. \tag{3.55}$$

Then we have the estimate

$$\|\partial^{\boldsymbol{\nu}} u(\boldsymbol{y})\|_{\mathcal{K}_1^s} \leq C \frac{\boldsymbol{\nu}!}{\boldsymbol{\rho}_{\boldsymbol{\nu}}^{\boldsymbol{\nu}}} \big(\exp\big(\|b(\boldsymbol{y})\|_{L^\infty}\big)^{2N_s+1} \Big(1 + \|b(\boldsymbol{y})\|_{\mathcal{W}_\infty^{s-1}}\Big)^{(s-1)N_s}.$$

Proof Let $\boldsymbol{\nu} \in \mathcal{F}$ with $\mathfrak{u} = \operatorname{supp}(\boldsymbol{\nu})$. By our assumption, it is clear that (with $\boldsymbol{\alpha} = 0$ in (3.55))

$$\Big\| \sum_{j \in \mathbb{N}} \rho_{\boldsymbol{\nu},j} |\psi_j| \Big\|_{L^\infty} \leq \kappa < \frac{\pi}{2}.$$

Consequently, if we fix the variable y_j with $j \notin \mathfrak{u}$, the function u of (3.17) is holomorphic on the domain $\mathcal{S}_{\mathfrak{u}}(\boldsymbol{\rho}_{\boldsymbol{\nu}})$, see Theorem 3.30. Hence, applying Cauchy's formula gives that

$$\begin{aligned}\|\partial^{\boldsymbol{\nu}} u(\boldsymbol{y})\|_{\mathcal{K}_1^s} &\leq \frac{\boldsymbol{\nu}!}{\boldsymbol{\rho}_{\boldsymbol{\nu}}^{\boldsymbol{\nu}}} \sup_{\boldsymbol{z}_{\mathfrak{u}} \in \mathcal{C}_{\mathfrak{u}}(\boldsymbol{y},\boldsymbol{\rho}_{\boldsymbol{\nu}})} \|u(\boldsymbol{z}_{\mathfrak{u}})\|_{\mathcal{K}_1^s} \\ &\leq C \frac{\boldsymbol{\nu}!}{\boldsymbol{\rho}_{\boldsymbol{\nu}}^{\boldsymbol{\nu}}} \sup_{\boldsymbol{z}_{\mathfrak{u}} \in \mathcal{C}_{\mathfrak{u}}(\boldsymbol{y},\boldsymbol{\rho}_{\boldsymbol{\nu}})} \frac{1}{\big(\rho(a(\boldsymbol{z}_{\mathfrak{u}}))\big)^{N_s+1}} \|a(\boldsymbol{z}_{\mathfrak{u}})\|_{\mathcal{W}_\infty^{s-1}}^{N_s},\end{aligned}$$

where $\mathcal{C}_{\mathfrak{u}}(\boldsymbol{y}, \boldsymbol{\rho}_{\boldsymbol{\nu}})$ is given as in (3.33). Notice that for $\boldsymbol{z}_{\mathfrak{u}} = (z_j)_{j \in \mathbb{N}} \in \mathcal{C}_{\mathfrak{u}}(\boldsymbol{y}, \boldsymbol{\rho}_{\boldsymbol{\nu}})$, we can write $z_j = y_j + \eta_j + \mathrm{i}\xi_j \in \mathcal{C}_{\boldsymbol{y},j}(\boldsymbol{\rho}_{\boldsymbol{\nu}})$ with $|\eta_j| \leq \rho_{\boldsymbol{\nu},j}$ and $|\xi_j| \leq \rho_{\boldsymbol{\nu},j}$ for $j \in \mathfrak{u}$. Hence, by (3.50), (3.51) and

$$\begin{aligned}\|a(\boldsymbol{z}_{\mathfrak{u}})\|_{\mathcal{W}_\infty^{s-1}} &\leq C \|a(\boldsymbol{z}_{\mathfrak{u}})\|_{L^\infty} \Big(1 + \|b(\boldsymbol{z}_{\mathfrak{u}})\|_{\mathcal{W}_\infty^{s-1}}\Big)^{s-1} \\ &= C \exp(\|b(\boldsymbol{y})\|_{L^\infty}) \\ &\quad \times \Bigg[1 + \sum_{|\boldsymbol{\alpha}| \leq s-1} \Big\| r_D^{|\boldsymbol{\alpha}|} \sum_{j \in \mathbb{N}} (y_j + \eta_j + \mathrm{i}\xi_j) D^{\boldsymbol{\alpha}} \psi_j \Big\|_{L^\infty}\Bigg]^{s-1} \\ &= C \exp(\|b(\boldsymbol{y})\|_{L^\infty}) \\ &\quad \times \Bigg[1 + \sum_{|\boldsymbol{\alpha}| \leq s-1} \Big\| 2 \sum_{j \in \mathfrak{u}} \rho_{\boldsymbol{\nu},j} | r_D^{|\boldsymbol{\alpha}|} D^{\boldsymbol{\alpha}} \psi_j | \Big\|_{L^\infty} + \|b(\boldsymbol{y})\|_{\mathcal{W}_\infty^{s-1}}\Bigg]^{s-1} \\ &\leq C \exp(\|b(\boldsymbol{y})\|_{L^\infty}) \Big(1 + 2\kappa + \|b(\boldsymbol{y})\|_{\mathcal{W}_\infty^{s-1}}\Big)^{s-1},\end{aligned}$$

we obtain the desired result. □

3.8.2 Sparsity of $K^s_\varkappa$-Norms of Wiener-Hermite PC Expansion Coefficients

To establish the sparsity, i.e., the weighted ℓ^2-summability and ℓ^p-summability of $K^s_\varkappa$-norms of Wiener-Hermite PC expansion coefficients we need the following assumption.

Assumption 3.33 *Let $s \in \mathbb{N}$. All functions ψ_j belong to $\mathcal{W}^{s-1}_\infty(D)$ and there exists a positive sequence $(\lambda_j)_{j\in\mathbb{N}}$ such that $\big(\exp(-\lambda_j^2)\big)_{j\in\mathbb{N}} \in \ell^1(\mathbb{N})$ and the series*

$$\sum_{j\in\mathbb{N}} \lambda_j \left| r_D^{|\boldsymbol{\alpha}|} D^{\boldsymbol{\alpha}} \psi_j \right|$$

converges in $L^\infty(D)$ for all $\boldsymbol{\alpha} \in \mathbb{N}_0^d$ with $|\boldsymbol{\alpha}| \le s-1$.

Lemma 3.34 *Suppose that Assumption 3.33 holds. Then $b(\boldsymbol{y})$ belongs to $\mathcal{W}^{s-1}_\infty(D)$ $\gamma - a.e.\ \boldsymbol{y} \in U$. Furthermore, $\mathbb{E}(\exp(k\|b(\boldsymbol{y})\|_{\mathcal{W}^{s-1}_\infty}))$ is finite for all $k \in [0,\infty)$.*

Proof Under Assumption 3.33, by Bachmayr et al. [9, Theorem 2.2.] we infer that for $\boldsymbol{\alpha} \in \mathbb{N}_0^d$, $|\boldsymbol{\alpha}| \le s-1$, the sequence

$$\left(\sum_{j=1}^N y_j r_D^{|\boldsymbol{\alpha}|} D^{\boldsymbol{\alpha}} \psi_j \right)_{N\in\mathbb{N}}$$

converges to some $\psi_{\boldsymbol{\alpha}}$ in L^∞ for $\gamma - a.e.\ \boldsymbol{y} \in U$ and $\mathbb{E}(\exp(k\|\psi_{\boldsymbol{\alpha}}(\boldsymbol{y})\|_{L^\infty}))$ is finite for all $k \in [0,\infty)$. Hence, for $\gamma - a.e.\ \boldsymbol{y} \in U$, the sequence $\big(\sum_{j=1}^N y_j \psi_j\big)_{N\in\mathbb{N}}$ is a Cauchy sequence in $\mathcal{W}^{s-1}_\infty(D)$. Since $\mathcal{W}^{s-1}_\infty(D)$ is a Banach space, the statement follows. □

Theorem 3.35 (General Case) *Let $s \in \mathbb{N}$, $s \ge 2$ and D be a bounded curvilinear polygonal domain. Let $f \in \mathcal{K}^{s-2}_{-1}(D)$ and Assumption 3.33 hold. Assume there exists a sequence*

$$\boldsymbol{\varrho} = (\varrho_j)_{j\in\mathbb{N}} \in (0,\infty)^\infty \ \ with\ (\varrho_j^{-1})_{j\in\mathbb{N}} \in \ell^q(\mathbb{N})$$

for some $0 < q < \infty$. Assume furthermore that, for each $\boldsymbol{\nu} \in \mathcal{F}$, there exists a sequence $\boldsymbol{\rho}_{\boldsymbol{\nu}} := (\rho_{\boldsymbol{\nu},j})_{j\in\mathbb{N}} \in [0,\infty)^\infty$ such that $\operatorname{supp}(\boldsymbol{\nu}) \subset \operatorname{supp}(\boldsymbol{\rho}_{\boldsymbol{\nu}})$,

$$\sup_{\boldsymbol{\nu}\in\mathcal{F}} \sum_{|\boldsymbol{\alpha}|\le s-1} \left\| \sum_{j\in\mathbb{N}} \rho_{\boldsymbol{\nu},j} |r_D^{|\boldsymbol{\alpha}|} D^{\boldsymbol{\alpha}} \psi_j| \right\|_{L^\infty} \le \kappa < \frac{\pi}{2}, \quad and \quad \sum_{\|\boldsymbol{\nu}\|_{\ell^\infty}\le r} \frac{\boldsymbol{\nu}!\boldsymbol{\varrho}^{2\boldsymbol{\nu}}}{\boldsymbol{\rho}_{\boldsymbol{\nu}}^{2\boldsymbol{\nu}}} < \infty$$

with $r \in \mathbb{N}$, $r > 2/q$.

Then

$$\sum_{\nu \in \mathcal{F}} \beta_\nu(r, \varrho) \|u_\nu\|^2_{\mathcal{K}^s_1} < \infty \quad with \quad \left(\beta_\nu(r, \varrho)^{-1/2}\right)_{\nu \in \mathcal{F}} \in \ell^q(\mathcal{F}),$$

where $\beta_\nu(r, \varrho)$ *is given in* (3.36). *Furthermore,*

$$(\|u_\nu\|_{\mathcal{K}^s_1})_{\nu \in \mathcal{F}} \in \ell^p(\mathcal{F}) \quad with \quad \frac{1}{p} = \frac{1}{q} + \frac{1}{2}.$$

Proof For each $\nu \in \mathcal{F}$ with $\mathrm{u} = \mathrm{supp}(\nu)$ and $y \in U$ such that $b(y) \in \mathcal{W}^{s-1}_\infty(D)$, Assumption 3.33 implies that the solution u of (3.17) is holomorphic in $\mathcal{S}_{\mathrm{u}}(\boldsymbol{\rho}_\nu)$ as a $\mathcal{K}^s_1(D) \cap V$-valued map, see Theorem 3.30.

We obtain from Lemmata 3.32 and 3.34

$$\begin{aligned}\int_U \|\partial^\nu u(y)\|^2_{\mathcal{K}^s_1} \, \mathrm{d}\gamma(y) &\le C \frac{\nu!}{\rho_\nu^{2\nu}} \int_U \left(\exp\left(\|b(y)\|_{L^\infty}\right)^{4N_s+2}\right. \\ &\quad \times \left(1 + \|b(y)\|_{\mathcal{W}^{s-1}_\infty}\right)^{2(s-1)N_s} \mathrm{d}\gamma(y) \\ &\le C \frac{\nu!}{\rho_\nu^{2\nu}} < \infty.\end{aligned}$$

This leads to

$$\begin{aligned}\sum_{\nu \in \mathcal{F}} \beta_\nu(r, \varrho) \|u_\nu\|^2_{\mathcal{K}^s_1} &= \sum_{\|\nu\|_{\ell^\infty} \le r} \frac{\varrho^{2\nu}}{\nu!} \int_U \|\partial^\nu u(y)\|^2_{\mathcal{K}^s_1} \, \mathrm{d}\gamma(y) \\ &\le C \sum_{\|\nu\|_{\ell^\infty} \le r} \frac{\nu! \varrho^{2\nu}}{\rho_\nu^{2\nu}} < \infty.\end{aligned}$$

The rest of the proof follows similarly to the proof of Theorem 3.13. □

Similarly to Corollaries 3.14 and 3.15 from Theorem 3.35 we obtain

Corollary 3.36 (The Case of Global Supports) *Let* $s \in \mathbb{N}$, $s \ge 2$ *and* D *be a bounded curvilinear, polygonal domain. Assume that for all* $j \in \mathbb{N}$ *holds* $\psi_j \in \mathcal{W}^{s-1}_\infty(D)$, *and that* $f \in \mathcal{K}^{s-2}_{-1}(D)$. *Assume further that there exists a sequence of positive numbers* $\boldsymbol{\lambda} = (\lambda_j)_{j \in \mathbb{N}}$ *such that*

$$\left(\lambda_j \|\psi_j\|_{\mathcal{W}^{s-1}_\infty}\right)_{j \in \mathbb{N}} \in \ell^1(\mathbb{N}) \quad and \quad (\lambda_j^{-1})_{j \in \mathbb{N}} \in \ell^q(\mathbb{N}),$$

for some $0 < q < \infty$. *Then we have* $(\|u_\nu\|_{\mathcal{K}^s_1})_{\nu \in \mathcal{F}} \in \ell^p(\mathcal{F})$ *with* $\frac{1}{p} = \frac{1}{q} + \frac{1}{2}$.

Corollary 3.37 (The Case of Disjoint Supports) *Let* $s \in \mathbb{N}$, $s \ge 2$ *and* $D \subset \mathbb{R}^d$ *with* $d \ge 2$ *be a bounded curvilinear polygonal domain. Assume that all the* *functions ψ_j belong to $\mathcal{W}_\infty^{s-1}(D)$ and have disjoint supports. Assume further that $f \in \mathcal{K}_{-1}^{s-2}(D)$ and that there exists a sequence of positive numbers $\boldsymbol{\lambda} = (\lambda_j)_{j\in\mathbb{N}}$ such that*

$$\left(\lambda_j \|\psi_j\|_{\mathcal{W}_\infty^{s-1}}\right)_{j\in\mathbb{N}} \in \ell^2(\mathbb{N}) \text{ and } (\lambda_j^{-1})_{j\in\mathbb{N}} \in \ell^q(\mathbb{N}),$$

for some $0 < q < \infty$. Then $(\|u_{\boldsymbol{\nu}}\|_{\mathcal{K}_1^s})_{\boldsymbol{\nu}\in\mathcal{F}} \in \ell^p(\mathcal{F})$ with $\frac{1}{p} = \frac{1}{q} + \frac{1}{2}$.

3.9 Bibliographical Remarks

In this section, we briefly recall some known related results in previous works on ℓ^p-summability and on weighted ℓ^2-summability of the generalized PC expansion coefficients of solutions to parametric divergence-form elliptic PDEs (3.17), as well as some applications to best n-term approximation.

A basic role in the approximation and numerical integration for parametric divergence-form elliptic PDEs (3.17) are generalized PC expansions for the dependence on the parametric variables. In [32, 37–39], based on the conditions $\left(\|\psi_j\|_{W_\infty^1}\right)_{j\in\mathbb{N}} \in \ell^p(\mathbb{N})$ for some $0 < p < 1$ on the affine expansion

$$a(\boldsymbol{y}) = \bar{a} + \sum_{j=1}^{\infty} y_j \psi_j, \quad \boldsymbol{y} \in [0,1]^\infty, \tag{3.56}$$

the authors have proven ℓ^p-summability of the coefficients in a Taylor or Legendre PC expansion and hence proposed adaptive best n-term rate optimal approximation methods of Galerkin and collocation type by choosing a set of n largest estimated terms in these expansions. To derive a fully discrete approximation, the best n-term approximants are then discretized by finite element methods. Some results on convergence rates of Galerkin approximation were proven in [71] for the log-Gaussian expansion (3.18), based on the summability $\left(j\|\psi_j\|_{W_\infty^1}\right)_{j\in\mathbb{N}} \in \ell^p(\mathbb{N})$ for some $0 < p < 1$. However, in these papers possible local support properties of the component functions ψ_j were not taken into account.

A different approach to studying summability that takes into account the support properties has been recently proposed in [10] for the affine-parametric case, in [9] for the log-exponential, parametric case, and in [8] for extension of both cases to higher-order Sobolev norms of the corresponding generalized PC expansion coefficients. This approach leads to significant improvements on the results on ℓ^p-summability and therefore, on best n-term semi-discrete and fully discrete approximations when the functions ψ_j have limited overlap, such as splines, finite elements or compactly supported wavelet bases. These approximation results provide a benchmark for convergence rates.

We present some results from [9] and [8] on ℓ^p-summability and weighted ℓ^2-summability of the Wiener-Hermite PC expansion coefficients of the solution to the parametric divergence-form elliptic PDEs (3.17)–(3.18) which were proven by real-variable bootstrapping arguments.

For convenience, we use the conventions:

$$W^1 := V,\quad W^2 := W,\quad H^{-1}(D) := V',\quad H^0(D) := L^2(D),\quad W^{0,\infty}(D) := L^\infty(D),$$

where we recall $W := \{v \in V \ :\ \Delta v \in L^2(D)\}$, is the space equipped with the norm $\|v\|_W := \|\Delta v\|_{L^2}$. The following theorem and lemma were proven in [9] for $i = 1$ and in [8] for $i = 2$.

Theorem 3.38 *Let $i = 1, 2$. Assume that the right side f in (3.17) belongs to $H^{i-2}(D)$, that the domain D has $C^{i-1,1}$ smoothness, that all functions ψ_j belong to $W^{i-1,\infty}(D)$. Assume that there exist a number $0 < q_i < \infty$ and a sequence $\varrho_i = (\varrho_{i;j})_{j\in\mathbb{N}}$ of positive numbers such that $(\varrho_{i;j}^{-1})_{j\in\mathbb{N}} \in \ell^{q_i}(\mathbb{N})$ and*

$$\sup_{|\alpha|\le i-1} \left\| \sum_{j\in\mathbb{N}} \varrho_{i;j} |D^\alpha \psi_j| \right\|_{L^\infty} < \infty. \tag{3.57}$$

Then we have for any $r \in \mathbb{N}$ the weighted ℓ^2-summability,

$$\sum_{\boldsymbol{\nu}\in\mathcal{F}} (\sigma_{i;\boldsymbol{\nu}} \|u_{\boldsymbol{\nu}}\|_{W^i})^2 < \infty \quad \text{and} \quad (\sigma_{i;\boldsymbol{\nu}}^{-1})_{\boldsymbol{\nu}\in\mathcal{F}} \in \ell^{q_i}(\mathcal{F}), \tag{3.58}$$

where

$$\sigma_{i;\boldsymbol{\nu}}^2 := \sum_{\|\boldsymbol{\nu}'\|_{\ell^\infty} \le r} \binom{\boldsymbol{\nu}}{\boldsymbol{\nu}'} \varrho_i^{2\boldsymbol{\nu}'}. \tag{3.59}$$

Furthermore,

$$(\|u_{\boldsymbol{\nu}}\|_{W^i})_{\boldsymbol{\nu}\in\mathcal{F}} \in \ell^{2q_i/(2+q_i)}(\mathcal{F}).$$

Notice that the assumption (3.57) which give the weighted ℓ^2-summability (3.58), already reflects the support properties of the component functions ψ_j.

For $\tau, \lambda \ge 0$, we define the family

$$p_{\boldsymbol{\nu}}(\tau, \lambda) := \prod_{j\in\mathbb{N}} (1 + \lambda \nu_j)^\tau, \quad \boldsymbol{\nu} \in \mathcal{F}, \tag{3.60}$$

with the abbreviation $p_{\boldsymbol{\nu}}(\tau) := p_{\boldsymbol{\nu}}(\tau, 1)$.

We make use of the following notation

$$\mathcal{F}_1 := \mathcal{F}, \quad \mathcal{F}_2 := \{\boldsymbol{\nu} \in \mathcal{F} : \nu_j \neq 1,\ j \in \mathbb{N}\}. \tag{3.61}$$

Lemma 3.39 *Let $0 < q < \infty$, $s = 1, 2$ and $\tau, \lambda \geq 0$. Let $\boldsymbol{\rho} = (\rho_j)_{j\in\mathbb{N}}$ be a sequence of positive numbers such $(\rho_j^{-1})_{j\in\mathbb{N}}$ belongs to $\ell^q(\mathbb{N})$. For $r \in \mathbb{N}$, define the family $(\sigma_{\boldsymbol{\nu}})_{\boldsymbol{\nu}\in\mathcal{F}}$ by*

$$\sigma_{\boldsymbol{\nu}}^2 := \sum_{\|\boldsymbol{\nu}'\|_{\ell^\infty} \leq r} \binom{\boldsymbol{\nu}}{\boldsymbol{\nu}'} \boldsymbol{\rho}^{2\boldsymbol{\nu}'}.$$

Then for any $r > \frac{2s(\tau+1)}{q}$, we have

$$\sum_{\boldsymbol{\nu}\in\mathcal{F}_s} p_{\boldsymbol{\nu}}(\tau, \lambda)\sigma_{\boldsymbol{\nu}}^{-q/s} < \infty.$$

This lemma has been proven in [43, Lemma 5.3]. Observe that for $s = 1$ and $\tau = 0$, an equivalent formulation of Lemma 3.39 is Lemma 3.11.

Theorem 3.38 and Lemma 3.39 directly imply the following corollary.

Corollary 3.40 *Under the assumptions of Theorem 3.38, let $s = 1, 2$ and $\tau, \lambda \geq 0$. Then we have that for any $r > \frac{2s(\tau+1)}{q}$,*

$$\sum_{\boldsymbol{\nu}\in\mathcal{F}_s} (\sigma_{i;\boldsymbol{\nu}} \|u_{\boldsymbol{\nu}}\|_{W^i})^2 < \infty \quad \text{and} \quad (p_{\boldsymbol{\nu}}(\tau, \lambda)\sigma_{i;\boldsymbol{\nu}}^{-1})_{\boldsymbol{\nu}\in\mathcal{F}_s} \in \ell^{q_i/s}(\mathcal{F}_s). \tag{3.62}$$

As commented in Sect. 3.6.2, in the case of disjoint or finitely overlapping supports the results on sparsity of Theorem 3.38 and Corollary 3.40 are stronger than those in Sects. 3.6.2 and 3.7.2. They play a basic role in best n-term approximation [8, 9] and linear approximation and quadrature [43] (see also [45]) of the solution to the parametric divergence-form elliptic PDEs (3.17)–(3.18).

Chapter 4
Sparsity for Holomorphic Functions

In Chap. 3 we introduced a concept of holomorphic (analytic) extensions of countably-parametric families $\{u(\boldsymbol{y}) : \boldsymbol{y} \in U\} \subset V$ in the separable Hilbert space V with respect to the parameter $\boldsymbol{y}$ into the Cartesian product $\mathcal{S}_{\mathrm{u}}(\boldsymbol{\rho})$ of strips in the complex domain (cp. (3.27)). We now introduce a refinement which is required for the ensuing results on rates of numerical approximation and integration of such families, based on sparsity (weighted ℓ^2-summability) and on Wiener-Hermite PC expansions of $\{u(\boldsymbol{y}) : \boldsymbol{y} \in U\}$: *quantified parametric holomorphy* of (complex extensions of) the parametric families $\{u(\boldsymbol{y}) : \boldsymbol{y} \in U\} \subset X$ for a separable Hilbert space X. Section 4.1 presents a definition of quantified holomorphy of families $\{u(\boldsymbol{y}) : \boldsymbol{y} \in U\}$ and discusses the sparsity of the Wiener-Hermite PC expansion coefficients of these families. In Sect. 4.2, we present the notion $(\boldsymbol{b}, \xi, \delta, X)$-holomorphy of composite functions. In Sect. 4.3, we analyze some examples of holomorphic functions which are solutions to certain parametric PDEs.

There are two basic steps in the approximations which we consider:

1. We truncate the countably-parametric family $\{u(\boldsymbol{y}) : \boldsymbol{y} \in U\} \subset X$ to a finite number $N \in \mathbb{N}$ of parameters. This step, which is sometimes also referred to as "dimension-truncation", of course implicitly depends on the enumeration of the coordinates $y_j \in \boldsymbol{y}$. *We assume throughout that this numbering is fixed by the indexing of the Parseval frame in Theorem 2.21* which frame is used as affine representation system to parametrize the uncertain input $a = \exp(b)$ of the PDE of interest. We emphasize that the finite dimension $N \in \mathbb{N}$ of the truncated parametric Wiener-Hermite PC expansion is a discretization parameter, and we will be interested in quantitative bounds on the error incurred by restricting $\{u(\boldsymbol{y}) : \boldsymbol{y} \in U\} \subset X$ to Wiener-Hermite PC expansions of the first N active variables only. We denote these restrictions by $\{u_N(\boldsymbol{y}) : \boldsymbol{y} \in U\}$.
2. The coefficients $u_{\boldsymbol{\nu}} \in X$ of the resulting, finite-parametric Wiener-Hermite PC expansion, can not be computed exactly, but must be numerically approximated. As is done in stochastic collocation and stochastic Galerkin algorithms, we

D. Dũng et al., *Analyticity and Sparsity in Uncertainty Quantification for PDEs with Gaussian Random Field Inputs*, Lecture Notes in Mathematics 2334,
https://doi.org/10.1007/978-3-031-38384-7_4

seek numerical approximations of u_ν in suitable, finite-dimensional subspaces $X_l \subset X$. Assuming the collection $(X_l)_{l\in\mathbb{N}} \subset X$ to be dense in X, any prescribed tolerance $\varepsilon > 0$ of approximation of $u_N(\mathbf{y})$ in $L^2(U, X; \gamma)$ can be met. For notational convenience, we also set $X_0 = \{0\}$.

In computational practice, however, given a target accuracy $\varepsilon \in (0, 1]$, one searches an allocation of $l : \mathcal{F} \times (0, 1] \to \mathbb{N} : (\boldsymbol{\nu}, \varepsilon) \mapsto l(\boldsymbol{\nu}, \varepsilon)$ of discretization levels along the "active" Wiener-Hermite PC expansion coefficients which ensures that the prescribed tolerance $\varepsilon \in (0, 1]$ is met with possibly minimal "computational budget". We propose and analyze the *a-priori construction of an allocation* l which ensures convergence rates of the corresponding collocation approximations which are independent of N (i.e. they are free from the "curse of dimensionality"). These approximations are based on "stochastic collocation", i.e. on sampling the parametric family $\{u(\mathbf{y}) : \mathbf{y} \in U\} \subset V$ in a collection of deterministic Gaussian coordinates in U. We prove, subsequently, dimension-independent convergence rates of the sparse collocation w.r.t. $\mathbf{y} \in U$ and w.r.t. the subspaces $X_l \subset X$ realize convergence rates which are free from the curse of dimensionality. These rates depend only on the summability (resp. sparsity) of the coefficients of the norm of the Wiener-Hermite PC expansion of the parametric family $\{u(\mathbf{y}) : \mathbf{y} \in U\}$ with respect to $\mathbf{y}$.

4.1 $(\boldsymbol{b}, \xi, \delta, X)$-Holomorphy and Sparsity

We introduce the concept of "$(\boldsymbol{b}, \xi, \delta, X)$-holomorphic functions", which constitutes a subset of $L^2(U, X; \gamma)$. As such these functions are typically not pointwise well defined for each $\mathbf{y} \in U$. In order to still define a suitable form of pointwise function evaluations to be used for numerical algorithms such as sampling at $\mathbf{y} \in U$ for "stochastic collocation" or for quadrature in "stochastic Galerkin" algorithms, we define them as $L^2(U, X; \gamma)$ limits of certain smooth (pointwise defined) functions, cp. Remark 4.4 and Example 6.9 ahead.

For $N \in \mathbb{N}$ and $\boldsymbol{\varrho} = (\varrho_j)_{j=1}^N \in (0, \infty)^N$ set (cp. (3.27))

$$\mathcal{S}(\boldsymbol{\varrho}) := \{z \in \mathbb{C}^N : |\mathfrak{Im} z_j| < \varrho_j \ \forall j\} \quad \text{and} \quad \mathcal{B}(\boldsymbol{\varrho}) := \{z \in \mathbb{C}^N : |z_j| < \varrho_j \ \forall j\}. \tag{4.1}$$

Definition 4.1 ($(\boldsymbol{b}, \xi, \delta, X)$-Holomorphy) Let X be a complex, separable Hilbert space, $\boldsymbol{b} = (b_j)_{j\in\mathbb{N}} \in (0, \infty)^\infty$ and $\xi > 0$, $\delta > 0$.

For $N \in \mathbb{N}$, $\boldsymbol{\varrho} \in (0, \infty)^N$ is called $(\boldsymbol{b}, \xi)$-*admissible* if

$$\sum_{j=1}^N b_j \varrho_j \le \xi. \tag{4.2}$$

A function $u \in L^2(U, X; \gamma)$ is called $(\boldsymbol{b}, \xi, \delta, X)$*-holomorphic* if

(i) for every $N \in \mathbb{N}$ there exists $u_N : \mathbb{R}^N \to X$, which, for every $(\boldsymbol{b}, \xi)$-admissible $\boldsymbol{\varrho} \in (0, \infty)^N$, admits a holomorphic extension (denoted again by u_N) from $\mathcal{S}(\boldsymbol{\varrho}) \to X$; furthermore, for all $N < M$

$$u_N(y_1, \ldots, y_N) = u_M(y_1, \ldots, y_N, 0, \ldots, 0) \qquad \forall (y_j)_{j=1}^N \in \mathbb{R}^N, \tag{4.3}$$

(ii) for every $N \in \mathbb{N}$ there exists $\varphi_N : \mathbb{R}^N \to \mathbb{R}_+$ such that $\|\varphi_N\|_{L^2(\mathbb{R}^N;\gamma_N)} \le \delta$ and

$$\sup_{\substack{\boldsymbol{\varrho} \in (0,\infty)^N \\ \text{is } (\boldsymbol{b},\xi)\text{-adm.}}} \sup_{\boldsymbol{z} \in \mathcal{B}(\boldsymbol{\varrho})} \|u_N(\boldsymbol{y} + \boldsymbol{z})\|_X \le \varphi_N(\boldsymbol{y}) \qquad \forall \boldsymbol{y} \in \mathbb{R}^N,$$

(iii) with $\tilde{u}_N : U \to X$ defined by $\tilde{u}_N(\boldsymbol{y}) := u_N(y_1, \ldots, y_N)$ for $\boldsymbol{y} \in U$ it holds

$$\lim_{N \to \infty} \|u - \tilde{u}_N\|_{L^2(U,X;\gamma)} = 0.$$

We interpret the definition of $(\boldsymbol{b}, \xi, \delta, X)$-holomorphy in the following remarks.

Remark 4.2 While the numerical value of $\xi > 0$ in Definition 4.1 of $(\boldsymbol{b}, \xi, \delta, X)$-holomorphy is of minor importance in the definition, the sequence $\boldsymbol{b}$ and the constant δ will crucially influence the magnitude of our upper bounds of the Wiener-Hermite PC expansion coefficients: The stronger the decay of $\boldsymbol{b}$, the larger we can choose the elements of the sequence $\boldsymbol{\varrho}$, so that $\boldsymbol{\varrho}$ satisfies (4.2). Hence stronger decay of $\boldsymbol{b}$ indicates larger domains of holomorphic extension. The constant δ is an upper bound of these extensions in the sense of item (ii). Importantly, the decay of $\boldsymbol{b}$ will determine the *sparsity* of the Wiener-Hermite PC expansion coefficients, while decreasing δ by a factor will roughly speaking translate to a decrease of all coefficients by the same *factor*.

Remark 4.3 Since $u_N \in L^2(\mathbb{R}^N, X; \gamma_N)$, the function $\tilde{u}_N$ in item (iii) belongs to $L^2(U, X; \gamma)$ by Fubini's theorem.

Remark 4.4 (Evaluation of Countably-Parametric Functions) In the following sections, for arbitrary $N \in \mathbb{N}$ and $(y_j)_{j=1}^N \in \mathbb{R}^N$ we will write

$$u(y_1, \ldots, y_N, 0, 0, \ldots) := u_N(y_1, \ldots, y_N). \tag{4.4}$$

This is well-defined due to (4.3). Note however that (4.4) should be considered as an abuse of notation, since pointwise evaluations of functions $u \in L^2(U, X; \gamma)$ are in general not well-defined.

Remark 4.5 The assumption of X being separable is not necessary in Definition 4.1: Every function $u_N : \mathbb{R}^N \to X$ as in Definition 4.1 is continuous since it allows a holomorphic extension. Hence for every $n, N \in \mathbb{N}$ the set,

$$A_{N,n} := \{u_N((y_j)_{j=1}^N) : y_j \in [-n, n] \, \forall j\} \subseteq X$$

is compact. Thus there is a countable set $X_{N,n} \subseteq X$ which is dense in $A_{N,n}$ for every $N, n \in \mathbb{N}$. Then $\bigcup_{n\in\mathbb{N}} A_{N,n}$ is contained in the (separable) closed span $\tilde{X}$ of

$$\bigcup_{N,n\in\mathbb{N}} X_{N,n} \subseteq X.$$

Since $\tilde{u}_N \in L^2(U, \tilde{X}; \gamma)$ for every $N \in \mathbb{N}$ we also have

$$u = \lim_{N\to\infty} u_N \in L^2(U, \tilde{X}; \gamma).$$

Hence, u is separably valued.

Lemma 4.6 *Let u be $(\boldsymbol{b}, \xi, \delta, X)$-holomorphic, let $N \in \mathbb{N}$ and $0 < \kappa < \xi < \infty$. Let u_N, φ_N be as in Definition 4.1. Then with $\boldsymbol{b}_N = (b_j)_{j=1}^N$ it holds for every $\boldsymbol{\nu} \in \mathbb{N}_0^N$*

$$\|\partial^{\boldsymbol{\nu}} u_N(\boldsymbol{y})\|_X \le \frac{\boldsymbol{\nu}!|\boldsymbol{\nu}|^{|\boldsymbol{\nu}|}\boldsymbol{b}_N^{\boldsymbol{\nu}}}{\kappa^{|\boldsymbol{\nu}|}\boldsymbol{\nu}^{\boldsymbol{\nu}}}\varphi_N(\boldsymbol{y}) \qquad \forall \boldsymbol{y} \in \mathbb{R}^N.$$

Proof For $\boldsymbol{\nu} \in \mathbb{N}_0^N$ fixed we choose $\boldsymbol{\varrho} = (\varrho_j)_{j=1}^N$ with $\varrho_j = \kappa \frac{\nu_j}{|\boldsymbol{\nu}|b_j}$ for $j \in \mathrm{supp}(\boldsymbol{\nu})$ and $\varrho_j = \frac{\xi-\kappa}{N b_j}$ for $j \notin \mathrm{supp}(\boldsymbol{\nu})$. Then

$$\sum_{j=1}^N \varrho_j b_j = \kappa \sum_{j\in\mathrm{supp}(\boldsymbol{\nu})} \frac{\nu_j}{|\boldsymbol{\nu}|} + \sum_{j\notin\mathrm{supp}(\boldsymbol{\nu})} \frac{\xi-\kappa}{N} \le \xi.$$

Hence $\boldsymbol{\varrho}$ is $(\boldsymbol{b}, \xi)$-admissible, i.e. there exists a holomorphic extension $u_N : \mathcal{S}(\boldsymbol{\varrho}) \to X$ as in Definition 4.1 (i)-(ii). Applying Cauchy's integral formula as in the proof of Lemma 3.9 we obtain the desired estimate. □

Let us recall the following. Let again X be a separable Hilbert space and $u \in L^2(U, X; \gamma)$. Then

$$L^2(U, X; \gamma) = L^2(U; \gamma) \otimes X$$

with Hilbertian tensor product, and u can be represented in a Wiener-Hermite PC expansion

$$u = \sum_{\boldsymbol{\nu} \in \mathcal{F}} u_{\boldsymbol{\nu}} H_{\boldsymbol{\nu}}, \tag{4.5}$$

where

$$u_{\boldsymbol{\nu}} = \int_U u(\boldsymbol{y}) H_{\boldsymbol{\nu}}(\boldsymbol{y}) \, \mathrm{d}\gamma(\boldsymbol{y})$$

are the Wiener-Hermite PC expansion coefficients. Also, there holds the Parseval-type identity

$$\|u\|^2_{L^2(U,X;\gamma)} = \sum_{\boldsymbol{\nu} \in \mathcal{F}} \|u_{\boldsymbol{\nu}}\|^2_X.$$

When u is $(\boldsymbol{b}, \xi, \delta, X)$-holomorphic, then we have for the functions $u_N : \mathbb{R}^N \to X$ in Definition 4.1

$$u_N = \sum_{\boldsymbol{\nu} \in \mathbb{N}_0^N} u_{N,\boldsymbol{\nu}} H_{\boldsymbol{\nu}},$$

where

$$u_{N,\boldsymbol{\nu}} = \int_{\mathbb{R}^N} u_N(\boldsymbol{y}) H_{\boldsymbol{\nu}}(\boldsymbol{y}) \, \mathrm{d}\gamma_N(\boldsymbol{y}).$$

In an analogous manner to (3.36), for $r \in \mathbb{N}$ and a finite sequence of nonnegative numbers $\boldsymbol{\varrho}_N = (\varrho_j)_{j=1}^N$, we define

$$\beta_{\boldsymbol{\nu}}(r, \boldsymbol{\varrho}_N) := \sum_{\boldsymbol{\nu}' \in \mathbb{N}_0^N : \, \|\boldsymbol{\nu}'\|_{\ell^\infty} \le r} \binom{\boldsymbol{\nu}}{\boldsymbol{\nu}'} \boldsymbol{\varrho}^{2\boldsymbol{\nu}'} = \prod_{j=1}^N \left(\sum_{\ell=0}^r \binom{\nu_j}{\ell} \varrho_j^{2\ell} \right), \quad \boldsymbol{\nu} \in \mathbb{N}_0^N. \tag{4.6}$$

Lemma 4.7 *Let u be $(\boldsymbol{b}, \xi, \delta, X)$-holomorphic, let $N \in \mathbb{N}$ and $\boldsymbol{\varrho}_N = (\varrho_j)_{j=1}^N \in [0, \infty)^N$.*

Then, for any fixed $r \in \mathbb{N}$, there holds the identity

$$\sum_{\boldsymbol{\nu} \in \mathbb{N}_0^N} \beta_{\boldsymbol{\nu}}(r, \boldsymbol{\varrho}_N) \|u_{N,\boldsymbol{\nu}}\|^2_X = \sum_{\{\boldsymbol{\nu} \in \mathbb{N}_0^N : \, \|\boldsymbol{\nu}\|_{\ell^\infty} \le r\}} \frac{\boldsymbol{\varrho}_N^{2\boldsymbol{\nu}}}{\boldsymbol{\nu}!} \int_{\mathbb{R}^N} \|\partial^{\boldsymbol{\nu}} u_N(\boldsymbol{y})\|^2_X \, \mathrm{d}\gamma_N(\boldsymbol{y}). \tag{4.7}$$

Proof From Lemma 4.6, for any $\boldsymbol{\nu} \in \mathbb{N}_0^N$, we have with $\boldsymbol{b}_N = (b_j)_{j=1}^N$

$$\begin{aligned}\int_{\mathbb{R}^N} \|\partial^{\boldsymbol{\nu}} u_N(\boldsymbol{y})\|_X^2 \,\mathrm{d}\gamma_N(\boldsymbol{y}) &\le \int_{\mathbb{R}^N} \left|\frac{\boldsymbol{\nu}!|\boldsymbol{\nu}|^{|\boldsymbol{\nu}|}\boldsymbol{b}_N^{\boldsymbol{\nu}}}{\kappa^{|\boldsymbol{\nu}|}\boldsymbol{\nu}^{\boldsymbol{\nu}}}\varphi_N(\boldsymbol{y})\right|^2 \mathrm{d}\gamma_N(\boldsymbol{y}) \\ &= \left(\frac{\boldsymbol{\nu}!|\boldsymbol{\nu}|^{|\boldsymbol{\nu}|}\boldsymbol{b}_N^{\boldsymbol{\nu}}}{\kappa^{|\boldsymbol{\nu}|}\boldsymbol{\nu}^{\boldsymbol{\nu}}}\right)^2 \int_{\mathbb{R}^N} |\varphi_N(\boldsymbol{y})|^2 \,\mathrm{d}\gamma_N(\boldsymbol{y}) < \infty \qquad (4.8)\end{aligned}$$

by our assumption. This condition allows us to integrate by parts as in the proof of [9, Theorem 3.3]. Following the argument there we obtain (4.7). □

Theorem 4.8 *Let u be $(\boldsymbol{b}, \xi, \delta, X)$-holomorphic for some $\boldsymbol{b} \in \ell^p(\mathbb{N})$ and for some $p \in (0, 1)$. Let $r \in \mathbb{N}$ be arbitrary.*

Then, with

$$\varrho_j := b_j^{p-1} \frac{\xi}{4\sqrt{r!}\|\boldsymbol{b}\|_{\ell^p}}, \quad j \in \mathbb{N}, \qquad (4.9)$$

and $\boldsymbol{\varrho}_N = (\varrho_j)_{j=1}^N$ it holds for all $N \in \mathbb{N}$,

$$\sum_{\boldsymbol{\nu}\in\mathbb{N}_0^N} \beta_{\boldsymbol{\nu}}(r, \boldsymbol{\varrho}_N)\|u_{N,\boldsymbol{\nu}}\|_X^2 \le \delta^2 C(\boldsymbol{b}) < \infty$$

$$\text{with } \|\beta_{\boldsymbol{\nu}}(r, \boldsymbol{\varrho}_N)^{-1/2}\|_{\ell^{p/(1-p)}(\mathbb{N}_0^N)} \le C'(\boldsymbol{b}, \xi) < \infty \qquad (4.10)$$

for some constants $C(\boldsymbol{b})$ and $C'(\boldsymbol{b}, \xi)$ depending on $\boldsymbol{b}$ and ξ, but independent of δ and $N \in \mathbb{N}$.

Furthermore, for every $N \in \mathbb{N}$ and every $q > 0$ there holds

$$(\|u_{N,\boldsymbol{\nu}}\|_X)_{\boldsymbol{\nu}\in\mathbb{N}_0^N} \in \ell^q(\mathbb{N}_0^N).$$

If $q \ge \frac{2p}{2-p}$ then there exists a constant $C > 0$ such that for all $N \in \mathbb{N}$ holds

$$\left\|(\|u_{N,\boldsymbol{\nu}}\|_X)_{\boldsymbol{\nu}\in\mathbb{N}_0^N}\right\|_{\ell^q(\mathbb{N}_0^N)} \le C < \infty.$$

Proof We have from (4.9)

$$\sum_{j\in\mathbb{N}} \varrho_j b_j = \frac{\xi}{4\sqrt{r!}\|\boldsymbol{b}\|_{\ell^p}} \sum_{j\in\mathbb{N}} b_j^p < \infty,$$

and $(\varrho_j^{-1})_{j\in\mathbb{N}} \in \ell^{p/(1-p)}(\mathbb{N})$. Set $\kappa := \xi/2 \in (0, \xi)$. Inserting (4.8) into (4.7) we obtain with $\boldsymbol{\varrho}_N = (\varrho_j)_{j=1}^N$

$$\sum_{\boldsymbol{\nu}\in\mathbb{N}_0^N} \beta_{\boldsymbol{\nu}}(r, \boldsymbol{\varrho}_N)\|u_{N,\boldsymbol{\nu}}\|_X^2 \le \delta^2 \sum_{\{\boldsymbol{\nu}\in\mathbb{N}_0^N : \|\boldsymbol{\nu}\|_{\ell^\infty}\le r\}} \left(\frac{(\boldsymbol{\nu}!)^{1/2}|\boldsymbol{\nu}|^{|\boldsymbol{\nu}|}\boldsymbol{\varrho}_N^{\boldsymbol{\nu}}\boldsymbol{b}_N^{\boldsymbol{\nu}}}{\kappa^{|\boldsymbol{\nu}|}\boldsymbol{\nu}^{\boldsymbol{\nu}}}\right)^2$$

$$\le \delta^2 \sum_{\{\boldsymbol{\nu}\in\mathbb{N}_0^N : \|\boldsymbol{\nu}\|_{\ell^\infty}\le r\}} \left(\frac{|\boldsymbol{\nu}|^{|\boldsymbol{\nu}|}\prod_{j=1}^N \left(\frac{b_j^p}{2\|\boldsymbol{b}\|_{\ell^p}}\right)^{\nu_j}}{\boldsymbol{\nu}^{\boldsymbol{\nu}}}\right)^2,$$

where we used $(\varrho_j b_j)^2 = b_j^{2p}\kappa/(2(r!))$ and the bound

$$\int_{\mathbb{R}^N} \varphi_N(\boldsymbol{y})^2 \,\mathrm{d}\gamma_N(\boldsymbol{y}) \le \delta^2$$

from Definition 4.1 (ii). With $\tilde{b}_j := b_j^p/(2\|\boldsymbol{b}\|_{\ell^p})$ the last term is bounded independent of N by $\delta^2 C(\boldsymbol{b})$ with

$$C(\boldsymbol{b}) := \left(\sum_{\boldsymbol{\nu}\in\mathcal{F}} \frac{|\boldsymbol{\nu}|^{|\boldsymbol{\nu}|}}{\boldsymbol{\nu}^{\boldsymbol{\nu}}}\tilde{\boldsymbol{b}}^{\boldsymbol{\nu}}\right)^2,$$

since the ℓ^1-norm is an upper bound of the ℓ^2-norm. As is well-known, the latter quantity is finite due to $\|\tilde{\boldsymbol{b}}\|_{\ell^1} < 1$, see, e.g., the argument in [37, Page 61].

Now introduce $\tilde{\varrho}_{N,j} := \varrho_j$ if $j \le N$ and $\tilde{\varrho}_{N,j} := \exp(j)$ otherwise. For any $q > 0$ we then have $(\tilde{\varrho}_{N,j}^{-1})_{j\in\mathbb{N}} \in \ell^q(\mathbb{N})$ and by Lemma 3.11 this implies

$$(\beta_{\boldsymbol{\nu}}(r, \tilde{\boldsymbol{\varrho}}_N)^{-1})_{\boldsymbol{\nu}\in\mathcal{F}} \in \ell^{q/2}(\mathcal{F})$$

as long as $r > 2/q$. Using $\beta_{\boldsymbol{\nu}}(r, \tilde{\boldsymbol{\varrho}}_N) = \beta_{(\nu_j)_{j=1}^N}(r, \boldsymbol{\varrho}_N)$ for all $\boldsymbol{\nu} \in \mathcal{F}$ with $\operatorname{supp}\boldsymbol{\nu} \subseteq \{1, \dots, N\}$ we conclude

$$(\beta_{\boldsymbol{\nu}}(r, \boldsymbol{\varrho}_N)^{-1})_{\boldsymbol{\nu}\in\mathbb{N}_0^N} \in \ell^{q/2}(\mathbb{N}_0^N)$$

for any $q > 0$. Now fix $q > 0$ (and choose $r \in \mathbb{N}$ so large that $2/q < r$). Then, by Hölder's inequality with $s := 2(q/2)/(1 + q/2)$, there holds

$$\sum_{\boldsymbol{\nu}\in\mathbb{N}_0^N} \|u_{N,\boldsymbol{\nu}}\|_X^s = \sum_{\boldsymbol{\nu}\in\mathbb{N}_0^N} \|u_{N,\boldsymbol{\nu}}\|_X^s \beta_{\boldsymbol{\nu}}(r, \boldsymbol{\varrho}_N)^{\frac{s}{2}} \beta_{\boldsymbol{\nu}}(r, \boldsymbol{\varrho}_N)^{-\frac{s}{2}}$$

$$\leq \Bigg(\sum_{\boldsymbol{\nu}\in\mathbb{N}_0^N} \|u_{N,\boldsymbol{\nu}}\|_X^2 \beta_{\boldsymbol{\nu}}(r, \boldsymbol{\varrho}_N)\Bigg)^{\frac{s}{2}} \Bigg(\sum_{\boldsymbol{\nu}\in\mathbb{N}_0^N} \beta_{\boldsymbol{\nu}}(r, \boldsymbol{\varrho}_N)^{-\frac{s}{2-s}}\Bigg)^{\frac{2-s}{2}},$$

which is finite since $s/(2-s) = q/2$. Thus we have shown

$$\forall q > 0, N \in \mathbb{N}: \quad (\|u_{N,\boldsymbol{\nu}}\|_X)_{\boldsymbol{\nu}\in\mathbb{N}_0^N} \in \ell^{q/(1+q/2)}(\mathbb{N}_0^N).$$

Finally, due to $(\varrho_j^{-1})_{j\in\mathbb{N}} \in \ell^{p/(1-p)}(\mathbb{N})$, Lemma 3.11 for all $N \in \mathbb{N}$ it holds

$$(\beta_{\boldsymbol{\nu}}(r, \boldsymbol{\varrho}_N)^{-1})_{\boldsymbol{\nu}\in\mathbb{N}_0^N} \in \ell^{p/(2(1-p))}(\mathbb{N}_0^N)$$

and there exists a constant $C'(\boldsymbol{b}, \xi)$ such that for all $N \in \mathbb{N}$ it holds

$$\left\|(\beta_{\boldsymbol{\nu}}(r, \boldsymbol{\varrho}_N)^{-1})_{\boldsymbol{\nu}}\right\|_{\ell^{p/(2(1-p))}(\mathbb{N}_0^N)} \leq C'(\boldsymbol{b}, \xi) < \infty.$$

This completes the proof. □

The following result states the sparsity of Wiener-Hermite PC expansion coefficients of $(\boldsymbol{b}, \xi, \delta, X)$-holomorphic maps.

Theorem 4.9 *Under the assumptions of Theorem 4.8 it holds*

$$\sum_{\boldsymbol{\nu}\in\mathcal{F}} \beta_{\boldsymbol{\nu}}(r, \boldsymbol{\varrho})\|u_{\boldsymbol{\nu}}\|_X^2 \leq \delta^2 C(\boldsymbol{b}) < \infty \quad \text{with} \quad \left(\beta_{\boldsymbol{\nu}}(r, \boldsymbol{\varrho})^{-1/2}\right)_{\boldsymbol{\nu}\in\mathcal{F}} \in \ell^{p/(1-p)}(\mathcal{F}), \tag{4.11}$$

where $C(\boldsymbol{b})$ is the same constant as in Theorem 4.8 and $\beta_{\boldsymbol{\nu}}(r, \boldsymbol{\varrho})$ is given in (3.36). *Furthermore,*

$$(\|u_{\boldsymbol{\nu}}\|_X)_{\boldsymbol{\nu}\in\mathcal{F}} \in \ell^{2p/(2-p)}(\mathcal{F}).$$

Proof Let $\tilde{u}_N \in L^2(U, X; \gamma)$ be as in Definition 4.1 and for $\boldsymbol{\nu} \in \mathcal{F}$ denote by

$$\tilde{u}_{N,\boldsymbol{\nu}} := \int_U \tilde{u}_N(\boldsymbol{y}) H_{\boldsymbol{\nu}}(\boldsymbol{y})\, \mathrm{d}\gamma(\boldsymbol{y}) \in X$$

the Wiener-Hermite PC expansion coefficient. By Fubini's theorem

$$\tilde{u}_{N,\boldsymbol{\nu}} = \int_U u_N((y_j)_{j=1}^N) \prod_{j=1}^N H_{\nu_j}(y_j)\, \mathrm{d}\gamma_N((y_j)_{j=1}^N) = u_{N,(\nu_j)_{j=1}^N}$$

for every $\boldsymbol{\nu} \in \mathcal{F}$ with $\operatorname{supp} \boldsymbol{\nu} \subseteq \{1, \ldots, N\}$. Furthermore, since $\tilde{u}_N$ is independent of the variables $(y_j)_{j=N+1}^\infty$ we have $\tilde{u}_{N,\boldsymbol{\nu}} = 0$ whenever $\operatorname{supp} \boldsymbol{\nu} \subsetneq \{1, \ldots, N\}$. Therefore Theorem 4.8 implies

$$\sum_{\boldsymbol{\nu} \in \mathcal{F}} \beta_{\boldsymbol{\nu}}(r, \boldsymbol{\varrho}) \|\tilde{u}_{N,\boldsymbol{\nu}}\|_X^2 \leq \frac{C(\boldsymbol{b})}{\delta^2} \qquad \forall N \in \mathbb{N}.$$

Now fix an arbitrary, finite set $\Lambda \subset \mathcal{F}$. Because of $\tilde{u}_N \to u \in L^2(U, X; \gamma)$ it holds

$$\lim_{N \to \infty} \tilde{u}_{N,\boldsymbol{\nu}} = u_{\boldsymbol{\nu}}$$

for all $\boldsymbol{\nu} \in \mathcal{F}$. Therefore

$$\sum_{\boldsymbol{\nu} \in \Lambda} \beta_{\boldsymbol{\nu}}(r, \boldsymbol{\varrho}) \|u_{\boldsymbol{\nu}}\|_X^2 = \lim_{N \to \infty} \sum_{\boldsymbol{\nu} \in \Lambda} \beta_{\boldsymbol{\nu}}(r, \boldsymbol{\varrho}) \|\tilde{u}_{N,\boldsymbol{\nu}}\|_X^2 \leq \frac{C(\boldsymbol{b})}{\delta^2}.$$

Since $\Lambda \subset \mathcal{F}$ was arbitrary, this shows that

$$\sum_{\boldsymbol{\nu} \in \mathcal{F}} \beta_{\boldsymbol{\nu}}(r, \boldsymbol{\varrho}) \|u_{\boldsymbol{\nu}}\|_X^2 \leq \delta^2 C(\boldsymbol{b}) < \infty.$$

Finally, due to $\boldsymbol{b} \in \ell^p(\mathbb{N})$, with

$$\varrho_j = b_j^{p-1} \frac{\xi}{4\sqrt{r!}\|\boldsymbol{b}\|_{\ell^p}}$$

as in Theorem 4.8 we have $(\varrho_j^{-1})_{j \in \mathbb{N}} \in \ell^{p/(1-p)}(\mathbb{N})$. By Lemma 3.11, it holds

$$(\beta_{\boldsymbol{\nu}}(r, \boldsymbol{\varrho})^{-1/2})_{\boldsymbol{\nu} \in \mathcal{F}} \in \ell^{p/(1-p)}(\mathcal{F}). \tag{4.12}$$

The relation (4.11) is proven. Hölder's inequality can be used to show that (4.12) gives

$$(\|u_{\boldsymbol{\nu}}\|_X)_{\boldsymbol{\nu} \in \mathcal{F}} \in \ell^{2p/(2-p)}(\mathcal{F})$$

(by a similar calculation as at the end of the proof of Theorem 4.8 with $q = p/(1-p)$). □

Remark 4.10 We establish the convergence rate of best n-term approximation of $(\boldsymbol{b}, \xi, \delta, X)$-holomorphic functions based on the ℓ^p-summability. Let u be $(\boldsymbol{b}, \xi, \delta, X)$-holomorphic for some $\boldsymbol{b} \in \ell^p(\mathbb{N})$ and some $p \in (0,1)$ as in Theorem 4.8. By Theorem 4.9 we then have $(\|u_{\boldsymbol{\nu}}\|_X)_{\boldsymbol{\nu}\in\mathcal{F}} \in \ell^{\frac{2p}{2-p}}$.

Let $\Lambda_n \subseteq \mathcal{F}$ be a set of cardinality $n \in \mathbb{N}$ containing n multiindices $\boldsymbol{\nu} \in \mathcal{F}$ such that $\|u_{\boldsymbol{\mu}}\|_X \le \|u_{\boldsymbol{\nu}}\|_X$ whenever $\boldsymbol{\nu} \in \Lambda_n$ and $\boldsymbol{\mu} \notin \Lambda_n$. Then, by Theorem 4.9, for the truncated the Wiener-Hermite PC expansion we have the error bound

$$\left\| u - \sum_{\boldsymbol{\nu}\in\Lambda_n} u_{\boldsymbol{\nu}} H_{\boldsymbol{\nu}} \right\|^2_{L^2(U,X;\gamma)} = \sum_{\boldsymbol{\nu}\in\mathcal{F}\setminus\Lambda_n} \|u_{\boldsymbol{\nu}}\|_X^2$$

$$\le \sup_{\boldsymbol{\nu}\in\mathcal{F}\setminus\Lambda_n} \|u_{\boldsymbol{\nu}}\|_X^{2-\frac{2p}{2-p}} \sum_{\boldsymbol{\mu}\in\mathcal{F}\setminus\Lambda_n} \|u_{\boldsymbol{\mu}}\|_X^{\frac{2p}{2-p}}.$$

For a nonnegative, monotonically decreasing sequence $(x_j)_{j\in\mathbb{N}} \in \ell^q(\mathbb{N})$ with $q > 0$ we have

$$x_n^q \le \frac{1}{n}\sum_{j=1}^n x_j^q$$

and thus

$$x_n \le n^{-\frac{1}{q}} \|(x_j)_{j\in\mathbb{N}}\|_{\ell^q(\mathbb{N})}.$$

With $q = \frac{2p}{2-p}$ this implies

$$\left(\sup_{\boldsymbol{\nu}\in\mathcal{F}\setminus\Lambda_n} \|u_{\boldsymbol{\nu}}\|_X\right)^{2-\frac{2p}{2-p}} \le \left(n^{-\frac{2-p}{2p}} \left(\sum_{\boldsymbol{\nu}\in\mathcal{F}} \|u_{\boldsymbol{\nu}}\|_X^{\frac{2p}{2-p}} \right)^{\frac{2-p}{2p}} \right)^{2-\frac{2p}{2-p}} = \mathcal{O}(n^{-\frac{2}{p}+2}).$$

Hence, by truncating the Wiener-Hermite PC expansion (4.5) after n largest terms yields the best n-term convergence rate

$$\left\| u - \sum_{\boldsymbol{\nu}\in\Lambda_n} u_{\boldsymbol{\nu}} H_{\boldsymbol{\nu}} \right\|_{L^2(U,X;\gamma)} = \mathcal{O}(n^{-\frac{1}{p}+1}) \qquad \text{as } n \to \infty. \tag{4.13}$$

4.2 $(\boldsymbol{b}, \xi, \delta, X)$-Holomorphy of Composite Functions

We now show that certain composite functions of the type

$$u(\boldsymbol{y}) = \mathcal{U}\Big(\exp\Big(\sum_{j\in\mathbb{N}} y_j\psi_j\Big)\Big) \tag{4.14}$$

are $(\boldsymbol{b}, \xi, \delta, X)$-holomorphic under certain conditions.

The significance of such functions is the following: if we think for example of $\mathcal{U}$ as the solution operator $\mathcal{S}$ in (3.5) (for a fixed f) which maps the diffusion coefficient $a \in L^\infty(D)$ to the solution $\mathcal{U}(a) \in H_0^1(D)$ of an elliptic PDE on some domain $D \subseteq \mathbb{R}^d$, then $\mathcal{U}\big(\exp\big(\sum_{j\in\mathbb{N}} y_j\psi_j\big)\big)$ is exactly the parametric solution discussed in Sects. 3.1–3.6. We explain this in more detail in Sect. 4.3.1. The presently developed, abstract setting allows, however, to *consider $\mathcal{U}$ as a solution operator of other, structurally similar PDEs with log-Gaussian random input data.* Furthermore, if $\mathcal{G}$ is another map with suitable holomorphy properties, the composition $\mathcal{G}\big(\mathcal{U}\big(\exp\big(\sum_{j\in\mathbb{N}} y_j\psi_j\big)\big)\big)$ is again of the general type $\tilde{\mathcal{U}}\big(\exp\big(\sum_{j\in\mathbb{N}} y_j\psi_j\big)\big)$ with $\tilde{\mathcal{U}} = \mathcal{G}\circ\mathcal{U}$.

This will allow to apply the ensuing results on convergence rates of deterministic collocation and quadrature algorithms to a wide range of PDEs with GRF inputs and functionals on their random solutions. As a particular case in point, we apply our results to posterior densities in Bayesian inversion, as we explain subsequently in Chap. 5. As a result, the concept of $(\boldsymbol{b}, \xi, \delta, X)$-holomorphy is fairly broad and covers a large range of parametric PDEs depending on log-Gaussian distributed data.

To formalize all of this, *we now provide sufficient conditions on the solution operator $\mathcal{U}$ and the sequence $(\psi_j)_{j\in\mathbb{N}}$ guaranteeing $(\boldsymbol{b}, \xi, \delta, X)$-holomorphy.* Let $d \in \mathbb{N}$, $D \subseteq \mathbb{R}^d$ be an open set and E a complex Banach space which is continuously embedded into $L^\infty(D;\mathbb{C})$, and finally let X be another complex Banach space. Additionally, suppose that there exists $C_E > 0$ such that for all $\psi_1, \psi_2 \in E$ and some $m \in \mathbb{N}$

$$\|\exp(\psi_1) - \exp(\psi_2)\|_E \le C_E\|\psi_1 - \psi_2\|_E \max\Big\{\exp\big(m\|\psi_1\|_E\big); \exp\big(m\|\psi_2\|_E\big)\Big\}. \tag{4.15}$$

This inequality covers in particular the Sobolev spaces $W^k_\infty(D;\mathbb{C})$, $k \in \mathbb{N}_0$, on bounded Lipschitz domains $D \subseteq \mathbb{R}^d$, but also the Kondrat'ev spaces $\mathcal{W}^k_\infty(D;\mathbb{C})$ on polygonal domains $D \subseteq \mathbb{R}^2$, cp. Lemma 3.28.

For a function $\psi \in E \subseteq L^\infty(D;\mathbb{C})$ we will write $\Re(\psi) \in L^\infty(D;\mathbb{R}) \subseteq L^\infty(D;\mathbb{C})$ to denote its real part and $\Im(\psi) \in L^\infty(D;\mathbb{R}) \subseteq L^\infty(D;\mathbb{C})$ its imaginary part so that $\psi = \Re(\psi) + \mathrm{i}\Im(\psi)$. Recall that the quantity $\rho(a)$ is defined in (3.25) for $a \in L^\infty(D;\mathbb{C})$.

Theorem 4.11 *Let* $0 < \delta < \delta_{\max}$, $K > 0$, $\eta > 0$ *and* $m \in \mathbb{N}$. *Let the inequality* (4.15) *hold for the space* E. *Assume that for an open set* $O \subseteq E$ *containing*

$$\{\exp(\psi) : \psi \in E, \ \|\Im(\psi)\|_E \leq \eta\},$$

it holds

(i) $\mathcal{U} : O \to X$ *is holomorphic,*
(ii) *for all* $a \in O$

$$\|\mathcal{U}(a)\|_X \leq \delta \left(\frac{1 + \|a\|_E}{\min\{1, \rho(a)\}} \right)^m,$$

(iii) *for all* $a, b \in O$

$$\|\mathcal{U}(a) - \mathcal{U}(b)\|_X \leq K \left(\frac{1 + \max\{\|a\|_E, \|b\|_E\}}{\min\{1, \rho(a), \rho(b)\}} \right)^m \|a - b\|_E,$$

(iv) $(\psi_j)_{j\in\mathbb{N}} \subseteq E \cap L^\infty(D)$ *and with* $b_j := \|\psi_j\|_E$ *it holds* $\boldsymbol{b} \in \ell^1(\mathbb{N})$.

Then there exists $\xi > 0$ *and for every* $\delta_{\max} > 0$ *there exists* $\tilde{C}$ *depending on* $\boldsymbol{b}$, $\delta_{\max}$, C_E *and* m *but independent of* $\delta \in (0, \delta_{\max})$, *such that with*

$$u_N\left((y_j)_{j=1}^N\right) = \mathcal{U}\left(\exp\left(\sum_{j=1}^N y_j \psi_j \right) \right) \qquad \forall (y_j)_{j=1}^N \in \mathbb{R}^N,$$

and $\tilde{u}_N(\boldsymbol{y}) = u_N(y_1, \ldots, y_N)$ *for* $\boldsymbol{y} \in U$, *the function*

$$u := \lim_{N\to\infty} \tilde{u}_N \in L^2(U, X; \gamma)$$

is well-defined and $(\boldsymbol{b}, \xi, \delta\tilde{C}, X)$*-holomorphic.*

Proof

Step 1. Choosing $\psi_2 \equiv 0$ in (4.15) with $\psi_1 = \psi$, we obtain

$$\|\exp(\psi)\|_E \leq C'_E \exp\left((m+1)\|\psi\|_E\right) \tag{4.16}$$

for some positive constant C'_E. Indeed,

$$\begin{aligned}
\|\exp(\psi_1)\|_E &\leq \|1\|_E + C_E\|\psi_1\|_E \exp\left(m\|1\|_E + m\|\psi_1\|_E\right) \\
&\leq C'_E(1 + \|\psi_1\|_E) \exp\left(m\|\psi_1\|_E\right) \\
&\leq C'_E \exp\left((m+1)\|\psi_1\|_E\right).
\end{aligned}$$

We show that $u_N \in L^2(\mathbb{R}^N, X; \gamma_N)$ for every $N \in \mathbb{N}$. To this end we recall that for any $s > 0$ (see, e.g., [72, Appendix B], [9, (38)] for a proof)

$$\int_{\mathbb{R}} \exp(s|y|)\, \mathrm{d}\gamma_1(y) \leq \exp\left(\frac{s^2}{2} + \frac{\sqrt{2}s}{\pi}\right). \tag{4.17}$$

Since E is continuously embedded into $L^\infty(D; \mathbb{C})$, there exists $C_0 > 0$ such that

$$\|\psi\|_{L^\infty(D)} \leq C_0 \|\psi\|_E \qquad \forall \psi \in E.$$

Using (ii), (4.16), and

$$\begin{aligned}
\frac{1}{\operatorname{ess\,inf}_{\boldsymbol{x}\in D}\left(\exp\left(\sum_{j=1}^N y_j \psi_j(\boldsymbol{x})\right)\right)} &\leq \left\| \exp\left(-\sum_{j=1}^N y_j \psi_j\right)\right\|_{L^\infty} \\
&\leq \exp\left(\left\|\sum_{j=1}^N y_j \psi_j\right\|_{L^\infty}\right) \\
&\leq \exp\left(C_0 \sum_{j=1}^N |y_j| \|\psi_j\|_E\right),
\end{aligned}$$

we obtain the bound

$$\begin{aligned}
\|u_N(\boldsymbol{y})\|_X &\leq \delta\left(1 + \left\|\exp\left(\sum_{j=1}^N y_j \psi_j\right)\right\|_E\right)^m \exp\left(C_0 m \sum_{j=1}^N |y_j| \|\psi_j\|_E\right) \\
&\leq \delta\left(1 + C'_E \exp\left((m+1)\sum_{j=1}^N |y_j|\, \|\psi_j\|_E\right)\right)^m \\
&\quad \times \exp\left(C_0 m \sum_{j=1}^N |y_j| \|\psi_j\|_E\right) \\
&\leq C_1 \exp\left((2 + C_0) m^2 \sum_{j=1}^N |y_j|\, \|\psi_j\|_E\right)
\end{aligned}$$

for some constant $C_1 > 0$ depending on δ, C_E and m. Hence, by (4.17) we have

$$\int_{\mathbb{R}^N} \|u_N(\boldsymbol{y})\|_X^2 \,\mathrm{d}\gamma_N(\boldsymbol{y}) \le C_1 \int_{\mathbb{R}^N} \exp\left((2+C_0)m^2)\sum_{j=1}^N |y_j|\|\psi_j\|_E\right) \mathrm{d}\gamma_N(\boldsymbol{y})$$

$$\le C_1 \exp\left(\frac{(2+C_0)^2 m^4}{2}\sum_{j=1}^N b_j^2 + \frac{\sqrt{2}(2+C_0)m^2}{\pi}\sum_{j=1}^N b_j\right)$$

$$< \infty.$$

Step 2. We show that $(\tilde{u}_N)_{N\in\mathbb{N}}$ which is defined as $\tilde{u}_N(\boldsymbol{y}) := u_N(y_1,\ldots,y_N)$ for $\boldsymbol{y} \in U$, is a Cauchy sequence in $L^2(U,X;\gamma)$. For any $N < M$ by (iii)

$$\|\tilde{u}_M - \tilde{u}_N\|_{L^2(U,X;\gamma)}^2$$

$$= \int_U \left\|\mathcal{U}\left(\exp\left(\sum_{j=1}^M y_j\psi_j\right)\right) - \mathcal{U}\left(\exp\left(\sum_{j=1}^N y_j\psi_j\right)\right)\right\|_X^2 \mathrm{d}\gamma(\boldsymbol{y})$$

$$\le K \int_U \left[\left(1 + \left\|\exp\left(\sum_{j=1}^M y_j\psi_j\right)\right\|_E + \left\|\exp\left(\sum_{j=1}^N y_j\psi_j\right)\right\|_E\right)^m\right.$$

$$\left.\times \exp\left(C_0 m\sum_{j=1}^M |y_j|\|\psi_j\|_E\right)\cdot\left\|\exp\left(\sum_{j=1}^M y_j\psi_j\right) - \exp\left(\sum_{j=1}^N y_j\psi_j\right)\right\|_E\right]\mathrm{d}\gamma(\boldsymbol{y}).$$

Using (4.16) again we can estimate

$$\|\tilde{u}_M - \tilde{u}_N\|_{L^2(U,X;\gamma)}^2$$

$$\le K \int_U \left[\left(1 + 2C_E' \exp\left((m+1)\sum_{j=1}^M |y_j|\,\|\psi_j\|_E\right)\right)^m\right.$$

$$\left.\times \exp\left(C_0 m\sum_{j=1}^M |y_j|\|\psi_j\|_E\right)\cdot\left\|\exp\left(\sum_{j=1}^M y_j\psi_j\right) - \exp\left(\sum_{j=1}^N y_j\psi_j\right)\right\|_E\right]\mathrm{d}\gamma(\boldsymbol{y})$$

$$\le C_2 \int_U \left[\exp\left((2+C_0)m^2)\sum_{j=1}^M |y_j|\,\|\psi_j\|_E\right)\right.$$

$$\left.\times \left\|\exp\left(\sum_{j=1}^M y_j\psi_j\right) - \exp\left(\sum_{j=1}^N y_j\psi_j\right)\right\|_E\right]\mathrm{d}\gamma(\boldsymbol{y})$$

for $C_2 > 0$ depending only on K, C_E, and m. Now, employing (4.15) we obtain

$$\left\| \exp\Big(\sum_{j=1}^{M} y_j\psi_j\Big) - \exp\Big(\sum_{j=1}^{N} y_j\psi_j\Big) \right\|_E$$

$$\leq C_E \sum_{j=N+1}^{M} |y_j| \|\psi_j\|_E \quad \exp\left(m \sum_{j=1}^{M} |y_j| \|\psi_j\|_E \right)$$

$$\leq C_E \sum_{j=N+1}^{M} |y_j| \|\psi_j\|_E \exp\left(m^2 \sum_{j=1}^{M} |y_j| \|\psi_j\|_E \right).$$

Therefore for a constant C_3 depending on C_E and δ (but independent of N), using $|y_j| \leq \exp(|y_j|)$,

$$\|\tilde{u}_M - \tilde{u}_N\|^2_{L^2(U,X;\gamma)}$$

$$\leq C_3 \sum_{j=N+1}^{M} \|\psi_j\|_E \int_{\mathbb{R}^M} |y_j| \exp\left((3 + C_0)m^2 \sum_{i=1}^{M} |y_i| \|\psi_i\|_E \right) \mathrm{d}\gamma_M((y_i)_{i=1}^M)$$

$$\leq C_3 \sum_{j=N+1}^{M} \|\psi_j\|_E \int_{\mathbb{R}^M} \exp\left(|y_j| + (3 + C_0)m^2 \sum_{i=1}^{M} |y_i| \|\psi_i\|_E \right) \mathrm{d}\gamma_M((y_i)_{i=1}^M)$$

$$\leq C_3 \Big(\sum_{j=N+1}^{M} b_j \Big)$$

$$\times \left(\exp\left(\frac{1}{2} + \frac{\sqrt{2}}{\pi} + \frac{(3 + C_0)^2 m^4}{2} \sum_{i=1}^{M} b_j^2 + \frac{\sqrt{2}(3 + C_0)m^2}{\pi} \sum_{j=1}^{M} b_j \right) \right),$$

where we used (4.17) and in the last inequality. Since $\boldsymbol{b} \in \ell^1(\mathbb{N})$ the last term is bounded by $C_4\Big(\sum_{j=N+1}^{\infty} b_j \Big)$ for a constant C_4 depending on C_E, K and $\boldsymbol{b}$ but independent of N, M. Due to $\boldsymbol{b} \in \ell^1(\mathbb{N})$, it also holds

$$\sum_{j=N+1}^{\infty} b_j \to 0 \text{ as } N \to \infty.$$

Since $N < M$ are arbitrary, we have shown that $(\tilde{u}_N)_{N\in\mathbb{N}}$ is a Cauchy sequence in the Banach space $L^2(U, X; \gamma)$. This implies that there is a function

$$u := \lim_{N\to\infty} \tilde{u}_N \in L^2(U, X; \gamma).$$

Step 3. To show that u is $(\boldsymbol{b}, \xi, \delta\tilde{C}, X)$ holomorphic, we provide constants $\xi > 0$ and $\tilde{C} > 0$ independent of δ so that u_N admits holomorphic extensions as in Definition 4.1. This concludes the proof.

Let $\xi := \pi/(4C_0)$. Fix $N \in \mathbb{N}$ and assume

$$\sum_{j=1}^{N} b_j \varrho_j < \xi$$

(i.e. $(\varrho_j)_{j=1}^N$ is $(\boldsymbol{b}, \delta_1)$-admissible). Then for $z_j = y_j + \mathrm{i}\zeta_j \in \mathbb{C}$ such that $|\Im(z_j)| = |\zeta_j| < \varrho_j$ for all j,

$$\rho\left(\exp\left(\sum_{j=1}^{N} z_j \psi_j(\boldsymbol{x})\right)\right) = \operatorname*{ess\,inf}_{\boldsymbol{x}\in D}\left(\exp\left(\sum_{j=1}^{N} y_j \psi_j(\boldsymbol{x})\right)\right)\cos\left(\sum_{j=1}^{N} \zeta_j \psi_j(\boldsymbol{x})\right).$$

Due to

$$\begin{aligned}\operatorname*{ess\,sup}_{\boldsymbol{x}\in D}\left|\sum_{j=1}^{N} \zeta_j \psi_j(\boldsymbol{x})\right| &\le \sum_{j=1}^{N} \varrho_j \|\psi_j\|_{L^\infty} \le \sum_{j=1}^{N} C_0 \varrho_j \|\psi_j\|_V \\ &= \sum_{j=1}^{N} C_0 \varrho_j b_j \le \frac{\pi}{4},\end{aligned}$$

we obtain for such $(z_j)_{j=1}^N$

$$\rho\left(\exp\left(\sum_{j=1}^{N} z_j \psi_j(\boldsymbol{x})\right)\right) \ge \exp\left(-\sum_{j=1}^{N} |y_j| \|\psi_j\|_{L^\infty}\right)\cos\left(\frac{\pi}{4}\right) > 0. \quad (4.18)$$

This shows that for every $\boldsymbol{\varrho} = (\varrho_j)_{j=1}^N \in (0,\infty)^N$ such that $\sum_{j=1}^N b_j \varrho_j < \xi$, it holds

$$\sum_{j=1}^{N} z_j \psi_j \in O \quad \forall \boldsymbol{z} \in \mathcal{S}(\boldsymbol{\varrho}).$$

Since $\mathcal{U} : O \to X$ is holomorphic, the function

$$u_N\left((y_j)_{j=1}^N\right) = \mathcal{U}\left(\exp\left(\sum_{j=1}^N y_j\psi_j\right)\right)$$

can be holomorphically extended to arguments $(z_j)_{j=1}^N \in \mathcal{S}(\boldsymbol{\varrho})$.

Finally we fix again $N \in \mathbb{N}$ and provide a function $\varphi_N \in L^2(U;\gamma)$ as in Definition 4.1. Fix $\boldsymbol{y} \in \mathbb{R}^N$ and $\boldsymbol{z} \in \mathcal{B}_{\boldsymbol{\varrho}}$ and set

$$a := \sum_{j=1}^N (y_j + z_j)\psi_j.$$

By (ii), (4.18) and because $b_j = \|\psi_j\|_E$ and

$$\sum_{j=1}^N b_j\varrho_j \le \xi,$$

we have that

$$\begin{aligned}
&\|u_N((y_j+z_j)_{j=1}^N)\|_X \\
&\le \delta\left(\frac{1+\|a\|_E}{\min\{1,\rho(a)\}}\right)^m \\
&\le \delta\left(\frac{1+C_E'\exp\left((m+1)\sum_{j=1}^N(|y_j|+|z_j|)\|\psi_j\|_E\right)}{\exp(-C_0\sum_{j=1}^N(|y_j|+|z_j|)\|\psi_j\|_E)\cos(\frac{\pi}{4})}\right)^m \\
&\le \delta\left(\frac{1+C_E'\exp\left((m+1)\sum_{j=1}^N|y_j|b_j\right)\exp((m+1)\xi)}{\exp(-C_0\sum_{j=1}^N|y_j|b_j)\exp(-C_0\xi)\cos(\frac{\pi}{4})}\right)^m \\
&\le \delta L\exp\left((2+C_0)m^2\sum_{j=1}^N|y_j|b_j\right)
\end{aligned}$$

for some L depending only on C_E, C_0 and m. Let us define the last quantity as $\varphi_N\left((y_j)_{j=1}^N\right)$. Then by (4.17) and because γ_N is a probability measure on $\mathbb{R}^N$,

$$\|\varphi_N\|_{L^2(\mathbb{R}^N;\gamma_N)} \le \delta L\exp\left(\sum_{j=1}^N\frac{(2+C_0)^2m^4b_j^2}{2}+(2+C_0)m^2\frac{\sqrt{2}b_j}{\pi}\right)$$

$$\leq \delta L \exp\left(\sum_{j\in\mathbb{N}} \frac{(2+C_0)^2 m^4 b_j^2}{2} + (2+C_0)m\frac{\sqrt{2}b_j}{\pi}\right)$$
$$\leq \delta\tilde{C}(\boldsymbol{b}, C_0, C_E, m),$$

for some constant $\tilde{C}(\boldsymbol{b}, C_0, C_E, m) \in (0, \infty)$ because $\boldsymbol{b} \in \ell^1(\mathbb{N})$. In all, we have shown that u satisfies $(\boldsymbol{b}, \xi, \delta\tilde{C}, X)$-holomorphy as in Definition 4.1. □

4.3 Examples of Holomorphic Data-to-Solution Maps

We revisit the example of linear elliptic divergence-form PDE with diffusion coefficient introduced in Chap. 3. Its coefficient-to-solution map S from (3.5) for a fixed $f \in X'$, gives rise to parametric maps which are parametric-holomorphic. This kind of function will, on the one hand, arise as generic model of Banach-space valued uncertain inputs of PDEs, and on the other hand as model of solution manifolds of PDEs. The connection is made through preservation of holomorphy under composition with inversion of boundedly invertible differential operators.

Let $f \in X'$ be given. If $A(a) \in \mathcal{L}_{\mathrm{is}}(X, X')$ is an isomorphism depending (locally) holomorphically on $a \in E$, then

$$\mathcal{U} : E \to X : a \mapsto (\mathrm{inv} \circ A(a))f$$

is also locally holomorphic as a function of $a \in E$. Here inv denotes the inversion map. This is a consequence of the fact that the map $\mathrm{inv} : \mathcal{L}_{\mathrm{is}}(X, X') \to \mathcal{L}_{\mathrm{is}}(X', X)$ is holomorphic, see e.g. [111, Example 1.2.38]. This argument can be used to show that the solution operator corresponding to the solution of certain PDEs is holomorphic in the parameter. We informally discuss this for some parametric PDEs and refer to [111, Chapters 1 and 5] for more details.

4.3.1 *Linear Elliptic Divergence-Form PDE with Parametric Diffusion Coefficient*

Let us again consider the model linear elliptic PDE

$$-\,\mathrm{div}(a\nabla\mathcal{U}(a)) = f \;\text{ in } D, \quad \mathcal{U}(a) = 0 \;\text{ on } \partial D, \tag{4.19}$$

where $d \in \mathbb{N}$, $D \subseteq \mathbb{R}^d$ is a bounded Lipschitz domain, $X := H_0^1(D; \mathbb{C})$, $f \in H^{-1}(D; \mathbb{C}) := (H_0^1(D; \mathbb{C}))'$ and $a \in E := L^\infty(D; \mathbb{C})$. Then the solution operator

$\mathcal{U} : O \to X$ maps the coefficient function a to the weak solution $\mathcal{U}(a)$, where

$$O := \{a \in L^\infty(D; \mathbb{C}) : \rho(a) > 0\},$$

with $\rho(a)$ defined in (3.25) for $a \in L^\infty(D; \mathbb{C})$. With $A(a)$ denoting the differential operator $-\operatorname{div}(a\nabla\cdot) \in L(X, X')$ we can also write $\mathcal{U}(a) = A(a)^{-1}f$. We now check assumptions (i)–(iii) of Theorem 4.11.

(i) As mentioned above, complex Fréchet differentiability (i.e. holomorphy) of $\mathcal{U} : O \to X$ is satisfied because the operation of inversion of linear operators is holomorphic on the set of boundedly invertible linear operators, A depends boundedly and linearly (thus holomorphically) on a, and therefore, the map

$$a \mapsto A(a)^{-1}f = \mathcal{U}(a)$$

is a composition of holomorphic functions. We refer once more to [111, Example 1.2.38] for more details.

(ii) For $a \in O$, it holds

$$\|\mathcal{U}(a)\|_X^2\, \rho(a) \le \left| \int_D \nabla\mathcal{U}(a)^\top a \overline{\nabla\mathcal{U}(a)}\, \mathrm{d}\boldsymbol{x} \right| = \left|\left\langle f, \overline{\mathcal{U}(a)} \right\rangle\right| \le \|f\|_{X'} \|\mathcal{U}(a)\|_X.$$

Here $\langle \cdot, \cdot \rangle$ denotes the dual product between X' and X. This gives the usual a-priori bound

$$\|\mathcal{U}(a)\|_X \le \frac{\|f\|_{X'}}{\rho(a)}. \tag{4.20}$$

(iii) For $a, b \in O$ and with $w := \mathcal{U}(a) - \mathcal{U}(b)$, we have that

$$\begin{aligned} \frac{\|w\|_X^2}{\rho(a)} &\le \left| \int_D \nabla w^\top a \overline{\nabla w}\, \mathrm{d}\boldsymbol{x} \right| \\ &= \left| \int_D \nabla\mathcal{U}(a)^\top a \overline{\nabla w}\, \mathrm{d}\boldsymbol{x} - \int_D \nabla\mathcal{U}(b)^\top b \overline{\nabla w}\, \mathrm{d}\boldsymbol{x} \right. \\ &\quad \left. - \int_D \nabla\mathcal{U}(b)^\top (a-b) \overline{\nabla w}\, \mathrm{d}\boldsymbol{x} \right| \\ &\le \|\mathcal{U}(b)\|_X \|w\|_X \|a-b\|_E \\ &\le \frac{\|f\|_{X'}}{\rho(b)} \|w\|_X \|a-b\|_E, \end{aligned}$$

and thus

$$\|\mathcal{U}(a) - \mathcal{U}(b)\|_X \le \|f\|_{X'} \frac{\|a\|_E}{\rho(b)} \|a-b\|_E. \tag{4.21}$$

Hence, if $(\psi_j)_{j\in\mathbb{N}} \subset E$ such that with $b_j := \|\psi_j\|_E$ it holds $\boldsymbol{b} \in \ell^1(\mathbb{N})$, then the solution

$$u(\boldsymbol{y}) = \lim_{N\to\infty} \mathcal{U}\left(\exp\left(\sum_{j=1}^{N} y_j\psi_j\right)\right) \in L^2(U, X; \gamma)$$

is well-defined and $(\boldsymbol{b}, \xi, \delta, X)$-holomorphic by Theorem 4.11.

This example can easily be generalized to spaces of higher-regularity, e.g., if $D \subseteq \mathbb{R}^d$ is a bounded C^{s-1} domain for some $s \in \mathbb{N}$, $s \geq 2$, then we may set $X := H_0^1(D;\mathbb{C}) \cap H^s(D;\mathbb{C})$ and $E := W_\infty^s(D;\mathbb{C})$ and repeat the above calculation.

4.3.2 Linear Parabolic PDE with Parametric Coefficient

Let $0 < T < \infty$ denote a finite time-horizon and let D be a bounded domain with Lipschitz boundary ∂D in $\mathbb{R}^d$. We define $I := (0, T)$ and consider the initial boundary value problem (IBVP for short) for the linear parabolic PDE

$$\begin{cases} \frac{\partial u(t,\boldsymbol{x})}{\partial t} - \operatorname{div}\left(a(\boldsymbol{x})\nabla u(t,\boldsymbol{x})\right) = f(t,\boldsymbol{x}), & (t,\boldsymbol{x}) \in I \times D, \\ u|_{\partial D\times I} = 0, & \\ u|_{t=0} = u_0(\boldsymbol{x}). & \end{cases} \tag{4.22}$$

In this section, we prove that the solution to this problem satisfies the assumptions of Theorem 4.11 for certain spaces E and X. We first review results on the existence and uniqueness of solutions to the Eq. (4.22). We refer to [97] and the references there for proofs and more detailed discussion.

We denote $V := H_0^1(D;\mathbb{C})$ and $V' := H^{-1}(D;\mathbb{C})$. The parabolic IBPV given by Eq. (4.22) is a well-posed operator equation in the intersection space of Bochner spaces (e.g. [97, Appendix], and e.g. [53, 110] for the definition of spaces)

$$X := L^2(I, V) \cap H^1(I, V') = \left(L^2(I) \otimes V\right) \cap \left(H^1(I) \otimes V'\right)$$

equipped with the sum norm

$$\|u\|_X := \left(\|u\|^2_{L^2(I,V)} + \|u\|^2_{H^1(I,V')}\right)^{1/2}, \qquad u \in X,$$

where

$$\|u\|^2_{L^2(I,V)} = \int_I \|u(t,\cdot)\|_V^2 \,\mathrm{d}t,$$

and

$$\|u\|^2_{H^1(I,V')} = \int_I \|\partial_t u(t,\cdot)\|^2_{V'} \, dt.$$

To state a space-time variational formulation and to specify the data space for (4.22), we introduce the test-function space

$$Y = L^2(I,V) \times L^2(D) = \left(L^2(I) \otimes V\right) \times L^2(D)$$

which we endow with the norm

$$\|v\|_Y = \left(\|v_1\|^2_{L^2(I,V)} + \|v_2\|^2_{L^2(D)}\right)^{1/2}, \qquad v = (v_1, v_2) \in Y.$$

Given a time-independent diffusion coefficient $a \in L^\infty(D;\mathbb{C})$ and $(f, u_0) \in Y'$, the continuous sesquilinear and antilinear forms corresponding to the parabolic problem (4.22) reads for $u \in X$ and $v = (v_1, v_2) \in Y$ as

$$B(u,v;a) := \int_I\int_D \partial_t u\, \overline{v_1}\, \mathrm{d}\boldsymbol{x}\, \mathrm{d}t + \int_I\int_D a\nabla u \cdot \overline{\nabla v_1}\, \mathrm{d}\boldsymbol{x}\, \mathrm{d}t + \int_D u_0\, \overline{v_2}\, \mathrm{d}\boldsymbol{x}$$

and

$$L(v) := \int_I \left\langle f(t,\cdot), v_1(t,\cdot)\right\rangle \mathrm{d}t + \int_D u_0\, \overline{v_2}\, \mathrm{d}\boldsymbol{x},$$

where $\langle\cdot,\cdot\rangle$ is the anti-duality pairing between V' and V. Then the space-time variational formulation of Eq. (4.22) is: Find $\mathcal{U}(a) \in X$ such that

$$B(\mathcal{U}(a), v; a) = L(v), \quad \forall v \in Y. \tag{4.23}$$

The existence and uniqueness of solution to the Eq. (4.23) was proved in [97] which reads as follows.

Proposition 4.12 *Assume that* $(f, u_0) \in Y'$ *and that*

$$0 < \rho(a) := \operatorname*{ess\,inf}_{\boldsymbol{x}\in D} \Re(a(\boldsymbol{x})) \le |a(\boldsymbol{x})| \le \|a\|_{L^\infty} < \infty, \qquad \boldsymbol{x} \in D. \tag{4.24}$$

Then the parabolic operator $\mathcal{B} \in \mathcal{L}(X, Y')$ *defined by*

$$(\mathcal{B}u)(v) = B(u, v; a),$$

is an isomorphism and $\mathcal{B}^{-1} : Y \to X$ *has the norm*

$$\|\mathcal{B}^{-1}\| \le \frac{1}{\beta(a)},$$

where

$$\beta(a) := \frac{\min\left(\rho(a)\|a\|_{L^\infty}^{-2}, \rho(a)\right)}{\sqrt{2\max(\rho(a)^{-2}, 1) + \vartheta^2}} \quad \textit{and} \quad \vartheta := \sup_{w\neq 0, w\in X} \frac{\|w(0,\cdot)\|_{L^2(D)}}{\|w\|_X}.$$

The constant ϑ depends only on T.

The data space for the Eq. (4.22) for complex-valued data is $E := L^\infty(D, \mathbb{C})$. With the set of admissible diffusion coefficients in the data space

$$O := \{a \in L^\infty(D, \mathbb{C}) \, : \, \rho(a) > 0\},$$

from the above proposition we immediately deduce that for given $(f, u_0) \in Y'$, the map

$$\mathcal{U} : O \to X : a \mapsto \mathcal{U}(a)$$

is well-defined.

Furthermore, there holds the a-priori estimate

$$\|\mathcal{U}(a)\|_X \le \frac{1}{\beta(a)}\left(\|f\|_{L^2(I,V')}^2 + \|u_0\|_{L^2}^2\right)^{1/2}. \tag{4.25}$$

This bound is a consequence of the following result which states that the data-to-solution map $a \to \mathcal{U}(a)$ is locally Lipschitz continuous.

Lemma 4.13 *Let $(f, u_0) \in Y'$. Assume that $\mathcal{U}(a)$ and $\mathcal{U}(b)$ be solutions to* (4.23) *with coefficients a, b satisfying* (4.24), *respectively.*

Then, with the function $\beta(\cdot)$ in variable a as in Proposition 4.12, we have

$$\|\mathcal{U}(a) - \mathcal{U}(b)\|_X \le \frac{1}{\beta(a)\beta(b)}\|a-b\|_{L_\infty}\left(\|f\|_{L^2(I,V')}^2 + \|u_0\|_{L^2}^2\right)^{1/2}.$$

Proof From (4.23) we find that for $w := \mathcal{U}(a) - \mathcal{U}(b)$,

$$\begin{aligned}\int_I\int_D \partial_t w\, \overline{v_1}\, \mathrm{d}\boldsymbol{x}\, \mathrm{d}t + \int_I\int_D a\nabla w\cdot\overline{\nabla v_1}\, \mathrm{d}\boldsymbol{x}\, \mathrm{d}t + \int_D w\big|_{t=0}\overline{v_2}\, \mathrm{d}\boldsymbol{x} \\ = -\int_I\int_D (a-b)\nabla\mathcal{U}(b)\cdot\overline{\nabla v_1}\, \mathrm{d}\boldsymbol{x}\, \mathrm{d}t.\end{aligned}$$

This is a parabolic equation in the variational form with $(\tilde{f}, 0) \in Y'$ where $\tilde{f} : L^2(I, V) \to \mathbb{C}$ is given by

$$\tilde{f}(v_1) := -\int_I\int_D (a-b)\nabla\mathcal{U}(b)\cdot\overline{\nabla v_1}\, \mathrm{d}\boldsymbol{x}\, \mathrm{d}t, \qquad v_1 \in L^2(I, V).$$

Now applying Proposition 4.12 we find

$$\|\mathcal{U}(a) - \mathcal{U}(b)\|_X \leq \frac{\|\tilde{f}\|_{L^2(I,V')}}{\beta(a)}. \tag{4.26}$$

We also have

$$\begin{aligned}\|\tilde{f}\|_{L^2(I,V')} &= \sup_{\|v_1\|_{L^2(I,V)}=1} |\tilde{f}(v_1)| \leq \|a-b\|_{L_\infty} \|\mathcal{U}(b)\|_{L^2(I,V)} \|v_1\|_{L^2(I,V)} \\ &\leq \|a-b\|_{L_\infty} \frac{1}{\beta(b)} \Big(\|f\|^2_{L^2(I,V')} + \|u_0\|^2_{L^2}\Big)^{1/2},\end{aligned}$$

where in the last estimate we used again Proposition 4.12. Inserting this into (4.26) we obtain the desired result. □

We are now in position to verify the assumptions (i)–(iii) of Theorem 4.11 for the data-to-solution map $a \mapsto \mathcal{U}(a)$ to the Eq. (4.22).

(i) For the first condition, it has been shown that the weak solution to the linear parabolic PDEs (4.22) depends holomorphically on the data $a \in O$ by the Ladyzhenskaya-Babuška-Brezzi theorem in Hilbert spaces over $\mathbb{C}$, see e.g. [37, Pages 26, 27].

(ii) Let $a \in O$. Using the elementary estimate $a + b \leq ab$ with $a, b \geq 2$, we get

$$\begin{aligned}\sqrt{2\max(\rho(a)^{-2},1) + \vartheta^2} &\leq \sqrt{2\max(\rho(a)^{-2},1) + \max(\vartheta^2, 2)} \\ &\leq \sqrt{2\max(\rho(a)^{-2},1)\max(\vartheta^2, 2)} \\ &\leq \max(\vartheta\sqrt{2}, 2)(\rho(a)^{-1} + 1).\end{aligned}$$

Hence, from (4.25) we can bound

$$\begin{aligned}\|\mathcal{U}(a)\|_X \leq \frac{C_0(\rho(a)^{-1}+1)}{\min\big(\rho(a)\|a\|^{-2}_{L^\infty}, \rho(a)\big)} &= \frac{C_0(1+\rho(a))}{\rho(a)^2 \min\big(\|a\|^{-2}_{L^\infty}, 1\big)} \\ &\leq \frac{C_0(1+\|a\|_{L^\infty})\|a\|^2_{L^\infty}}{\min\big(\rho(a)^4, 1\big)} \\ &\leq C_0\bigg(\frac{1+\|a\|_{L^\infty}}{\min\big(\rho(a), 1\big)}\bigg)^4,\end{aligned} \tag{4.27}$$

where

$$C_0 = \max(\vartheta\sqrt{2}, 2)\big(\|f\|^2_{L^2(I,V')} + \|u_0\|^2_{L^2}\big)^{1/2}.$$

(iii) The third assumption follows from Lemma 4.13 and the part (ii), i.e., for $a, b \in O$ holds

$$\|\mathcal{U}(a) - \mathcal{U}(b)\|_X \le C\left(\frac{1 + \|a\|_{L^\infty}}{\min\left(\rho(a), 1\right)}\right)^4 \left(\frac{1 + \|b\|_{L^\infty}}{\min\left(\rho(b), 1\right)}\right)^4 \|a - b\|_{L_\infty}, \tag{4.28}$$

for some $C > 0$ depending on f, u_0 and T.

In conclusion, if $(\psi_j)_{j\in\mathbb{N}} \subset L^\infty(D)$ such that with $b_j := \|\psi_j\|_{L^\infty}$ it holds $\boldsymbol{b} \in \ell^1(\mathbb{N})$, then the solution

$$u(\boldsymbol{y}) = \lim_{N\to\infty} \mathcal{U}\left(\exp\left(\sum_{j=1}^N y_j\psi_j\right)\right) \tag{4.29}$$

belonging to $L^2(U, X; \gamma)$ is well-defined and $(\boldsymbol{b}, \xi, \delta, X)$-holomorphic by Theorem 4.11.

We continue studying the holomorphy of the solution map to the Eq. (4.22) in function space of higher-regularity. Denote by $H^1(I, L^2(D))$ the space of all functions $v(t, \boldsymbol{x}) \in L^2(I, L^2(D))$ such that the norm

$$\|v\|_{H^1(I,L^2)} := \left(\|v\|^2_{L(I,L^2)} + \|\partial_t v\|^2_{L^2(I,L^2)}\right)^{1/2}$$

is finite. We put

$$Z := L^2(I, W) \cap H^1(I, L^2(D)), \quad W := \left\{v \in V : \Delta v \in L^2(D)\right\},$$

and

$$\|v\|_Z := \left(\|v\|^2_{H^1(I,L^2)} + \|v\|^2_{L^2(I,W)}\right)^{1/2}.$$

In the following the constants C and C' may change their values from line to line.

Lemma 4.14 *Assume that $a \in W^1_\infty(D) \cap O$ and $f \in L^2(I, L^2(D))$ and $u_0 \in V$. Suppose further that $\mathcal{U}(a) \in X$ is the weak solution to the Eq. (4.22). Then $\mathcal{U}(a) \in L^2(I, W) \cap H^1(I, L^2(D))$. Furthermore,*

$$\|\partial_t \mathcal{U}(a)\|_{L^2(I,L^2)} \le \left(\frac{1 + \|a\|_{W^1_\infty}}{\min(\rho(a), 1)}\right)^4 \left(\|u_0\|^2_V + \|f\|^2_{L^2(I,L^2)}\right)^{1/2},$$

and

$$\|\Delta\mathcal{U}(a)\|_{L^2(I,L^2)} \le C\left(\frac{1 + \|a\|_{W^1_\infty}}{\min(\rho(a), 1)}\right)^5 \left(\|u_0\|^2_V + \|f\|^2_{L^2(I,L^2)}\right)^{1/2},$$

where $C > 0$ independent of f and u_0. Therefore,

$$\|\mathcal{U}(a)\|_Z \leq C\left(\frac{1+\|a\|_{W^1_\infty}}{\min(\rho(a), 1)}\right)^5 \left(\|u_0\|_V^2 + \|f\|^2_{L^2(I,L^2)}\right)^{1/2}.$$

Proof The argument follows along the lines of, e.g., [53, Section 7.1.3] by separation of variables. Let $(\omega_k)_{k\in\mathbb{N}} \subset V$ be an orthogonal basis which is orthonormal basis of $L^2(D)$, [eigenbasis in polygon generally not smooth], see, e.g. [53, Page 353]. Let further, for $m \in \mathbb{N}$,

$$\mathcal{U}_m(a) = \sum_{k=1}^{m} d_m^k(t)\omega_k \in V_m$$

be a Galerkin approximation to $\mathcal{U}(a)$ on $V_m := \operatorname{span}\{\omega_k,\ k = 1, \ldots, m\}$.

Then we have

$$\partial_t \mathcal{U}_m(a) = \sum_{k=1}^{m} \frac{d}{dt} d_m^k(t)\omega_k \in V_m.$$

Multiplying both sides with $\partial_t \mathcal{U}_m(a)$ we get

$$\int_D \partial_t \mathcal{U}_m(a)\partial_t \overline{\mathcal{U}_m(a)}\,\mathrm{d}\boldsymbol{x} + \int_D a\nabla\mathcal{U}_m(a)\cdot\partial_t\overline{\nabla\mathcal{U}_m(a)}\,\mathrm{d}\boldsymbol{x} = \int_D f\partial_t\overline{\mathcal{U}_m(a)}\,\mathrm{d}\boldsymbol{x}.$$

The conjugate equation is given by

$$\int_D \partial_t \mathcal{U}_m(a)\partial_t \overline{\mathcal{U}_m(a)}\,\mathrm{d}\boldsymbol{x} + \int_D \bar{a}\,\overline{\nabla\mathcal{U}_m(a)}\cdot\partial_t\nabla\mathcal{U}_m(a)\,\mathrm{d}\boldsymbol{x} = \int_D \bar{f}\partial_t\mathcal{U}_m(a)\,\mathrm{d}\boldsymbol{x}.$$

Consequently we obtain

$$\begin{aligned} 2\|\partial_t\mathcal{U}_m(a)\|^2_{L^2} + \frac{\mathrm{d}}{\mathrm{d}t}\int_D \Re(a)|\nabla\mathcal{U}_m(a)|^2\,\mathrm{d}\boldsymbol{x} &= \int_D f\partial_t\overline{\mathcal{U}_m(a)}\,\mathrm{d}\boldsymbol{x} \\ &\quad + \int_D \bar{f}\partial_t\mathcal{U}_m(a)\,\mathrm{d}\boldsymbol{x}. \end{aligned}$$

Integrating both sides with respect to t on I and using the Cauchy-Schwarz inequality we arrive at

$$\begin{aligned} &2\|\partial_t\mathcal{U}_m(a)\|^2_{L^2(I,L^2)} + \int_D \Re(a)\left|\nabla\mathcal{U}_m(a)\big|_{t=T}\right|^2\mathrm{d}\boldsymbol{x} \\ &\leq \int_D \Re(a)\left|\nabla\mathcal{U}_m(a)\big|_{t=0}\right|^2\mathrm{d}\boldsymbol{x} + \|f\|^2_{L^2(I,L^2)} + \|\partial_t\mathcal{U}_m(a)\|^2_{L^2(I,L^2)}, \end{aligned}$$

which implies

$$\begin{aligned}\|\partial_t \mathcal{U}_m(a)\|^2_{L^2(I,L^2)} &\le \int_D \Re(a)\left|\nabla \mathcal{U}_m(a)\big|_{t=0}\right|^2 \mathrm{d}\boldsymbol{x} + \|f\|^2_{L^2(I,L^2)} \\ &\le \|a\|_{L^\infty} \left\|\nabla \mathcal{U}_m(a)\big|_{t=0}\right\|^2_{L^2} + \|f\|^2_{L^2(I,L^2)} \\ &\le \|a\|_{L^\infty}\|u_0\|^2_V + \|f\|^2_{L^2(I,L^2)},\end{aligned} \tag{4.30}$$

where we used the bounds $\|\nabla \mathcal{U}_m(a)\big|_{t=0}\|_{L^2} \le \|u_0\|_V$, see [53, Page 362].

Passing to limits we deduce that

$$\begin{aligned}\|\partial_t \mathcal{U}(a)\|_{L^2(I,L^2)} &\le \left(\|a\|_{L^\infty}\|u_0\|^2_V + \|f\|^2_{L^2(I,L^2)}\right)^{1/2} \\ &\le (\|a\|_{L^\infty} + 1)^{1/2}\left(\|u_0\|^2_V + \|f\|^2_{L^2(I,L^2)}\right)^{1/2} \\ &\le C\left(\frac{1 + \|a\|_{W^1_\infty}}{\min(\rho(a), 1)}\right)^4 \left(\|u_0\|^2_V + \|f\|^2_{L^2(I,L^2)}\right)^{1/2}.\end{aligned}$$

We also have from (4.25) and (4.27) that

$$\begin{aligned}\|\mathcal{U}(a)\|_{L^2(I,L^2)} \le C\|\mathcal{U}(a)\|_{L^2(I,V)} &\le \frac{C}{\beta(a)}\left(\|f\|^2_{L^2(I,V')} + \|u_0\|^2_{L^2}\right)^{1/2} \\ &\le \frac{C}{\beta(a)}\left(\|u_0\|^2_V + \|f\|^2_{L^2(I,L^2)}\right)^{1/2} \\ &\le C\left(\frac{1 + \|a\|_{L^\infty}}{\min(\rho(a), 1)}\right)^4 \left(\|u_0\|^2_V + \|f\|^2_{L^2(I,L^2)}\right)^{1/2} \\ &\le C\left(\frac{1 + \|a\|_{W^1_\infty}}{\min(\rho(a), 1)}\right)^4 \left(\|u_0\|^2_V + \|f\|^2_{L^2(I,L^2)}\right)^{1/2}.\end{aligned} \tag{4.31}$$

We now estimate $\|\Delta \mathcal{U}(a)\|_{L^2(I,L^2)}$. From the identity (valid in $L^2(I, L^2(D))$)

$$-\Delta \mathcal{U}(a) = \frac{1}{a}\left[\nabla a \cdot \nabla \mathcal{U}(a) + f - \partial_t \mathcal{U}(a)\right],$$

and (4.30), (4.31) we obtain that

$$\begin{aligned}\|\Delta \mathcal{U}(a)\|_{L^2(I,L^2)} \le \frac{1}{\rho(a)}\Big[&\|a\|_{W^1_\infty}\|\mathcal{U}(a)\|_{L^2(I,V)} + \|f\|_{L^2(I,L^2)} \\ &+ \|\partial_t \mathcal{U}(a)\|_{L^2(I,L^2)}\Big]\end{aligned}$$

$$\le C\frac{\|a\|_{W^1_\infty}}{\rho(a)}\left(\frac{1+\|a\|_{L^\infty}}{\min(\rho(a),1)}\right)^4\left(\|u_0\|_V^2+\|f\|_{L^2(I,L^2)}^2\right)^{1/2}$$

$$\le C\left(\frac{1+\|a\|_{W^1_\infty}}{\min(\rho(a),1)}\right)^5\left(\|u_0\|_V^2+\|f\|_{L^2(I,L^2)}^2\right)^{1/2},$$

with $C>0$ independent of f and u_0. Combining this and (4.30), (4.31), the desired result follows. □

Lemma 4.15 *Assume $f\in L^2(I,L^2(D))$ and $u_0\in V$. Let $\mathcal{U}(a)$ and $\mathcal{U}(b)$ be the solutions to* (4.23) *with $a,b\in W^1_\infty(D)\cap O$, respectively. Then we have*

$$\|\mathcal{U}(a)-\mathcal{U}(b)\|_Z\le C'\left(\frac{1+\|a\|_{W^1_\infty}}{\min(\rho(a),1)}\right)^5\left(\frac{1+\|b\|_{W^1_\infty}}{\min(\rho(b),1)}\right)^5\|a-b\|_{W^1_\infty},$$

with $C'>0$ depending on f and u_0.

Proof Denote $w:=\mathcal{U}(a)-\mathcal{U}(b)$. Then w is the solution to the equation

$$\begin{cases}\partial_t w-\operatorname{div}\left(a\nabla w\right)=\nabla(a-b)\cdot\nabla\mathcal{U}(b)+(a-b)\Delta\mathcal{U}(b),\\ w|_{\partial D\times I}=0,\\ w|_{t=0}=0.\end{cases}\tag{4.32}$$

Hence

$$-\Delta w=\frac{1}{a}\left[\nabla a\cdot\nabla w+\nabla(a-b)\cdot\nabla\mathcal{U}(b)+(a-b)\Delta\mathcal{U}(b)-\partial_t w\right]$$

which leads to

$$\|\Delta w\|_{L^2(I,L^2)}\le\frac{1}{\rho(a)}\Big[\|a\|_{W^1_\infty}\|w\|_{L^2(I,V)}+\|\partial_t w\|_{L^2(I,L^2)}$$

$$+\|a-b\|_{W^1_\infty}\left(\|\mathcal{U}(b)\|_{L^2(I,W)}+\|\mathcal{U}(b)\|_{L^2(I,V)}\right)\Big].$$

Lemma 4.14 gives that

$$\|\partial_t w\|_{L^2(I,L^2)}\le\left(\frac{1+\|a\|_{W^1_\infty}}{\min(\rho(a),1)}\right)^5\Big(\|\nabla(a-b)\cdot\nabla\mathcal{U}(b)\|_{L^2(I,L^2)}^2$$

$$+\|(a-b)\Delta\mathcal{U}(b)\|_{L^2(I,L^2)}^2\Big)^{1/2}$$

$$\le\left(\frac{1+\|a\|_{W^1_\infty}}{\min(\rho(a),1)}\right)^5\|a-b\|_{W^1_\infty}\Big(\|\mathcal{U}(b)\|_{L^2(I,W)}^2$$

$$+\|\mathcal{U}(b)\|_{L^2(I,V)}^2\Big)^{1/2},$$

and

$$\|\mathcal{U}(b)\|_{L^2(I,W)} + \|\mathcal{U}(b)\|_{L^2(I,V)} \le C\left(\frac{1+\|b\|_{W^1_\infty}}{\min(\rho(b),1)}\right)^5 \left(\|u_0\|_V^2 + \|f\|^2_{L^2(I,L^2)}\right)^{1/2},$$

which implies

$$\begin{aligned}
&\|a-b\|_{W^1_\infty}\left(\|\mathcal{U}(b)\|_{L^2(I,W)} + \|\mathcal{U}(b)\|_{L^2(I,V)}\right) + \|\partial_t w\|_{L^2(I,L^2)} \\
&\le C'\left(\frac{1+\|a\|_{W^1_\infty}}{\min(\rho(a),1)}\right)^5 \left(\frac{1+\|b\|_{W^1_\infty}}{\min(\rho(b),1)}\right)^5 \|a-b\|_{W^1_\infty}.
\end{aligned}$$

We also have

$$\begin{aligned}
\|w\|_{L^2(I,V)} &\le \frac{1}{\beta(a)\beta(b)}\|a-b\|_{L_\infty} \\
&\le C'\left(\frac{1+\|a\|_{L^\infty}}{\min(\rho(a),1)}\right)^4 \left(\frac{1+\|b\|_{L^\infty}}{\min(\rho(b),1)}\right)^4 \|a-b\|_{W^1_\infty},
\end{aligned}$$

see (4.28). Hence

$$\|\Delta w\|_{L^2(I,L^2)} \le C'\left(\frac{1+\|a\|_{W^1_\infty}}{\min(\rho(a),1)}\right)^5 \left(\frac{1+\|b\|_{W^1_\infty}}{\min(\rho(b),1)}\right)^5 \|a-b\|_{W^1_\infty}. \tag{4.33}$$

Since the terms $\|\partial_t w\|_{L^2(I,L^2)}$ and $\|w\|_{L^2(I,L^2)}$ are also bounded by the right side of (4.33), we arrive at

$$\begin{aligned}
\|w\|_Z &= \left(\|\Delta w\|^2_{L^2(I,L^2)} + \|\partial_t w\|^2_{L^2(I,L^2)} + \|w\|^2_{L^2(I,L^2)}\right)^{1/2} \\
&\le C'\left(\frac{1+\|a\|_{W^1_\infty}}{\min(\rho(a),1)}\right)^5 \left(\frac{1+\|b\|_{W^1_\infty}}{\min(\rho(b),1)}\right)^5 \|a-b\|_{W^1_\infty}
\end{aligned}$$

which is the claim. □

From Lemma 4.15, by the same argument as in the proof of [70, Proposition 4.5] we can verify that the solution map $a \mapsto \mathcal{U}(a)$ from $W^1_\infty(D) \cap O$ to Z is holomorphic. If we assume further that $(\psi_j)_{j\in\mathbb{N}} \subseteq W^1_\infty(D)$ and with $b_j := \|\psi_j\|_{W^1_\infty}$, it holds $\boldsymbol{b} \in \ell^1(\mathbb{N})$ and all the conditions in Theorem 4.11 are satisfied. Therefore, $u(\boldsymbol{y})$ given by the formula (4.29) is $(\boldsymbol{b}, \xi, \delta, Z)$-holomorphic with appropriate ξ and δ.

Remark 4.16 For $s > 1$, let

$$Z^s := \bigcap_{k=0}^{s} H^k(I, H^{2s-2k}(D))$$

with the norm

$$\|v\|_{Z^s} = \left(\sum_{k=0}^{s} \left\| \frac{\mathrm{d}^k v}{\mathrm{d}t^k} \right\|^2_{L^2(I, H^{2s-2k})} \right)^{1/2}.$$

Assume that $a \in W^{2s-1}_\infty(D) \cap O$. At present we do not know whether the solution map $a \mapsto \mathcal{U}(a)$ from $W^{2s-1}_\infty(D) \cap O$ to Z^s is holomorphic. To obtain the holomorphy of the solution map, we need a result similar to that in Lemma 4.15. In order for this to hold, higher-order regularity and compatibility of the data for Eq. (4.32) is required, i.e,

$$g_0 = 0 \in V, \quad g_1 = h(0) - Lg_0 \in V, \ldots, g_s = \frac{\mathrm{d}^{s-1} h}{\mathrm{d}t^{s-1}}(0) - Lg_{s-1} \in V,$$

where

$$h = \nabla(a-b) \cdot \nabla \mathcal{U}(b) + (a-b)\Delta \mathcal{U}(b), \quad L = \partial_t \cdot - \operatorname{div}\big(a\nabla \cdot\big).$$

See e.g. [110, Theorem 27.2]. It is known that without such compatibility, the solution will develop spatial singularities at the corners and edges of D, and temporal singularities as $t \downarrow 0$; see e.g. [81].

In general the compatibility condition does not hold when we only assume that

$$u_0 \in H^{2s-1}(D) \cap V \text{ and } \frac{\mathrm{d}^k f}{\mathrm{d}t^k} \in L^2(I, H^{2s-2k-2}(D))$$

for $k = 0, \ldots, s-1$.

4.3.3 *Linear Elastostatics with Log-Gaussian Modulus of Elasticity*

We illustrate the foregoing abstract setting of Sect. 4.1 for another class of boundary value problems. In computational mechanics, one is interested in the numerical approximation of deformations of elastic bodies. We refer to e.g. [108] for an accessible exposition of the mathematical foundations and assumptions. In *linearized elastostatics* one is concerned with small (in a suitable sense, see [108] for details) deformations.

We consider an elastic body occupying the domain $D \subset \mathbb{R}^d$, $d = 2, 3$ (the physically relevant case naturally is $d = 3$, we include $d = 2$ to cover the so-called model of "plane-strain" which is widely used in engineering, and has governing equations with the same mathematical structure). In the linear theory, small deformations of the elastic body occupying D, subject to, e.g., body forces

$f : D \to \mathbb{R}^d$ such as gravity are modeled in terms of the *displacement field* $\boldsymbol{u} : D \to \mathbb{R}^d$, describing the displacement of a *material point* $\boldsymbol{x} \in D$ (see [108] for a discussion of axiomatics related to this mathematical concept). Importantly, unlike the scale model problem considered up to this point, modeling now involves vector fields of data (e.g., $\boldsymbol{f}$) and solution (i.e., $\boldsymbol{u}$).

Governing equations for the mathematical model of linearly elastic deformation, subject to homogeneous Dirichlet boundary conditions on ∂D, read: to find $\boldsymbol{u} : D \to \mathbb{R}^d$ such that

$$\begin{aligned} \operatorname{div}\boldsymbol{\sigma}[\boldsymbol{u}] + \boldsymbol{f} &= 0 \quad \text{in } D, \\ \boldsymbol{u} &= 0 \quad \text{on } \partial D. \end{aligned} \tag{4.34}$$

Here $\boldsymbol{\sigma} : D \to \mathbb{R}^{d\times d}_{\text{sym}}$ is a symmetric matrix function, the so-called *stress tensor*. It depends on the displacement field u via the so-called (linearized) *strain tensor* $\boldsymbol{\epsilon}[\boldsymbol{u}] : D \to \mathbb{R}^{d\times d}_{\text{sym}}$, which is given by

$$\boldsymbol{\epsilon}[\boldsymbol{u}] := \frac{1}{2}\left(\operatorname{grad}\boldsymbol{u} + (\operatorname{grad}\boldsymbol{u})^\top\right), \quad (\boldsymbol{\epsilon}[\boldsymbol{u}])_{ij} := \frac{1}{2}(\partial_j u_i + \partial_i u_j), i, j = 1, \ldots, d. \tag{4.35}$$

In the linearized theory, the tensors $\boldsymbol{\sigma}$ and $\boldsymbol{\epsilon}$ in (4.34), (4.35) are related by the linear constitutive stress-strain relation ("Hooke's law")

$$\boldsymbol{\sigma} = \mathtt{A}\boldsymbol{\epsilon}. \tag{4.36}$$

In (4.36), $\mathtt{A}$ is a fourth order tensor field, i.e.

$$\mathtt{A} = \{\mathtt{A}_{ijkl} : i, j, k, l = 1, \ldots, d\},$$

with certain symmetries that must hold among its d^4 components independent of the particular material constituting the elastic body (see, e.g., [108] for details). Thus, (4.36) reads in components as $\sigma_{ij} = \mathtt{A}_{ijkl}\epsilon_{kl}$ with summation over repeated indices implied. Let us now fix $d = 3$. Symmetry implies that ϵ and σ are characterized by 6 components. If, in addition, the material constituting the elastic body is *isotropic*, the tensor $\mathtt{A}$ can in fact be characterized by only two independent coefficient functions. We adopt here the *Poisson ratio*, denoted ν, and the modulus of elasticity $\mathtt{E}$. With these two parameters, the stress-strain law (4.36) can be expressed in the component form

$$\begin{pmatrix} \sigma_{11} \\ \sigma_{22} \\ \sigma_{33} \\ \sigma_{12} \\ \sigma_{13} \\ \sigma_{23} \end{pmatrix} = \frac{\mathtt{E}}{(1+\nu)(1-2\nu)}$$

$$\times \begin{pmatrix} 1-\nu & \nu & \nu & 0 & 0 & 0 \\ \nu & 1-\nu & \nu & 0 & 0 & 0 \\ \nu & \nu & 1-\nu & 0 & 0 & 0 \\ 0 & 0 & 0 & 1-2\nu & 0 & 0 \\ 0 & 0 & 0 & 0 & 1-2\nu & 0 \\ 0 & 0 & 0 & 0 & 0 & 1-2\nu \end{pmatrix} \begin{pmatrix} \epsilon_{11} \\ \epsilon_{22} \\ \epsilon_{33} \\ \epsilon_{12} \\ \epsilon_{13} \\ \epsilon_{23} \end{pmatrix}. \tag{4.37}$$

We see from (4.37) that for isotropic elastic materials, the tensor $\mathtt{A}$ is proportional to the modulus $\mathtt{E} > 0$, with the Poisson ratio $\nu \in [0, 1/2)$. We remark that for common materials, $\nu \uparrow 1/2$ arises in the so-called *incompressible limit*. In that case, (4.34) can be described by the Stokes equations.

With the constitutive law (4.36), we may cast the governing Eq. (4.34) into the so-called "primal", or "displacement-formulation": find $\boldsymbol{u} : D \to \mathbb{R}^d$ such that

$$-\operatorname{div}(\mathtt{A}\epsilon[\boldsymbol{u}]) = \boldsymbol{f} \quad \text{in } D, \qquad \boldsymbol{u}|_{\partial D} = 0. \tag{4.38}$$

This form is structurally identical to the scalar diffusion problem (3.1).

Accordingly, we fix $\nu \in [0, 1/2)$ and model uncertainty in the elastic modulus $\mathtt{E} > 0$ in (4.37) by a log-Gaussian random field

$$\mathtt{E}(\boldsymbol{y})(\boldsymbol{x}) := \exp(b(\boldsymbol{y}))(\boldsymbol{x}), \quad \boldsymbol{x} \in D, \ \ \boldsymbol{y} \in U. \tag{4.39}$$

Here, $b(\boldsymbol{y})$ is a Gaussian series representation of the GRF $b(Y(\omega))$ as discussed in Sect. 2.5. The log-Gaussian ansatz $\mathtt{E} = \exp(b)$ ensures

$$E_{\min}(\boldsymbol{y}) := \operatorname*{ess\,inf}_{\boldsymbol{x}\in D} \mathtt{E}(\boldsymbol{y})(\boldsymbol{x}) > 0 \qquad \gamma\text{-a.e. } \boldsymbol{y} \in U,$$

i.e., the γ-almost sure positivity of (realizations of) the elastic modulus $\mathtt{E}$. Denoting the 3×3 matrix relating the stress and strain components in (4.37) also by $\mathtt{A}$ (this slight abuse of notation should, however, not cause confusion in the following), we record that for $0 \le \nu < 1/2$, the matrix $\mathtt{A}$ is invertible:

$$\mathtt{A}^{-1} = \frac{1}{\mathtt{E}} \begin{pmatrix} 1 & -\nu & -\nu & 0 & 0 & 0 \\ -\nu & 1 & -\nu & 0 & 0 & 0 \\ -\nu & -\nu & 1 & 0 & 0 & 0 \\ 0 & 0 & 0 & 1+\nu & 0 & 0 \\ 0 & 0 & 0 & 0 & 1+\nu & 0 \\ 0 & 0 & 0 & 0 & 0 & 1+\nu \end{pmatrix}. \tag{4.40}$$

It readily follows from this explicit expression that due to

$$\mathtt{E}^{-1}(\boldsymbol{y})(\boldsymbol{x}) = \exp(-b(\boldsymbol{y})(\boldsymbol{x})),$$

by the Gerschgorin theorem invertibility holds for γ-a.e. $\boldsymbol{y} \in U$. Also, the components of $\mathtt{A}^{-1}$ are GRFs (which are, however, fully correlated for deterministic ν).

Occasionally, instead of the constants $\mathtt{E}$ and ν, one finds the (equivalent) so-called *Lamé-constants* λ, μ. They are related to $\mathtt{E}$ and ν by

$$\lambda = \frac{\mathtt{E}\nu}{(1+\nu)(1-2\nu)}, \quad \mu = \frac{\mathtt{E}}{2(1+\nu)}. \tag{4.41}$$

For GRF models (4.39) of $\mathtt{E}$, (4.41) shows that for each fixed $\nu \in (0, 1/2)$, also the Lamé-constants are GRFs *which are fully correlated*. This implies, in particular, that "large" realizations of the GRF (4.39) do not cause so-called "volume locking" in the equilibrium Eq. (4.34): this effect is related to the elastic material described by the constitutive Eq. (4.36) being nearly incompressible. Incompressibility here arises as either $\nu \uparrow 1/2$ at fixed $\mathtt{E}$ or, equivalently, as $\lambda \to \infty$ at fixed μ.

Parametric weak solutions of (4.38) with (4.39) are within the scope of the abstract theory developed up to this point. To see this, we provide a variational formulation of (4.38). Assuming for convenience homogeneous Dirichlet boundary conditions, we multiply (4.38) by a test displacement field $\boldsymbol{v} \in X := V^d$ with $V := H_0^1(D)$, and integrate by parts, to obtain the weak formulation: find $\boldsymbol{u} \in X$ such that, for all $\boldsymbol{v} \in X$ holds (in the matrix-vector notation (4.37))

$$\int_D \boldsymbol{\epsilon}[\boldsymbol{v}] \cdot \mathtt{A}\boldsymbol{\epsilon}[\boldsymbol{u}]\,\mathrm{d}\boldsymbol{x} = 2\mu(\boldsymbol{\epsilon}[\boldsymbol{u}], \boldsymbol{\epsilon}[\boldsymbol{v}]) + \lambda(\operatorname{div}\boldsymbol{u}, \operatorname{div}\boldsymbol{v}) = (\boldsymbol{f}, \boldsymbol{v}). \tag{4.42}$$

The variational form (4.42) suggests that, as $\lambda \to \infty$ for fixed μ, the "volume-preservation" constraint $\|\operatorname{div}\boldsymbol{u}\|_{L^2} = 0$ is imposed for $\boldsymbol{v} = \boldsymbol{u}$ in (4.42).

Unique solvability of (4.42) follows upon verifying coercivity of the corresponding bilinear form on the left-hand side of (4.42). It follows from (4.37) and (4.40) that

$$\forall \boldsymbol{v} \in H^1(D)^d: \quad \mathtt{E}c_{\min}(\nu)\|\boldsymbol{\epsilon}[\boldsymbol{v}]\|_{L^2}^2 \le \int_D \boldsymbol{\epsilon}[\boldsymbol{v}] \cdot \mathtt{A}\boldsymbol{\epsilon}[\boldsymbol{v}]\,\mathrm{d}\boldsymbol{x} \le \mathtt{E}c_{\max}(\nu)\|\boldsymbol{\epsilon}[\boldsymbol{v}]\|_{L^2}^2.$$

Here, the constants $c_{\min}, c_{\max}$ are positive and bounded for $0 < \nu < 1/2$ and independent of $\mathtt{E}$.

For the log-Gaussian model (4.39) of the elastic modulus $\mathtt{E}$, the relations (4.41) show in particular, that *the volume-locking effect arises as in the deterministic setting only if* $\nu \simeq 1/2$, *independent of the realization of* $\mathtt{E}(\boldsymbol{y})$. Let us consider well-posedness of the variational formulation (4.42), for log-Gaussian, parametric elastic modulus $\mathtt{E}(\boldsymbol{y})$ as in (4.39). To this end, with $\mathtt{A}_1$ denoting the matrix $\mathtt{A}$ in (4.37) with $\mathtt{E} = 1$, we introduce in (4.42) the parametric bilinear forms

$$b(\boldsymbol{u}, \boldsymbol{v}; \boldsymbol{y}) := \mathtt{E}(\boldsymbol{y}) \int_D \boldsymbol{\epsilon}[\boldsymbol{v}] \cdot \mathtt{A}_1\boldsymbol{\epsilon}[\boldsymbol{u}]\,\mathrm{d}\boldsymbol{x}$$

$$= \frac{\mathrm{E}(y)}{1+\nu}\left((\epsilon[u], \epsilon[v]) + \frac{\nu}{1-2\nu}(\operatorname{div} u, \operatorname{div} v) \right).$$

Let us verify continuity and coercivity of the parametric bilinear forms

$$\{b(\cdot, \cdot; y) : X \times X \to \mathbb{R} : y \in U\}, \tag{4.43}$$

where we recall that $U := \mathbb{R}^\infty$. With A_1 as defined above, we write for arbitrary $v \in X = H_0^1(D)^d$, $d = 2, 3$, and for all $y \in U_0 \subset U$ where the set U_0 is as in (3.21),

$$\begin{aligned}
b(v, v; y) &= \int_D \epsilon[v] \cdot (\mathrm{A}\epsilon[v])\, \mathrm{d}x = \int_D E(y)\, (\epsilon[v] \cdot (\mathrm{A}_1\epsilon[v]))\, \mathrm{d}x \\
&\geq c(\nu) \int_D E(y) \|\epsilon[v]\|_2^2\, \mathrm{d}x \\
&\geq c(\nu) \exp(-\|b(y)\|_{L^\infty}) \int_D \|\epsilon[v]\|_2^2\, \mathrm{d}x \\
&\geq \frac{c(\nu)}{2} a_{\min}(y) |v|_{H^1}^2 \\
&\geq C_P \frac{c(\nu)}{2} a_{\min}(y) \|v\|_{H^1}^2 .
\end{aligned}$$

Here, in the last two steps we employed the first Korn's inequality, and the Poincaré inequality, respectively. The lower bound $E(y) \geq \exp(-\|b(y)\|_{L^\infty})$ is identical to (3.20) in the scalar diffusion problem.

In a similar fashion, continuity of the bilinear forms (4.43) may be established: there exists a constant $c'(\nu) > 0$ such that

$$\forall u, v \in X,\ \forall y \in U_0 : \quad |b(u, v; y)| \leq c'(\nu) \exp(\|b(y)\|_{L^\infty}) \|u\|_{H^1} \|v\|_{H^1} .$$

With continuity and coercivity of the parametric forms (4.43) verified for $y \in U_0$, the Lax-Milgram lemma ensures for given $f \in L^2(D)^d$ the existence of the parametric solution family

$$\{u(y) \in X : b(u, v; y) = (f, v)\ \forall v \in X,\ y \in U_0\}. \tag{4.44}$$

Similar to the scalar case discussed in Proposition 3.7, the following result on almost everywhere existence and measurability holds.

Proposition 4.17 *Under Assumption 3.6, $\gamma(U_0) = 1$. For all $k \in \mathbb{N}$ there holds, with $\mathbb{E}(\cdot)$ denoting expectation with respect to γ,*

$$\mathbb{E}\left(\exp(k\|b(\cdot)\|_{L^\infty})\right) < \infty.$$

The parametric solution family (4.44) *of the parametric elliptic boundary value problem* (4.42) *with log-Gaussian modulus* $E(y)$ *as in* (4.39) *is in* $L^k(U, V; \gamma)$ *for every finite* $k \in \mathbb{N}$.

For the parametric solution family (4.44), analytic continuations into complex parameter domains, and parametric regularity results may be developed in analogy to the development in Sects. 3.7 and 3.8. The key result for bootstrapping to higher order regularity is, in the case of smooth boundaries ∂D, classical elliptic regularity for linear, Agmon-Douglis-Nirenberg elliptic systems which comprise (4.38). In the polygonal (for $d = 2$) or polyhedral ($d = 3$) case, weighted regularity shifts in Kondrat'ev type spaces are available in [62, Theorem 5.2] (for $d = 2$) and in [103] (for both, $d = 2, 3$).

4.3.4 Maxwell Equations with Log-Gaussian Permittivity

Similar models are available for time-harmonic, electromagnetic waves in dielectric media with uncertain conductivity. We refer to [76], where log-Gaussian models are employed. There, also the parametric regularity analysis of the parametric electric and magnetic fields is discussed, albeit by real-variable methods. The setting in [76] is, however, so that the presently developed, complex variable methods can be brought to bear on it. We refrain from developing the details.

4.3.5 Linear Parametric Elliptic Systems and Transmission Problems

In Sect. 3.8.1, Theorem 3.29 we obtained parameter-explicit elliptic regularity shifts for a scalar, linear second order parametric elliptic divergence-form PDE in polygonal domain $D \subset \mathbb{R}^2$. A key feature of these estimates in the subsequent analysis of sparsity of gpc expansions was the *polynomial dependence on the parameter in the bounds on parametric solutions in corner-weighted Sobolev spaces of Kondrat'ev type*. Such a-priori bounds are not limited to the particular setting considered in Sect. 3.8.1, but hold for rather general, linear elliptic PDEs in smooth domains $D \subset \mathbb{R}^d$ of space dimension $d \geq 2$, with parametric differential and boundary operators of general integer order. In particular, for example, for linear, anisotropic elastostatics in $\mathbb{R}^3$, for parametric fourth order PDEs in $\mathbb{R}^2$ which arise in dimensionally reduced models of elastic continua (plates, shells, etc.). We refer to [80] for statements of results and proofs.

In the results in Sect. 3.8.1, we admitted inhomogeneous coefficients which are regular in all of D. In many applications, *transmission problems* with parametric, inhomogeneous coefficients with are piecewise regular on a given, fixed (i.e. non-parametric) partition of D is of interest. Also in these cases, corresponding a-priori estimates of parametric solution families with norm bounds which are polynomial with respect to the parameters hold. We refer to [95] for such results, in smooth domains D, with smooth interfaces.

Chapter 5
Parametric Posterior Analyticity and Sparsity in BIPs

We have investigated the parametric analyticity of the *forward solution maps* of linear PDEs with uncertain parametric inputs which typically arise from GRF models for these inputs. We have also provided an analysis of sparsity in the Wiener-Hermite PC expansion of the corresponding parametric solution families.

We now explore the notion of parametric holomorphy in the context of BIPs for linear PDEs. For these PDEs we adopt the Bayesian setting as outlined, e.g., in [48] and the references there. This Bayesian setting is briefly recapitulated in Sect. 5.1. With a suitable version of Bayes' theorem, the main result is a (short) proof of parametric $(\boldsymbol{b}, \xi, \delta, \mathbb{C})$-holomorphy of the Bayesian posterior density for unbounded parameter ranges. This implies sparsity of the coefficients in Wiener-Hermite PC expansions of the Bayesian posterior density, which can be leveraged to obtain higher-order approximation rates that are free from the curse of dimensionality for various deterministic approximation methods of the Bayesian expectations, for several classes of function space priors modelled by product measures on the parameter sequences $\boldsymbol{y}$. In particular, the construction of Gaussian priors described in Sect. 2.2 is applicable. Concerning related previous works, we remark the following. In [98] holomorphy for a bounded parameter domain (in connection with uniform prior measure) has been addressed by complex variable arguments in the same fashion. In [96], MC and QMC integration has been analyzed by real-variable arguments for such Gaussian priors. In [68], corresponding results have been obtained also for so-called *Besov priors*, again by real-variable arguments for the parametric posterior. Since the presently developed, quantified parametric holomorphy results are independent of the particular measure placed upon the unbounded parameter domain $\mathbb{R}^\infty$. The sparsity and approximation rate bounds for the parametric deterministic posterior densities will imply approximate rate bounds also for prior constructions beyond the Gaussian ones.

D. Dũng et al., *Analyticity and Sparsity in Uncertainty Quantification for PDEs with Gaussian Random Field Inputs*, Lecture Notes in Mathematics 2334,
https://doi.org/10.1007/978-3-031-38384-7_5

5.1 Formulation and Well-Posedness

With E and X denoting separable Banach and Hilbert spaces over $\mathbb{C}$, respectively, we consider a forward solution map $\mathcal{U} : E \to X$ and an observation map $\mathcal{O} : X \to \mathbb{R}^m$. In the context of the previous sections, $\mathcal{U}$ could denote again the map which associates with a diffusion coefficient $a \in E := L^\infty(D; \mathbb{C})$ the solution $\mathcal{U}(a) \in X := H_0^1(D; \mathbb{C})$ of the Eq. (5.7) below. We assume the map $\mathcal{U}$ to be Borel measurable.

The inverse problem consists in determining the (expected value of an) uncertain input datum $a \in E$ from noisy observation data $\mathfrak{d} \in \mathbb{R}^m$. Here, the observation noise $\eta \in \mathbb{R}^m$ is assumed additive centered Gaussian, i.e., the observation data $\mathfrak{d}$ for input a is

$$\mathfrak{d} = \mathcal{O} \circ \mathcal{U}(a) + \eta,$$

where $\eta \sim \mathcal{N}(0, \mathbf{\Gamma})$. We assume the observation noise covariance $\mathbf{\Gamma} \in \mathbb{R}^{m\times m}$ is symmetric positive definite.

In the so-called Bayesian setting of the inverse problem, one assumes that the uncertain input a is modelled as RV which is distributed according to a prior measure π_0 on E. Then, under suitable conditions, which are made precise in Theorem 5.2 below, the posterior distribution $\pi(\cdot|\mathfrak{d})$ on the conditioned RV $\mathcal{U}|\mathfrak{d}$ is absolutely continuous w.r.t. the prior measure π_0 on E and there holds Bayes' theorem in the form

$$\frac{d\pi(\cdot|\mathfrak{d})}{d\pi_0}(a) = \frac{1}{Z}\Theta(a). \tag{5.1}$$

In (5.1), the posterior density Θ and the normalization constant Z are given by

$$\begin{aligned} \Theta(a) &= \exp(-\Phi(\mathfrak{d}; a)), \quad \Phi(\mathfrak{d}; a) = \frac{1}{2}\|\mathbf{\Gamma}^{-1/2}(\mathfrak{d} - \mathcal{O}(\mathcal{U}(a)))\|_2^2, \\ Z &= \mathbb{E}_{\pi_0}[\Theta(\cdot)]. \end{aligned} \tag{5.2}$$

Additional conditions ensure that the posterior measure $\pi(\cdot|\mathfrak{d})$ is well-defined and that (5.1) holds according to the following result from [48].

Proposition 5.1 *Assume that $\mathcal{O} \circ \mathcal{U} : E \to \mathbb{R}^m$ is continuous and that $\pi_0(E) = 1$. Then the posterior $\pi(\cdot|\mathfrak{d})$ is absolutely continuous with respect to π_0, and* (5.2) *holds.*

The condition $\pi_0(E) = 1$ can in fact be weakened to $\pi_0(E) > 0$ (e.g. [48, Theorem 3.4]).

The solution of the BIP amounts to the evaluation of the posterior expectation $\mathbb{E}_{\mu^{\mathfrak{d}}}[\cdot]$ of a continuous linear map $\phi : X \to Q$ of the map $\mathcal{U}(a)$, where Q is a suitable Hilbert space over $\mathbb{C}$. Solving the Bayesian inverse problem is thus closely related to the numerical approximation of the posterior expectation

$$\mathbb{E}_{\pi(\cdot|\mathfrak{d})}[\phi(\mathcal{U}(\cdot))] \in Q.$$

For computational purposes, and to facilitate Wiener-Hermite PC approximation of the density Θ in (5.1), one parametrizes the input data $a = a(\boldsymbol{y}) \in E$ by a Gaussian series as discussed in Sect. 2.5. Inserting into $\Theta(a)$ in (5.1), (5.2) this results in a countably-parametric density $U \ni \boldsymbol{y} \mapsto \Theta(a(\boldsymbol{y}))$, for $\boldsymbol{y} \in U$, and the Gaussian reference measure π_0 on E in (5.1) is pushed forward into a countable product γ of the sequence of Gaussian measures $\{\gamma_{1,n}\}_{n\in\mathbb{N}}$ on $\mathbb{R}$: using (5.1) and choosing a Gaussian prior (e.g. [48, Section 2.4] or [68, 85])

$$\pi_0 = \gamma = \bigotimes_{j\in\mathbb{N}} \gamma_{1,n}$$

on U (see Example 2.17), the Bayesian estimate, i.e., the posterior expectation, can then be written as a (countably) iterated integral [48, 96, 98] with respect to the product GM γ, i.e.

$$\mathbb{E}_{\pi(\cdot|\mathfrak{d})}[\phi(\mathcal{U}(a(\cdot)))] = \frac{1}{Z}\int_U \phi(\mathcal{U}(a(\boldsymbol{y})))\Theta(a(\boldsymbol{y}))\,\mathrm{d}\gamma(\boldsymbol{y}) \in Q,$$
$$Z = \int_U \Theta(a(\boldsymbol{y}))\,\mathrm{d}\gamma(\boldsymbol{y}) \in \mathbb{R}. \tag{5.3}$$

The parametric density $U \to \mathbb{R}$ in (5.3) which arises in Bayesian PDE inversion under Gaussian prior and also under more general, so-called Besov prior measures on U, see, e.g. [48, Section 2.3], [68, 85]. The parametric density

$$\boldsymbol{y} \mapsto \phi(\mathcal{U}(a(\boldsymbol{y})))\Theta(a(\boldsymbol{y}))$$

inherits sparsity from the forward map $\boldsymbol{y} \mapsto \mathcal{U}(a(\boldsymbol{y}))$, whose sparsity is expressed as before in terms of ℓ^p-summability and weighted ℓ^2-summability of Wiener-Hermite PC expansion coefficients. We employ the parametric holomorphy of the forward map $a \mapsto \mathcal{U}(a)$ to quantify the sparsity of the parametric posterior densities $\boldsymbol{y} \mapsto \Theta(a(\boldsymbol{y}))$ and $\boldsymbol{y} \mapsto \phi(\mathcal{U}(a(\boldsymbol{y})))\Theta(a(\boldsymbol{y}))$ in (5.3).

5.2 Posterior Parametric Holomorphy

With a Gaussian series in the data space E, for the resulting parametric data-to-solution map

$$u : U \to X : \boldsymbol{y} \mapsto \mathcal{U}(a(\boldsymbol{y})),$$

we now prove that under certain conditions both, the corresponding parametric posterior density

$$\boldsymbol{y} \mapsto \exp\Big(-(\mathfrak{d} - \mathcal{O}(u(\boldsymbol{y})))^\top \boldsymbol{\Gamma}^{-1}(\mathfrak{d} - \mathcal{O}(u(\boldsymbol{y})))\Big) \tag{5.4}$$

in (5.2), and the integrand

$$\boldsymbol{y} \mapsto \phi(u(\boldsymbol{y})) \exp\Big(-(\mathfrak{d} - \mathcal{O}(u(\boldsymbol{y})))^\top \boldsymbol{\Gamma}^{-1}(\mathfrak{d} - \mathcal{O}(u(\boldsymbol{y})))\Big) \tag{5.5}$$

in (5.3) are $(\boldsymbol{b}, \xi, \delta, \mathbb{C})$-holomorphic and $(\boldsymbol{b}, \xi, \delta, Q)$-holomorphic, respectively.

Theorem 5.2 *Let $r > 0$. Assume that the map $u : U \to X$ is $(\boldsymbol{b}, \xi, \delta, X)$-holomorphic with constant functions $\varphi_N \equiv r$, $N \in \mathbb{N}$, in Definition 4.1. Let the observation noise covariance matrix $\boldsymbol{\Gamma} \in \mathbb{R}^{m\times m}$ be symmetric positive definite.*

Then, for any bounded linear quantity of interest $\phi \in L(X, Q)$, and for any observable $\mathcal{O} \in (X')^m$ with arbitrary, finite m, the function in (5.4) *is $(\boldsymbol{b}, \xi, \delta, \mathbb{C})$-holomorphic and the function in* (5.5) *is $(\boldsymbol{b}, \xi, \delta, Q)$-holomorphic.*

Proof We only show the statement for the parametric integrand in (5.5), as the argument for the posterior density in (5.4) is completely analogous.

Consider the map

$$\Xi : \{v \in X : \|v\|_X \le r\} \to Q : v \mapsto \phi(v) \exp(-(\mathfrak{d} - \mathcal{O}(v))\boldsymbol{\Gamma}^{-1}(\mathfrak{d} - \mathcal{O}(v))).$$

This function is well-defined. We have $|\mathcal{O}(v)| \le \|\mathcal{O}\|_{X'} r$ and $|\phi(v)| \le \|\phi\|_{L(X;Q)} r$ for all $v \in X$ with $\|v\|_X \le r$. Since $\exp : \mathbb{C} \to \mathbb{C}$ is Lipschitz continuous on compact subsets of $\mathbb{C}$ and since $\phi \in L(X; Q)$ is bounded linear map (and thus Lipschitz continuous), we find that

$$\sup_{\|v\|_X \le r} \|\Xi(v)\|_Q =: \tilde{r} < \infty$$

and that

$$\Xi : \{v \in X : \|v\|_X \le r\} \to \mathbb{C}$$

is Lipschitz continuous with some Lipschitz constant $L > 0$.

Let us recall that the $(\boldsymbol{b}, \xi, \delta, X)$-holomorphy of $u : U \to X$, implies the existence of (continuous) functions $u_N \in L^2(\mathbb{R}^N, X; \gamma_N)$ such that with $\tilde{u}_N(\boldsymbol{y}) = u_N(y_1, \ldots, y_N)$ it holds $\lim_{N\to\infty} \tilde{u}_N = u$ in the sense of $L^2(U, X; \gamma)$. Furthermore, if

$$\sum_{j=1}^N b_j \varrho_j \le \delta$$

(i.e. $\boldsymbol{\varrho} = (\varrho_j)_{j=1}^N$ is $(\boldsymbol{b}, \xi)$-admissible in the sense of Definition 4.1), then u_N allows a holomorphic extension

$$u_N : \mathcal{S}_{\boldsymbol{\varrho}} \to X$$

such that for all $\boldsymbol{y} \in \mathbb{R}^N$

$$\sup_{\boldsymbol{z} \in \mathcal{B}_{\boldsymbol{\varrho}}} \|u_N(\boldsymbol{y} + \boldsymbol{z})\|_X \leq \varphi_N(\boldsymbol{y}) = r \qquad \forall \boldsymbol{y} \in \mathbb{R}^N, \tag{5.6}$$

see (4.1) for the definition of $\mathcal{S}_{\boldsymbol{\varrho}}$ and $\mathcal{B}_{\boldsymbol{\varrho}}$.

We want to show that $f(\boldsymbol{y}) := \Xi(u(\boldsymbol{y}))$ is well-defined in $L^2(U, Q; \gamma)$, and given as the limit of the functions

$$\tilde{f}_N(\boldsymbol{y}) = f_N((y_j)_{j=1}^N)$$

for all $\boldsymbol{y} \in U$ and $N \in \mathbb{N}$, where

$$f_N((y_j)_{j=1}^N) = \Xi(u_N((y_j)_{j=1}^N).$$

Note at first that $f_N : \mathbb{R}^N \to Q$ is well-defined. In the case

$$\sum_{j=1}^N b_j \varrho_j \leq \delta,$$

f_N allows a holomorphic extension $f_N : \mathcal{S}_{\boldsymbol{\varrho}} \to X$ given through $\Xi \circ u_N$. Using (5.6), this extension satisfies for any $N \in \mathbb{N}$ and any $(\boldsymbol{b}, \xi)$-admissible $\varrho \in (0, \infty)^N$

$$\sup_{\boldsymbol{z} \in \mathcal{B}_{\boldsymbol{\varrho}}} |f_N(\boldsymbol{y} + \boldsymbol{z})| \leq \sup_{\|v\|_X \leq r} |\Xi(v)| = \tilde{r} \qquad \forall \boldsymbol{y} \in \mathbb{R}^N.$$

This shows assumptions (i)–(ii) of Definition 4.1 for $f_N : \mathbb{R}^N \to Q$.

Finally we show assumption (iii) of Definition 4.1. By assumption it holds $\lim_{N\to\infty} \tilde{u}_N = u$ in the sense of $L^2(U, X; \gamma)$. Thus for $f = \Xi \circ u$ and with $f_N = \Xi \circ u_N$

$$\begin{aligned} \int_U \|f(\boldsymbol{y}) - f_N(\boldsymbol{y})\|_Q^2 \,\mathrm{d}\gamma(\boldsymbol{y}) &= \int_U \|\Xi(u(\boldsymbol{y})) - \Xi(u_N(\boldsymbol{y}))\|_Q^2 \,\mathrm{d}\gamma(\boldsymbol{y}) \\ &\leq L^2 \int_U \|u(\boldsymbol{y}) - u_N(\boldsymbol{y})\|_X^2 \,\mathrm{d}\gamma(\boldsymbol{y}), \end{aligned}$$

which tends to 0 as $N \to \infty$. Here we used that L is a Lipschitz constant of Ξ. □

Let us now discuss which functions satisfy the requirements of Theorem 5.2. Additional to $(\boldsymbol{b}, \xi, \delta, X)$-holomorphy, we had to assume boundedness of the holomorphic extensions in Definition 4.1. For functions of the type as in Theorem 4.11 $u(\boldsymbol{y}) = \lim_{N\to\infty} \mathcal{U}\left(\exp\left(\sum_{j=1}^N y_j\psi_j\right)\right)$, the following result gives sufficient conditions such that the assumptions of Theorem 5.2 are satisfied for the forward map.

Corollary 5.3 *Assume that $\mathcal{U} : O \to X$ and $(\psi_j)_{j\in\mathbb{N}} \subset E$ satisfy Assumptions (i), (iii) and (iv) of Theorem 4.11 and additionally for some $r > 0$*

(ii) $\|\mathcal{U}(a)\|_X \leq r$ for all $a \in O$.

Then

$$u(\boldsymbol{y}) = \lim_{N\to\infty} \mathcal{U}\left(\exp\left(\sum_{j=1}^N y_j\psi_j\right)\right) \in L^2(U, X; \gamma)$$

is $(\boldsymbol{b}, \xi, \delta, X)$-holomorphic with constant functions $\varphi_N \equiv r$, $N \in \mathbb{N}$, in Definition 4.1.

Proof By Theorem 4.11, u is $(\boldsymbol{b}, \xi, \delta, X)$-holomorphic. Recalling the construction of $\varphi_N : \mathbb{R}^N \to \mathbb{R}$ in Step 3 of the proof of Theorem 4.11, we observe that φ_N can be chosen as $\varphi_N \equiv r$. □

5.3 Example: Parametric Diffusion Coefficient

We revisit the example of the diffusion equation with parametric log-Gaussian coefficient as introduced in Sect. 3.5 and used in Sect. 4.3.1. With the Lipschitz continuity of the data-to-solution map established in Sect. 4.3.1, we verify the well-posedness of the corresponding BIP.

We fix the dimension $d \in \mathbb{N}$ of the physical domain $D \subseteq \mathbb{R}^d$, being a bounded Lipschitz domain, and choose $E = L^\infty(D; \mathbb{C})$ and $X = H_0^1(D; \mathbb{C})$. We assume that $f \in X'$ and $a_0 \in E$ with

$$\rho(a_0) > 0.$$

For

$$a \in O := \{a \in E : \rho(a) > 0\},$$

let $\mathcal{U}(a)$ be the solution to the equation

$$-\operatorname{div}((a_0 + a)\nabla\mathcal{U}(a)) = f \text{ in } D, \quad \mathcal{U}(a) = 0 \text{ on } \partial D, \tag{5.7}$$

for some fixed $f \in X'$.

Due to

$$\rho(a_0 + a) \geq \rho(a_0) > 0,$$

for every $a \in O$, as in (4.20) we find that $\mathcal{U}(a)$ is well-defined and it holds

$$\|\mathcal{U}(a)\|_X \leq \frac{\|f\|_{X'}}{\rho(a_0)} =: r \qquad \forall a \in O.$$

This shows assumption (ii) in Corollary 5.3. Slightly adjusting the arguments in Sect. 4.3.1 one observes that $\mathcal{U} : O \to X$ satisfies assumptions (i) and (iii) in Theorem 4.11. Fix a representation system $(\psi_j)_{j\in\mathbb{N}} \subseteq V$ such that with $b_j := \|\psi_j\|_E$ it holds $(b_j)_{j\in\mathbb{N}} \in \ell^1(\mathbb{N})$. Then Corollary 5.3 implies that the forward map

$$u(\boldsymbol{y}) = \lim_{N\to\infty} \mathcal{U}\Big(\exp\Big(\sum_{j=1}^{N} y_j\psi_j\Big)\Big)$$

satisfies the assumptions of Theorem 5.2. Theorem 5.2 in turn implies that the posterior density for this model is $(\boldsymbol{b}, \xi, \delta, X)$-holomorphic. We shall prove in Chap. 6 that sparse-grid quadratures can be constructed which achieve higher order convergence for the integrands in (5.4) and (5.5), with the convergence rate being a decreasing function of $p \in (0, 4/5)$ such that $\boldsymbol{b} \in \ell^p(\mathbb{N})$, see Theorem 6.16. Furthermore, Theorem 4.9 implies a certain sparsity for the family of Wiener-Hermite PC expansion coefficients of the parametric maps in (5.4) and (5.5).

Chapter 6
Smolyak Sparse-Grid Interpolation and Quadrature

Theorem 4.9 shows that if v is $(\boldsymbol{b}, \xi, \delta, X)$-holomorphic for some $\boldsymbol{b} \in \ell^p(\mathbb{N})$ and some $p \in (0, 1)$, then $(\|v_{\boldsymbol{\nu}}\|_X)_{\boldsymbol{\nu}\in\mathcal{F}} \in \ell^{2p/(2-p)}(\mathcal{F})$. In Remark 4.10, based on this summability of the Wiener-Hermite PC expansion coefficients, we derived the convergence rate of best n-term approximation as in (4.13). This approximation is not linear since the approximant is taken accordingly to the N largest terms $\|v_{\boldsymbol{\nu}}\|_X$. To construct a linear approximation which gives the same convergence rate it is suitable to use the stronger weighted ℓ^2-summability result (4.11) in Theorem 4.9.

In Theorem 4.9 of Chap. 4, we have obtained the weighted ℓ^2-summability

$$\sum_{\boldsymbol{\nu}\in\mathcal{F}} \beta_{\boldsymbol{\nu}}(r, \boldsymbol{\varrho})\|u_{\boldsymbol{\nu}}\|_X^2 < \infty \quad \text{with} \quad \left(\beta_{\boldsymbol{\nu}}(r, \boldsymbol{\varrho})^{-1/2}\right)_{\boldsymbol{\nu}\in\mathcal{F}} \in \ell^{p/(1-p)}(\mathcal{F}), \tag{6.1}$$

for the norms of the Wiener-Hermite PC expansion coefficients of $(\boldsymbol{b}, \xi, \delta, X)$-holomorphic functions u if $\boldsymbol{b} \in \ell^p(\mathbb{N})$ for some $0 < p < 1$. In Sect. 4.2 and Chap. 5 we saw that solutions to certain parametric PDEs as well as posterior densities satisfy $(\boldsymbol{b}, \xi, \delta, X)$-holomorphy.

The goal of this section is in a constructive way to sharpen and improve these results in a form more suitable for numerical implementation by using some ideas from [43, 45, 113]. We shall construct a new weight family $(c_{\boldsymbol{\nu}})_{\boldsymbol{\nu}\in\mathcal{F}}$ based on $(\beta_{\boldsymbol{\nu}}(r, \boldsymbol{\varrho}))_{\boldsymbol{\nu}\in\mathcal{F}}$, such that (6.1) with $\beta_{\boldsymbol{\nu}}(r, \boldsymbol{\varrho})$ replaced by $c_{\boldsymbol{\nu}}$, and its generalization of the form (3.43) for $\sigma_{\boldsymbol{\nu}} = c_{\boldsymbol{\nu}}^{1/2}$ hold. Once a suitable family $(c_{\boldsymbol{\nu}})_{\boldsymbol{\nu}\in\mathcal{F}}$ has been identified, we obtain a multiindex set $\Lambda_\varepsilon \subseteq \mathcal{F}$ for $\varepsilon > 0$ via

$$\Lambda_\varepsilon := \{\boldsymbol{\nu} \in \mathcal{F} : c_{\boldsymbol{\nu}}^{-1} \geq \varepsilon\}. \tag{6.2}$$

The set Λ_ε will then serve as an index set to define interpolation operators $\mathbf{I}_{\Lambda_\varepsilon}$ and quadrature operators $\mathbf{Q}_{\Lambda_\varepsilon}$. As the sequence $(c_{\boldsymbol{\nu}})_{\boldsymbol{\nu}\in\mathcal{F}}$ is used to construct sets

D. Dũng et al., *Analyticity and Sparsity in Uncertainty Quantification for PDEs with Gaussian Random Field Inputs*, Lecture Notes in Mathematics 2334,
https://doi.org/10.1007/978-3-031-38384-7_6

of multiindices, it should possess certain features, including each $c_{\boldsymbol{\nu}}$ to be easily computable for $\boldsymbol{\nu} \in \mathcal{F}$, and for the resulting numerical algorithm to be efficient.

6.1 Smolyak Sparse-Grid Interpolation and Quadrature

6.1.1 Smolyak Sparse-Grid Interpolation

Recall that for every $n \in \mathbb{N}_0$ we denote by $(\chi_{n,j})_{j=0}^n \subseteq \mathbb{R}$ the Gauss-Hermite points in one dimension (in particular, $\chi_{0,0} = 0$), that is, the roots of Hermite polynomial H_{n+1}. Let

$$I_n : C^0(\mathbb{R}) \to C^0(\mathbb{R})$$

be the univariate polynomial Lagrange interpolation operator defined by

$$(I_n u)(y) := \sum_{j=0}^{n} u(\chi_{n,j}) \prod_{\substack{i=0 \\ i \neq j}}^{n} \frac{y - \chi_{n,i}}{\chi_{n,j} - \chi_{n,i}}, \qquad y \in \mathbb{R},$$

with the convention that $I_{-1} : C^0(\mathbb{R}) \to C^0(\mathbb{R})$ is defined as the constant 0 operator.

For any multi-index $\boldsymbol{\nu} \in \mathcal{F}$, introduce the tensorized operators $\mathbf{I}_{\boldsymbol{\nu}}$ by

$$\mathbf{I}_{\mathbf{0}} u := u((\chi_{0,0})_{j \in \mathbb{N}}),$$

and for $\boldsymbol{\nu} \neq \mathbf{0}$ via

$$\mathbf{I}_{\boldsymbol{\nu}} := \bigotimes_{j \in \mathbb{N}} I_{\nu_j}, \tag{6.3}$$

i.e.,

$$\mathbf{I}_{\boldsymbol{\nu}} u(\boldsymbol{y}) = \sum_{\{\boldsymbol{\mu} \in \mathcal{F} : \boldsymbol{\mu} \leq \boldsymbol{\nu}\}} u((\chi_{\nu_j,\mu_j})_{j \in \mathbb{N}}) \prod_{j \in \mathbb{N}} \prod_{\substack{i=0 \\ i \neq \mu_j}}^{\nu_j} \frac{y_j - \chi_{\nu_j,i}}{\chi_{\nu_j,\mu_j} - \chi_{\nu_j,i}}, \qquad \boldsymbol{y} \in U.$$

The operator $\mathbf{I}_{\boldsymbol{\nu}}$ can thus be applied to functions u which are pointwise defined at each $(\chi_{\nu_j,\mu_j})_{j \in \mathbb{N}} \in U$. Via Remark 4.4, we can apply it in particular to $(\boldsymbol{b}, \xi, \delta, X)$-holomorphic functions. Observe that the product over $j \in \mathbb{N}$ in (6.3) is a finite product, since for every j with $\nu_j = 0$, the inner product over $i \in \{0, \ldots, \mu_j - 1, \mu_j + 1, \ldots, \nu_j\}$ is over an empty set, and therefore equal to one by convention.

For a finite set $\Lambda \subseteq \mathcal{F}$

$$\mathbf{I}_\Lambda := \sum_{\boldsymbol{\nu}\in\Lambda} \bigotimes_{j\in\mathbb{N}} (I_{\nu_j} - I_{\nu_j - 1}). \tag{6.4}$$

Expanding all tensor product operators, we get

$$\mathbf{I}_\Lambda = \sum_{\boldsymbol{\nu}\in\Lambda} \sigma_{\Lambda;\boldsymbol{\nu}} \mathbf{I}_{\boldsymbol{\nu}} \qquad \text{where} \qquad \sigma_{\Lambda;\boldsymbol{\nu}} := \sum_{\{\boldsymbol{e}\in\{0,1\}^\infty : \boldsymbol{\nu}+\boldsymbol{e}\in\Lambda\}} (-1)^{|\boldsymbol{e}|}. \tag{6.5}$$

Definition 6.1 An index set $\Lambda \subseteq \mathcal{F}$ is called *downward closed*, if it is finite and if for every $\boldsymbol{\nu} \in \Lambda$ it holds $\boldsymbol{\mu} \in \Lambda$ whenever $\boldsymbol{\mu} \le \boldsymbol{\nu}$. Here, the ordering "$\le$" between two indices $\boldsymbol{\mu} = (\mu_j)_{j\in\mathbb{N}}$ and $\boldsymbol{\nu} = (\nu_j)_{j\in\mathbb{N}}$ in $\mathcal{F}$ expresses that for all $j \in \mathbb{N}$ holds $\mu_j \le \nu_j$.

As is well-known, $\mathbf{I}_\Lambda$ possesses the following crucial property, see for example [111, Lemma 1.3.3].

Lemma 6.2 *Let $\Lambda \subseteq \mathcal{F}$ be downward closed. Then $\mathbf{I}_\Lambda f = f$ for all $f \in \operatorname{span}\{\boldsymbol{y}^{\boldsymbol{\nu}} : \boldsymbol{\nu} \in \Lambda\}$.*

The reason to choose the collocation points $(\chi_{n,j})_{j=0}^n$ as the Gauss-Hermite points, is that it was recently shown that the interpolation operators I_n then satisfy the following stability estimate, see [52, Lemma 3.13].

Lemma 6.3 *For every $n \in \mathbb{N}_0$ and every $m \in \mathbb{N}$ it holds*

$$\|I_n(H_m)\|_{L^2(\mathbb{R};\gamma_1)} \le 4\sqrt{2m-1}.$$

With the presently adopted normalization of the GM γ_1, it holds $H_0 \equiv 1$ and therefore $I_n(H_0) = H_0$ for all $n \in \mathbb{N}_0$ (since the interpolation operator I_n exactly reproduces all polynomials of degree $n \in \mathbb{N}_0$). Hence

$$\|I_n(H_0)\|_{L^2(\mathbb{R};\gamma_1)} = \|H_0\|_{L^2(\mathbb{R};\gamma_1)} = 1$$

for all $n \in \mathbb{N}_0$. Noting that $4\sqrt{2m-1} \le (1+m)^2$ for all $m \in \mathbb{N}$, we get

$$\|I_n(H_m)\|_{L^2(\mathbb{R};\gamma_1)} \le (1+m)^2 \qquad \forall n,\ m \in \mathbb{N}_0.$$

Consequently

$$\|\mathbf{I}_{\boldsymbol{\nu}}(H_{\boldsymbol{\mu}})\|_{L^2(U;\gamma)} = \prod_{j\in\mathbb{N}} \|I_{\nu_j}(H_{\mu_j})\|_{L^2(\mathbb{R};\gamma_1)} \le \prod_{j\in\mathbb{N}} (1+\mu_j)^2 \qquad \forall \boldsymbol{\nu},\ \boldsymbol{\mu} \in \mathcal{F}. \tag{6.6}$$

Recall that for $\boldsymbol{\nu} \in \mathcal{F}$ and $\tau \geq 0$, we denote

$$p_{\boldsymbol{\nu}}(\tau) := \prod_{j\in\mathbb{N}}(1+\nu_j)^\tau.$$

If $\nu_j > \mu_j$ then $(I_{\nu_j} - I_{\nu_j-1})H_{\mu_j} = 0$. Thus,

$$\bigotimes_{j\in\mathbb{N}}(I_{\nu_j} - I_{\nu_j-1})H_{\boldsymbol{\mu}} = 0,$$

whenever there exists $j \in \mathbb{N}$ such that $\nu_j > \mu_j$. Hence, for any downward closed set Λ, it holds

$$\|\mathbf{I}_\Lambda(H_{\boldsymbol{\mu}})\|_{L^2(U;\gamma)} \leq p_{\boldsymbol{\mu}}(3). \tag{6.7}$$

Indeed,

$$\begin{aligned}\|\mathbf{I}_\Lambda(H_{\boldsymbol{\mu}})\|_{L^2(U;\gamma)} &\leq \sum_{\{\boldsymbol{\nu}\in\Lambda\,:\,\boldsymbol{\nu}\leq\boldsymbol{\mu}\}} p_{\boldsymbol{\mu}}(2) \leq |\{\boldsymbol{\nu}\in\Lambda\,:\,\boldsymbol{\nu}\leq\boldsymbol{\mu}\}|\,p_{\boldsymbol{\mu}}(2)\\ &= \prod_{j\in\mathbb{N}}(1+\mu_j)p_{\boldsymbol{\mu}}(2) = p_{\boldsymbol{\mu}}(3).\end{aligned}$$

6.1.2 Smolyak Sparse-Grid Quadrature

Analogous to I_n we introduce univariate polynomial quadrature operators via

$$Q_n u := \sum_{j=0}^{n} u(\chi_{n,j})\omega_{n,j}, \qquad \omega_{n,j} := \int_{\mathbb{R}} \prod_{i\neq j} \frac{y-\chi_{n,i}}{\chi_{n,j}-\chi_{n,i}}\,\mathrm{d}\gamma_1(y).$$

Furthermore, we define

$$\mathbf{Q}_0 u := u((\chi_{0,0})_{j\in\mathbb{N}}),$$

and for $\boldsymbol{\nu} \neq \mathbf{0}$,

$$\mathbf{Q}_{\boldsymbol{\nu}} := \bigotimes_{j\in\mathbb{N}} Q_{\nu_j},$$

i.e.,

$$\mathbf{Q}_{\boldsymbol{\nu}} u = \sum_{\{\boldsymbol{\mu} \in \mathcal{F} : \boldsymbol{\mu} \le \boldsymbol{\nu}\}} u((\chi_{\nu_j,\mu_j})_{j\in\mathbb{N}}) \prod_{j\in\mathbb{N}} \omega_{\nu_j,\mu_j}.$$

Finally for a finite downward closed $\Lambda \subseteq \mathcal{F}$ with $\sigma_{\Lambda;\boldsymbol{\nu}}$ as in (6.5),

$$\mathbf{Q}_\Lambda := \sum_{\boldsymbol{\nu}\in\Lambda} \sigma_{\Lambda;\boldsymbol{\nu}} Q_{\boldsymbol{\nu}}.$$

Again we emphasize that the above formulas are meaningful as long as point evaluations of u at each $(\chi_{\nu_j,\mu_j})_{j\in\mathbb{N}}$ are well defined, $\boldsymbol{\nu} \in \mathcal{F}$, $\boldsymbol{\mu} \le \boldsymbol{\nu}$. Also note that

$$\mathbf{Q}_\Lambda f = \int_U \mathbf{I}_\Lambda f(\boldsymbol{y}) \, \mathrm{d}\gamma(\boldsymbol{y}). \tag{6.8}$$

Recall that the set $\mathcal{F}_2$ is defined by

$$\mathcal{F}_2 := \{\boldsymbol{\nu} \in \mathcal{F} \, : \, \nu_j \neq 1 \, \forall j\}. \tag{6.9}$$

We thus have $\mathcal{F}_2 \subsetneq \mathcal{F}$. Similar to Lemma 6.2 we have the following lemma, which can be proven completely analogous to [111, Lemma 1.3.16] (also see [113, Remark 4.2]).

Lemma 6.4 *Let $\Lambda \subseteq \mathcal{F}$ be downward closed. Then*

$$Q_\Lambda v = \int_U v(\boldsymbol{y}) \, \mathrm{d}\gamma(\boldsymbol{y})$$

for all $v \in$ span$\{\boldsymbol{y}^{\boldsymbol{\nu}} \, : \, \boldsymbol{\nu} \in \Lambda \cup (\mathcal{F}\backslash\mathcal{F}_2)\}$.

With (6.6) it holds

$$\begin{aligned} |\mathbf{Q}_{\boldsymbol{\nu}}(H_{\boldsymbol{\mu}})| &= \left| \int_U \mathbf{I}_{\boldsymbol{\nu}}(H_{\boldsymbol{\mu}})(\boldsymbol{y}) \, \mathrm{d}\gamma(\boldsymbol{y}) \right| \le \|\mathbf{I}_{\boldsymbol{\nu}}(H_{\boldsymbol{\mu}})\|_{L^2(U;\gamma)} \\ &\le \prod_{j\in\mathbb{N}} (1+\mu_j)^2 \qquad \forall \boldsymbol{\nu}, \, \boldsymbol{\mu} \in \mathcal{F}, \end{aligned}$$

and similarly, using (6.7), we have the bound

$$|\mathbf{Q}_\Lambda(H_{\boldsymbol{\mu}})| \le p_{\boldsymbol{\mu}}(3). \tag{6.10}$$

6.2 Multiindex Sets

In this section, we first recall some arguments from [43, 45, 113] which allow to bound the number of required function evaluations in the interpolation an quadrature algorithm. Subsequently, a construction of a suitable family $(c_{k,\boldsymbol{\nu}})_{\boldsymbol{\nu}\in\mathcal{F}}$ is provided for $k \in \{1, 2\}$. The index k determines whether the family will be used for a sparse-grid interpolation ($k = 1$) or a Smolyak-type sparse-grid quadrature ($k = 2$) algorithm. Finally, it is shown that the multiindex sets $\Lambda_{k,\varepsilon}$ as in (6.2) based on $(c_{k,\boldsymbol{\nu}})_{\boldsymbol{\nu}\in\mathcal{F}}$, guarantee algebraic convergence rates for certain truncated Wiener-Hermite PC expansions. This will be exploited to verify convergence rates for interpolation in Sect. 6.3 and for quadrature in Sect. 6.4.

6.2.1 *Number of Function Evaluations*

In order to obtain a convergence rate in terms of the number of evaluations of u, we need to determine the number of interpolation points used by the operator $\mathbf{I}_\Lambda$ or $\mathbf{Q}_\Lambda$. Since the discussion of $\mathbf{Q}_\Lambda$ is very similar, we concentrate here on $\mathbf{I}_\Lambda$.

Computing the interpolant $\mathbf{I}_{\boldsymbol{\nu}} u$ in (6.3) requires knowledge of the function values of u at each point in

$$\{(\chi_{\nu_j,\mu_j})_{j\in\mathbb{N}} \,:\, \boldsymbol{\mu} \le \boldsymbol{\nu}\}.$$

The cardinality of this set is bounded by $\prod_{j\in\mathbb{N}}(1+\nu_j) = p_{\boldsymbol{\nu}}(1)$. Denote by

$$\mathrm{pts}(\Lambda) := \{(\chi_{\nu_j,\mu_j})_{j\in\mathbb{N}} \,:\, \boldsymbol{\mu} \le \boldsymbol{\nu},\ \boldsymbol{\nu} \in \Lambda\} \tag{6.11}$$

the set of interpolation points defining the interpolation operator $\mathbf{I}_\Lambda$ (i.e., $|\mathrm{pts}(\Lambda)|$ is the number of function evaluations of u required to compute $\mathbf{I}_\Lambda u$). By (6.5) we obtain the bound

$$|\mathrm{pts}(\Lambda)| \le \sum_{\{\boldsymbol{\nu}\in\Lambda\,:\,\sigma_{\Lambda,\boldsymbol{\nu}}\neq 0\}} \prod_{j\in\mathbb{N}}(1+\nu_j) = \sum_{\{\boldsymbol{\nu}\in\Lambda\,:\,\sigma_{\Lambda,\boldsymbol{\nu}}\neq 0\}} p_{\boldsymbol{\nu}}(1). \tag{6.12}$$

6.2.2 *Construction of $(c_{k,\boldsymbol{\nu}})_{\boldsymbol{\nu}\in\mathcal{F}}$*

We are now in position to construct $(c_{k,\boldsymbol{\nu}})_{\boldsymbol{\nu}\in\mathcal{F}}$. As mentioned above, we distinguish between the cases $k = 1$ and $k = 2$, which correspond to polynomial interpolation or quadrature. Note that in the next lemma we define $c_{k,\boldsymbol{\nu}}$ for all $\boldsymbol{\nu} \in \mathcal{F}$, but the estimate provided in the lemma merely holds for $\boldsymbol{\nu} \in \mathcal{F}_k$, $k \in \{1, 2\}$, where $\mathcal{F}_1 := \mathcal{F}$

and $\mathcal{F}_2$ is defined in (6.9). Throughout what follows, empty products shall equal 1 by convention.

Lemma 6.5 *Assume that* $\tau > 0$, $k \in \{1, 2\}$ *and* $r > \max\{\tau, k\}$. *Let* $\boldsymbol{\varrho} \in (0, \infty)^\infty$ *be such that* $\varrho_j \to \infty$ *as* $j \to \infty$.

Then there exist $K > 0$ *and* $C_0 > 0$ *such that*

$$c_{k,\boldsymbol{\nu}} := \prod_{j \in \mathrm{supp}(\boldsymbol{\nu})} \max\left\{1, K\varrho_j\right\}^{2k} \nu_j^{r-\tau}, \quad \boldsymbol{\nu} \in \mathcal{F}, \tag{6.13}$$

satisfies

$$C_0 c_{k,\boldsymbol{\nu}} p_{\boldsymbol{\nu}}(\tau) \le \beta_{\boldsymbol{\nu}}(r, \boldsymbol{\varrho}) \quad \forall \boldsymbol{\nu} \in \mathcal{F}_k \tag{6.14}$$

with $\beta_{\boldsymbol{\nu}}(r, \boldsymbol{\varrho})$ *as in* (3.36).

Proof

Step 1. Fix $\boldsymbol{\nu} \in \mathcal{F}_k$, then $j \in \mathrm{supp}(\boldsymbol{\nu})$ implies $\nu_j \ge k$ and thus $\min\{r, \nu_j\} \ge k$ since $r > k$ by assumption. With $s := \min\{r, \nu_j\} \le \nu_j$, for all $j \in \mathbb{N}$ holds

$$\begin{aligned}\binom{\nu_j}{s} &= \frac{\nu_j!}{(\nu_j - s)!s!} \ge \frac{1}{s!}(\nu_j - s + 1)^s \ge \nu_j^s \frac{1}{s!s^s} \ge \nu_j^s \frac{1}{r!r^r} \\ &= \nu_j^{\min\{\nu_j, r\}} \frac{1}{r!r^r} \ge \nu_j^r \frac{1}{r!r^{2r}}.\end{aligned}$$

Furthermore, if $j \in \mathrm{supp}(\boldsymbol{\nu})$, then due to $s = \min\{\nu_j, r\} \ge k$, with $\varrho_0 := \min\{1, \min_{j \in \mathbb{N}} \varrho_j\}$ we have

$$\varrho_0^{2r} \le \min\{1, \varrho_j\}^{2r} \le \varrho_j^{2(s-k)}.$$

Thus

$$\varrho_j^{\min\{\nu_j, r\}} \ge \varrho_0^{2r} \varrho_j^{2k}$$

for all $j \in \mathbb{N}$. In all, we conclude

$$\begin{aligned}\beta_{\boldsymbol{\nu}}(r, \boldsymbol{\varrho}) &= \prod_{j \in \mathbb{N}} \left(\sum_{l=0}^{r} \binom{\nu_j}{l} \varrho_j^{2l} \right) \ge \prod_{j \in \mathrm{supp}(\boldsymbol{\nu})} \binom{\nu_j}{\min\{\nu_j, r\}} \varrho_j^{2\min\{\nu_j, r\}} \\ &\ge \prod_{j \in \mathrm{supp}(\boldsymbol{\nu})} \frac{\varrho_0^{2r}}{r!r^{2r}} \varrho_j^{2k} \nu_j^r.\end{aligned} \tag{6.15}$$

Since $\boldsymbol{\nu} \in \mathcal{F}_k$ was arbitrary, this estimate holds for all $\boldsymbol{\nu} \in \mathcal{F}_k$.

Step 2. Denote $\hat{\varrho}_j := \max\{1, K\varrho_j\}$, where $K > 0$ is still at our disposal. We have

$$p_{\boldsymbol{\nu}}(\tau) \leq \prod_{j \in \mathrm{supp}(\boldsymbol{\nu})} 2^\tau \nu_j^\tau$$

and thus

$$c_{k,\boldsymbol{\nu}} p_{\boldsymbol{\nu}}(\tau) \leq \prod_{j \in \mathrm{supp}(\boldsymbol{\nu})} 2^\tau \hat{\varrho}_j^{2k} \nu_j^r. \tag{6.16}$$

Again, this estimate holds for any $\boldsymbol{\nu} \in \mathcal{F}_k$.

With $\varrho_0 := \min\{1, \min_{j \in \mathbb{N}} \varrho_j\}$ denote

$$C_b := \left(\frac{\varrho_0^{2r}}{r!r^{2r}}\right)^{1/(2k)} \qquad \text{and} \qquad C_c := (2^\tau)^{1/(2k)}.$$

Set

$$K := \frac{C_b}{C_c}, \qquad \tilde{\varrho}_j = K\varrho_j$$

for all $j \in \mathbb{N}$. Then

$$C_b \varrho_j = C_c \tilde{\varrho}_j = C_c \hat{\varrho}_j \begin{cases} 1 & \text{if } K\varrho_j \geq 1, \\ K\varrho_j & \text{if } K\varrho_j < 1. \end{cases}$$

Let

$$C_0 := \prod_{\{j \in \mathbb{N} : K\varrho_j < 1\}} (K\varrho_j)^{2k}$$

and note that this product is over a finite number of indices, since $\varrho_j \to \infty$ as $j \to \infty$. Then for any $\boldsymbol{\nu} \in \mathcal{F}_k$

$$\prod_{j \in \mathrm{supp}(\boldsymbol{\nu})} C_c \tilde{\varrho}_j \geq C_0^{\frac{1}{2k}} \prod_{j \in \mathrm{supp}(\boldsymbol{\nu})} C_c \hat{\varrho}_j.$$

With (6.15) and (6.16) we thus obtain for every $\boldsymbol{\nu} \in \mathcal{F}_k$,

$$\begin{aligned}\beta_{\boldsymbol{\nu}}(r, \boldsymbol{\varrho}) &\geq \prod_{j \in \operatorname{supp}(\boldsymbol{\nu})} (C_b \varrho_j)^{2k} \nu_j^r = \prod_{j \in \operatorname{supp}(\boldsymbol{\nu})} \left(C_c \tilde{\varrho}_j\right)^{2k} \nu_j^r \\ &\geq C_0 \prod_{j \in \operatorname{supp}(\boldsymbol{\nu})} (C_c \hat{\varrho}_j)^{2k} \nu_j^r \geq C_0 c_{k,\boldsymbol{\nu}} p_{\boldsymbol{\nu}}(\tau).\end{aligned}$$

□

6.2.3 Summability Properties of the Collection $(c_{k,\boldsymbol{\nu}})_{\boldsymbol{\nu} \in \mathcal{F}}$

First we discuss the summability of the collection $(c_{k,\boldsymbol{\nu}})_{\boldsymbol{\nu} \in \mathcal{F}}$. We will require the following lemma which is a modification of [43, Lemma 6.2].

Lemma 6.6 *Let $\theta \geq 0$. Let further $k \in \{1, 2\}$, $\tau > 0$, $r > \max\{k, \tau\}$ and $q > 0$ be such that $(r - \tau)q/(2k) - \theta > 1$. Assume that $(\varrho_j)_{j \in \mathbb{N}} \in (0, \infty)^\infty$ satisfies $(\varrho_j^{-1})_{j \in \mathbb{N}} \in \ell^q(\mathbb{N})$. Then with $(c_{k,\boldsymbol{\nu}})_{\boldsymbol{\nu} \in \mathcal{F}}$ as in Lemma 6.5 it holds*

$$\sum_{\boldsymbol{\nu} \in \mathcal{F}} p_{\boldsymbol{\nu}}(\theta) c_{k,\boldsymbol{\nu}}^{-\frac{q}{2k}} < \infty.$$

Proof This lemma can be proven in the same way as the proof of [43, Lemma 6.2]. We provide a proof for completeness. With $\hat{\varrho}_j := \max\{1, K\varrho_j\}$ it holds $(\hat{\varrho}_j^{-1})_{j \in \mathbb{N}} \in \ell^q(\mathbb{N})$. By definition of $c_{k,\boldsymbol{\nu}}$, factorizing, we get

$$\begin{aligned}\sum_{\boldsymbol{\nu} \in \mathcal{F}} p_{\boldsymbol{\nu}}(\theta) c_{k,\boldsymbol{\nu}}^{-\frac{q}{2k}} &= \sum_{\boldsymbol{\nu} \in \mathcal{F}} \prod_{j \in \operatorname{supp}(\boldsymbol{\nu})} (1 + \nu_j)^\theta \left(\hat{\varrho}_j^{2k} \nu_j^{r-\tau}\right)^{-\frac{q}{2k}} \\ &\leq \prod_{j \in \mathbb{N}} \left(2^\theta \hat{\varrho}_j^{-q} \sum_{n \in \mathbb{N}} n^{\frac{-q(r-\tau)}{2k}} n^\theta\right).\end{aligned}$$

The sum over n equals some finite constant C since by assumption $q(r-\tau)/2k-\theta > 1$. Using the inequality $\log(1 + x) \leq x$ for all $x > 0$, we get

$$\sum_{\boldsymbol{\nu} \in \mathcal{F}} c_{k,\boldsymbol{\nu}}^{-\frac{q}{2k}} \leq \prod_{j \in \mathbb{N}} \left(1 + C\hat{\varrho}_j^{-q}\right) = \exp\left(\sum_{j \in \mathbb{N}} \log(1 + C\hat{\varrho}_j^{-q})\right) \leq \exp\left(\sum_{j \in \mathbb{N}} C\hat{\varrho}_j^{-q}\right),$$

which is finite since $(\hat{\varrho}_j^{-1}) \in \ell^q(\mathbb{N})$. □

Based on (6.2), for $\varepsilon > 0$ and $k \in \{1, 2\}$ let

$$\Lambda_{k,\varepsilon} := \{\boldsymbol{\nu} \in \mathcal{F} : c_{k,\boldsymbol{\nu}}^{-1} \geq \varepsilon\} \subseteq \mathcal{F}. \tag{6.17}$$

The summability shown in Lemma 6.6 implies algebraic convergence rates of the tail sum as provided by the following proposition. This is well-known and follows by Stechkin's lemma [102] which itself is a simple consequence of Hölder's inequality.

Proposition 6.7 *Let $k \in \{1, 2\}$, $\tau > 0$, and $q > 0$. Let $(\varrho_j^{-1})_{j\in\mathbb{N}} \in \ell^q(\mathbb{N})$ and $r > \max\{k, \tau\}$, $(r - \tau)q/(2k) > 2$. Assume that $(a_{\boldsymbol{\nu}})_{\boldsymbol{\nu}\in\mathcal{F}} \in [0, \infty)^\infty$ is such that*

$$\sum_{\boldsymbol{\nu}\in\mathcal{F}} \beta_{\boldsymbol{\nu}}(r, \boldsymbol{\varrho}) a_{\boldsymbol{\nu}}^2 < \infty. \tag{6.18}$$

Then there exists a constant C solely depending on $(c_{k,\boldsymbol{\nu}})_{\boldsymbol{\nu}\in\mathcal{F}}$ in (6.13) such that for all $\varepsilon > 0$ it holds that

$$\sum_{\boldsymbol{\nu}\in\mathcal{F}_k\backslash\Lambda_{k,\varepsilon}} p_{\boldsymbol{\nu}}(\tau) a_{\boldsymbol{\nu}} \leq C \left(\sum_{\boldsymbol{\nu}\in\mathcal{F}} \beta_{\boldsymbol{\nu}}(r, \boldsymbol{\varrho}) a_{\boldsymbol{\nu}}^2\right)^{\frac{1}{2}} \varepsilon^{\frac{1}{2}-\frac{q}{4k}},$$

and

$$|\mathrm{pts}(\Lambda_{k,\epsilon})| \leq C\varepsilon^{-\frac{q}{2k}}. \tag{6.19}$$

Proof We estimate

$$\sum_{\boldsymbol{\nu}\in\mathcal{F}_k\backslash\Lambda_{k,\varepsilon}} p_{\boldsymbol{\nu}}(\tau) a_{\boldsymbol{\nu}} \leq \left(\sum_{\boldsymbol{\nu}\in\mathcal{F}_k\backslash\Lambda_{k,\varepsilon}} p_{\boldsymbol{\nu}}(\tau)^2 a_{\boldsymbol{\nu}}^2 c_{k,\boldsymbol{\nu}}\right)^{1/2} \left(\sum_{\boldsymbol{\nu}\in\mathcal{F}_k\backslash\Lambda_{k,\varepsilon}} c_{k,\boldsymbol{\nu}}^{-1}\right)^{1/2}.$$

The first sum is finite by (6.18) and because $C_0 p_{\boldsymbol{\nu}}(\tau)^2 c_{k,\boldsymbol{\nu}} \leq \beta_{\boldsymbol{\nu}}(r, \boldsymbol{\varrho})$ according to (6.14). By Lemma 6.6 and (6.17) we obtain

$$\sum_{\boldsymbol{\nu}\in\mathcal{F}_k\backslash\Lambda_{k,\varepsilon}} c_{k,\boldsymbol{\nu}}^{-1} = \sum_{c_{k,\boldsymbol{\nu}}^{-1}<\varepsilon} c_{k,\boldsymbol{\nu}}^{-\frac{q}{2k}} c_{k,\boldsymbol{\nu}}^{-1+\frac{q}{2k}} \leq C\varepsilon^{1-\frac{q}{2k}}$$

which proves the first statement. Moreover, using (6.12)

$$|\mathrm{pts}(\Lambda_{k,\epsilon})| = \sum_{\boldsymbol{\nu}\in\Lambda_{k,\epsilon}} p_{\boldsymbol{\nu}}(1) = \sum_{c_{k,\boldsymbol{\nu}}^{-1}\geq\varepsilon} p_{\boldsymbol{\nu}}(1) c_{k,\boldsymbol{\nu}}^{-\frac{q}{2k}} c_{k,\boldsymbol{\nu}}^{\frac{q}{2k}} \leq \varepsilon^{-\frac{q}{2k}} \sum_{\boldsymbol{\nu}\in\mathcal{F}_k} p_{\boldsymbol{\nu}}(1) c_{k,\boldsymbol{\nu}}^{-\frac{q}{2k}} \leq C\varepsilon^{-\frac{q}{2k}}$$

again by Lemma 6.6 and (6.17). □

6.2.4 Computing Λ_ε

Having identified appropriate sequences $(c_{\boldsymbol{\nu}})_{\boldsymbol{\nu}\in\mathcal{F}}$, in order to be able to implement the Smolyak sparse-grid interpolation operator $\mathbf{I}_{\Lambda_\varepsilon}$ and the Smolyak sparse-grid quadrature operator $\mathbf{Q}_{\Lambda_\varepsilon}$, in practice it remains to compute the sets Λ_ε in (6.2). We now recall Algorithm 2 in [111, Sec. 3.1.3] which achieves this in $O(|\Lambda_\varepsilon|)$ work and memory. For the convenience of the reader we recall the main statement regarding the algorithm's complexity below in Lemma 6.8. Additionally, we point to [19, Alg. 4.13] which presents an alternative approach—a recursive algorithm that also achieves linear computational complexity.

In the following denote $\boldsymbol{e}_j := (\delta_{ij})_{j\in\mathbb{N}} \in \mathbb{N}_0^\infty$.

Algorithm 1 Lambda(ε, $(c_{\boldsymbol{\nu}})_{\boldsymbol{\nu}\in\mathcal{F}}$))

$\boldsymbol{\nu} \leftarrow \mathbf{0}$
if $c_{\boldsymbol{\nu}}^{-1} < \varepsilon$ **then**
    $\Lambda \leftarrow \emptyset$
    return Λ
else
    $\Lambda \leftarrow \{\boldsymbol{\nu}\}$
while True **do**
    $d \leftarrow 1$
    while $c_{\boldsymbol{\nu}+\boldsymbol{e}_d}^{-1} < \varepsilon$ **do**
        if $\nu_d \neq 0$ **then** ▷ Reject $\boldsymbol{\nu}+\boldsymbol{e}_d$ where $\nu_d \neq 0$
            $\nu_d \leftarrow 0$
            $d \leftarrow d+1$
        else if $\boldsymbol{\nu} \neq \mathbf{0}$ **then** ▷ Reject $\boldsymbol{\nu}+\boldsymbol{e}_d$ where $\nu_d = 0$
            $d = \min\{j \in \mathbb{N} : \nu_j \neq 0\}$
        else ▷ Reject $\boldsymbol{e}_d \Rightarrow$ stop algorithm
            return Λ
    $\boldsymbol{\nu} \leftarrow \boldsymbol{\nu}+\boldsymbol{e}_d$
    $\Lambda \leftarrow \Lambda \cup \{\boldsymbol{\nu}\}$

The algorithm is of linear complexity in the following sense [111, 3.1.12]:

Lemma 6.8 *Let $(c_{\boldsymbol{\nu}}^{-1})_{\boldsymbol{\nu}\in\mathcal{F}} \subseteq [0,\infty)$ be a null-sequence such that (i) $\boldsymbol{\mu} \leq \boldsymbol{\nu}$ implies $c_{\boldsymbol{\mu}}^{-1} \geq c_{\boldsymbol{\nu}}^{-1}$ and (ii) if $\boldsymbol{\nu} \in \mathcal{F}$ and for some $i < j$ it holds $\nu_i = \nu_j = 0$, then $c_{\boldsymbol{\nu}+\boldsymbol{e}_i}^{-1} \geq c_{\boldsymbol{\nu}+\boldsymbol{e}_j}^{-1}$.*

Then for any $\varepsilon > 0$, Algorithm 1 terminates and returns Λ_ε in (6.2). Moreover each line of Algorithm 1 is executed at most $4|\Lambda_\varepsilon|+1$ times.

6.3 Interpolation Convergence Rate

If X is a Hilbert space, then the Wiener-Hermite PC expansion of $u : U \to X$ converges in general only in $L^2(U, X; \gamma)$. As mentioned before this creates some subtleties when working with interpolation and quadrature operators based on pointwise evaluations of the target function. To demonstrate this, we recall the following example from [40], which *does not satisfy* $(\boldsymbol{b}, \xi, \delta, \mathbb{C})$*-holomorphy*, since Definition 4.1 (iii) does not hold.

Example 6.9 Define $u : U \to \mathbb{C}$ pointwise by

$$u(\boldsymbol{y}) := \begin{cases} 1 & \text{if } |\{j \in \mathbb{N} : y_j \neq 0\}| < \infty, \\ 0 & \text{otherwise.} \end{cases}$$

Then u vanishes on the complement of the γ-null set

$$\bigcup_{n \in \mathbb{N}} \mathbb{R}^n \times \{0\}^\infty.$$

Consequently u is equal to the constant zero function in the sense of $L^2(U; \gamma)$. Hence there holds the expansion $u = \sum_{\boldsymbol{\nu} \in \mathcal{F}} 0 \cdot H_{\boldsymbol{\nu}}$ with convergence in $L^2(U; \gamma)$. Now let $\Lambda \subseteq \mathcal{F}$ be nonempty, finite and downward closed. As explained in Sect. 6.1.1, the interpolation operator $\mathbf{I}_\Lambda$ reproduces all polynomials in $\text{span}\{\boldsymbol{y}^{\boldsymbol{\nu}} : \boldsymbol{\nu} \in \Lambda\}$. Since any point $(\chi_{\nu_j, \mu_j})_{j \in \mathbb{N}}$ with $\mu_j \le \nu_j$ is zero in all but finitely many coordinates (due to $\chi_{0,0} = 0$), we observe that

$$\mathbf{I}_\Lambda u \equiv 1 \neq 0 \equiv \sum_{\boldsymbol{\nu} \in \mathcal{F}} 0 \cdot \mathbf{I}_\Lambda H_{\boldsymbol{\nu}}.$$

This is due to the fact that $u = \sum_{\boldsymbol{\nu} \in \mathcal{F}} 0 \cdot H_{\boldsymbol{\nu}}$ only holds in the $L^2(U; \gamma)$ sense, and interpolation or quadrature (which require pointwise evaluation of the function) are not meaningful for $L^2(U; \gamma)$ functions.

The above example shows that if

$$u = \sum_{\boldsymbol{\nu} \in \mathcal{F}} u_{\boldsymbol{\nu}} H_{\boldsymbol{\nu}} \in L^2(U; \gamma)$$

with Wiener-Hermite PC expansion coefficients $(u_{\boldsymbol{\nu}})_{\boldsymbol{\nu} \in \mathcal{F}} \subset \mathbb{R}$, then the formal equalities

$$\mathbf{I}_\Lambda u = \sum_{\boldsymbol{\nu} \in \mathcal{F}} u_{\boldsymbol{\nu}} \mathbf{I}_\Lambda H_{\boldsymbol{\nu}},$$

and

$$\mathbf{Q}_\Lambda u = \sum_{\boldsymbol{\nu} \in \mathcal{F}} u_{\boldsymbol{\nu}} \mathbf{Q}_\Lambda H_{\boldsymbol{\nu}}$$

do in general not hold in $L^2(U;\gamma)$. Our definition of $(\boldsymbol{b},\xi,\delta,X)$-holomorphy allows to circumvent this by interpolating not u itself but the approximations u_N to u which are pointwise defined and only depend on finitely many variables, cp. Definition 4.1.

Our analysis starts with the following result about pointwise convergence. For $k \in \{1,2\}$ and $N \in \mathbb{N}$ we introduce the notation

$$\mathcal{F}_k^N := \{\boldsymbol{\nu} \in \mathcal{F}_k \, : \, \mathrm{supp}(\boldsymbol{\nu}) \subseteq \{1,\ldots,N\}\}.$$

These sets thus contain multiindices $\boldsymbol{\nu}$ for which $\nu_j = 0$ for all $j > N$.

Lemma 6.10 *Let u be $(\boldsymbol{b},\xi,\delta,X)$-holomorphic for some $\boldsymbol{b} \in (0,\infty)^\infty$. Let $N \in \mathbb{N}$, and let $\tilde{u}_N : U \to X$ be as in Definition 4.1. For $\boldsymbol{\nu} \in \mathcal{F}$ define*

$$\tilde{u}_{N,\boldsymbol{\nu}} := \int_U \tilde{u}_N(\boldsymbol{y}) H_{\boldsymbol{\nu}}(\boldsymbol{y}) \, \mathrm{d}\gamma(\boldsymbol{y}).$$

Then,

$$\tilde{u}_N(\boldsymbol{y}) = \sum_{\boldsymbol{\nu} \in \mathcal{F}_1^N} \tilde{u}_{N,\boldsymbol{\nu}} H_{\boldsymbol{\nu}}(\boldsymbol{y}) \tag{6.20}$$

with the equality and pointwise absolute convergence in X for all $\boldsymbol{y} \in U$.

Proof From the Cramér bound

$$|\tilde{H}_n(x)| < 2^{n/2}\sqrt{n!}\exp(x^2/2),$$

see [74], and where $\tilde{H}_n(x/\sqrt{2}) := 2^{n/2}\sqrt{n!}H_n(x)$, see [1, Page 787], we have for all $n \in \mathbb{N}_0$

$$\sup_{x \in \mathbb{R}} \exp(-x^2/4)|H_n(x)| \leq 1. \tag{6.21}$$

By Theorem 4.8 $(\tilde{u}_{N,\boldsymbol{\nu}})_{\boldsymbol{\nu}\in\mathcal{F}} \in \ell^1(\mathcal{F})$. Note that for $\boldsymbol{\nu} \in \mathcal{F}_1^N$

$$\tilde{u}_{N,\boldsymbol{\nu}} = \int_U \tilde{u}_N(\boldsymbol{y}) H_{\boldsymbol{\nu}}(\boldsymbol{y}) \, \mathrm{d}\gamma(\boldsymbol{y}) = \int_{\mathbb{R}^N} u_N(y_1,\ldots,y_N) \prod_{j=1}^N H_{\nu_j}(y_j) \, \mathrm{d}\gamma_N((y_j)_{j=1}^N)$$

and thus $\tilde{u}_{N,\boldsymbol{\nu}}$ coincides with the Wiener-Hermite PC expansion coefficient of u_N w.r.t. the multiindex $(\nu_j)_{j=1}^N \in \mathbb{N}_0^N$. The summability of the collection

$$\left(\|u_{N,\boldsymbol{\nu}}\|_X \| \prod_{j=1}^N H_{\nu_j}(y_j) \|_{L^2(\mathbb{R}^N;\gamma_N)} \right)_{\boldsymbol{\nu} \in \mathcal{F}_1^N}$$

now implies in particular,

$$u_N((y_j)_{j=1}^N) = \sum_{\nu \in \mathcal{F}_1^N} u_{N,\nu} \prod_{j=1}^N H_{\nu_j}(y_j)$$

in the sense of $L^2(\mathbb{R}^N; \gamma_N)$.

Due to (6.21) and $(\|u_{N,\nu}\|_X)_{\nu \in \mathcal{F}_1^N} \in \ell^1(\mathcal{F}_1^N)$ we can define a continuous function

$$\hat{u}_N : (y_j)_{j=1}^N \mapsto \sum_{\nu \in \mathbb{N}_0^N} u_{N,\nu} \prod_{j=1}^N H_{\nu_j}(y_j) \tag{6.22}$$

on $\mathbb{R}^N$. By (6.21), for every fixed $(y_j)_{j=1}^N \in \mathbb{R}^N$ we have the uniform bound $|\prod_{j=1}^N H_{\nu_j}(y_j)| \le \prod_{j=1}^N \exp(\frac{y_j^2}{4})$ independent of $\nu \in \mathcal{F}_1^N$. The summability of $(\|u_{N,\nu}\|_X)_{\nu \in \mathcal{F}_1^N}$ implies the absolute convergence of the series in (6.22) for every fixed $(y_j)_{j=1}^N \in \mathbb{R}^N$.

Since they have the same Wiener-Hermite PC expansion, it holds $\hat{u}_N = u_N$ in the sense of $L^2(\mathbb{R}^N; \gamma_N)$.

By Definition 4.1 the function $u : \mathbb{R}^N \to X$ is in particular continuous (it even allows a holomorphic extension to some subset of $\mathbb{C}^N$ containing $\mathbb{R}^N$). Now $\hat{u}_N$, $u_N : \mathbb{R}^N \to X$ are two continuous functions which are equal in the sense of $L^2(\mathbb{R}^N; \gamma_N)$. Thus they coincide pointwise and it holds in X for every $\boldsymbol{y} \in U$,

$$\tilde{u}_N(\boldsymbol{y}) = u_N((y_j)_{j=1}^N) = \sum_{\nu \in \mathcal{F}_1^N} \tilde{u}_{N,\nu} H_\nu(\boldsymbol{y}).$$

□

The result on the pointwise absolute convergence in Lemma 6.10 is not sufficient for establishing convergence of the interpolation approximation in the space $L^2(U, X; \gamma)$. To this end, we need the result on convergence in the space $L^2(U, X; \gamma)$ in the following lemma.

Lemma 6.11 *Let u be $(\boldsymbol{b}, \xi, \delta, X)$-holomorphic for some $\boldsymbol{b} \in (0, \infty)^\infty$. Let $N \in \mathbb{N}$, and let $\tilde{u}_N : U \to X$ be as in Definition 4.1 and $\tilde{u}_{N,\nu}$ as in Lemma 6.10. Let $\Lambda \subset \mathcal{F}_1$ be a finite, downward closed set.*

Then we have

$$\boldsymbol{I}_\Lambda \tilde{u}_N = \sum_{\nu \in \mathcal{F}_1^N} \tilde{u}_{N;\nu} \boldsymbol{I}_\Lambda H_\nu \tag{6.23}$$

with equality and unconditional convergence in the space $L^2(U, X; \gamma)$.

Proof For a function $v : U \to X$ we have

$$\mathbf{I}_\Lambda v(\boldsymbol{y}) = \sum_{\boldsymbol{\nu}\in\Lambda} \sigma_{\Lambda;\boldsymbol{\nu}} \sum_{\boldsymbol{\mu}\in\mathcal{F},\boldsymbol{\mu}\le\boldsymbol{\nu}} v(\chi_{\boldsymbol{\nu},\boldsymbol{\mu}}) L_{\boldsymbol{\nu},\boldsymbol{\mu}}(\boldsymbol{y}), \tag{6.24}$$

where $\sigma_{\Lambda;\boldsymbol{\nu}}$ is defined in (6.5) and recall, $\chi_{\boldsymbol{\nu},\boldsymbol{\mu}} = (\chi_{\nu_j,\mu_j})_{j\in\mathbb{N}}$ and

$$L_{\boldsymbol{\nu},\boldsymbol{\mu}}(\boldsymbol{y}) := \prod_{j\in\mathbb{N}} \prod_{\substack{i=0\\ i\ne\mu_j}}^{\nu_j} \frac{y_j - \chi_{\nu_j,i}}{\chi_{\nu_j,\mu_j} - \chi_{\nu_j,i}}, \quad \boldsymbol{y}\in U. \tag{6.25}$$

Since in a Banach space absolute convergence implies unconditional convergence, from Lemma 6.10 it follows that for any $\boldsymbol{y}\in U$,

$$\tilde{u}_N(\boldsymbol{y}) = \sum_{\boldsymbol{\nu}\in\mathcal{F}_1^N} \tilde{u}_{N,\boldsymbol{\nu}} H_{\boldsymbol{\nu}}(\boldsymbol{y}) \tag{6.26}$$

with equality and unconditional convergence in X. Let $\{F_n\}_{n\in\mathbb{N}} \subset \mathcal{F}_1^N$ be any sequence of finite sets in $\mathcal{F}_1^N$ exhausting $\mathcal{F}_1^N$. Then

$$\forall \boldsymbol{y}\in U:\ \tilde{u}_N^{(n)}(\boldsymbol{y}) := \sum_{\boldsymbol{\nu}\in F_n} \tilde{u}_{N,\boldsymbol{\nu}} H_{\boldsymbol{\nu}}(\boldsymbol{y}) \to \tilde{u}_N(\boldsymbol{y}),\ n\to\infty, \tag{6.27}$$

with convergence in the space X. Notice that the functions $\mathbf{I}_\Lambda \tilde{u}_N$ and $\sum_{\boldsymbol{\nu}\in F_n} \tilde{u}_{N,\boldsymbol{\nu}} \mathbf{I}_\Lambda H_{\boldsymbol{\nu}}$ belong to the space $L^2(U,X;\gamma)$. Hence we have that

$$\begin{aligned}
\Big\| \mathbf{I}_\Lambda \tilde{u}_N - \sum_{\boldsymbol{\nu}\in F_n} \tilde{u}_{N,\boldsymbol{\nu}} \mathbf{I}_\Lambda H_{\boldsymbol{\nu}} \Big\|_{L^2(U,X;\gamma)} &= \big\| \mathbf{I}_\Lambda \tilde{u}_N - \mathbf{I}_\Lambda \tilde{u}_N^{(n)} \big\|_{L^2(U,X;\gamma)} \\
&= \big\| \mathbf{I}_\Lambda \big(\tilde{u}_N - \tilde{u}_N^{(n)}\big) \big\|_{L^2(U,X;\gamma)} \\
&\le \sum_{\boldsymbol{\nu}\in\Lambda} |\sigma_{\Lambda;\boldsymbol{\nu}}| \sum_{\boldsymbol{\mu}\in\mathcal{F},\boldsymbol{\mu}\le\boldsymbol{\nu}} \big\| \tilde{u}_N(\chi_{\boldsymbol{\nu},\boldsymbol{\mu}}) - \tilde{u}_N^{(n)}(\chi_{\boldsymbol{\nu},\boldsymbol{\mu}}) \big\|_X \\
&\quad \times \int_U |L_{\boldsymbol{\nu},\boldsymbol{\mu}}(\boldsymbol{y})| \, \mathrm{d}\gamma(\boldsymbol{y}).
\end{aligned} \tag{6.28}$$

Observe that $L_{\boldsymbol{\nu},\boldsymbol{\mu}}$ is a polynomial of order $|\boldsymbol{\nu}|$. Since $\{\boldsymbol{\mu}\in\mathcal{F} : \boldsymbol{\mu}\le\boldsymbol{\nu}\}$ and Λ are finite sets, we can choose $C := C(\Lambda) > 0$ so that

$$\int_U |L_{\boldsymbol{\nu},\boldsymbol{\mu}}(\boldsymbol{y})| \, \mathrm{d}\gamma(\boldsymbol{y}) \le C$$

for all $\boldsymbol{\mu}\le\boldsymbol{\nu}$ and $\boldsymbol{\nu}\in\Lambda$, and, moreover, by using (6.27) we can choose n_0 so that

$$\|\tilde{u}_N(\chi_{\nu,\mu}) - \tilde{u}_N^{(n)}(\chi_{\nu,\mu})\|_X \leq \varepsilon$$

for all $n \geq n_0$ and $\mu \leq \nu$, $\nu \in \Lambda$. Consequently, we have that for all $n \geq n_0$,

$$\left\| \mathbf{I}_\Lambda \tilde{u}_N - \sum_{\nu \in F_n} \tilde{u}_{N,\nu} \mathbf{I}_\Lambda H_\nu \right\|_{L^2(U,X;\gamma)} \leq C \sum_{\nu \in \Lambda} |\sigma_{\Lambda;\nu}| \sum_{\mu \leq \nu} \varepsilon = C\varepsilon \sum_{\nu \in \Lambda} |\sigma_{\Lambda;\nu}| p_\nu(1). \tag{6.29}$$

Hence we derive the convergence in the space $L^2(U, X; \gamma)$ of the sequence $\sum_{\nu \in F_n} \tilde{u}_{N,\nu} \mathbf{I}_\Lambda H_\nu$ to $\mathbf{I}_\Lambda \tilde{u}_N$ $(n \to \infty)$ for any sequence of finite sets $\{F_n\}_{n\in\mathbb{N}} \subset \mathcal{F}_1^N$ exhausting $\mathcal{F}_1^N$. This proves the lemma. □

Remark 6.12 Under the assumption of Lemma 6.11, in a similar way, we can prove that for every $\boldsymbol{y} \in U$

$$\mathbf{I}_\Lambda \tilde{u}_N(\boldsymbol{y}) = \sum_{\nu \in \mathcal{F}_1^N} \tilde{u}_{N;\nu} \mathbf{I}_\Lambda H_\nu(\boldsymbol{y}) \tag{6.30}$$

with the equality and unconditional convergence in the space X.

We arrive at the following convergence rate result, which improves the convergence rate in [52] (in terms of the number of function evaluations) by a factor 2 (for the case when the elements of the representation system are supported globally in D). Additionally, we provide an explicit construction of suitable index sets. Recall that pointwise evaluations of a $(\boldsymbol{b}, \xi, \delta, X)$-holomorphic functions are understood in the sense of Remark. 4.4.

Theorem 6.13 *Let u be $(\boldsymbol{b}, \xi, \delta, X)$-holomorphic for some $\boldsymbol{b} \in \ell^p(\mathbb{N})$ and some $p \in (0, 2/3)$. Let $(c_{1,\nu})_{\nu \in \mathcal{F}}$ be as in Lemma 6.5 with $\boldsymbol{\varrho}$ as in Theorem 4.8.*

Then there exist $C > 0$ and, for every $n \in \mathbb{N}$, $\varepsilon_n > 0$ such that $|\mathrm{pts}(\Lambda_{1,\varepsilon_n})| \leq n$ (with $\Lambda_{1,\varepsilon_n}$ as in (6.17)) and

$$\|u - \boldsymbol{I}_{\Lambda_{1,\varepsilon_n}} u\|_{L^2(U,X;\gamma)} \leq C n^{-\frac{1}{p}+\frac{3}{2}}.$$

Proof For $\varepsilon > 0$ small enough and satisfying $|\Lambda_{1,\varepsilon}| > 0$, take $N \in \mathbb{N}$ with

$$N \geq \max\{j \in \mathrm{supp}(\boldsymbol{\nu}) : \boldsymbol{\nu} \in \Lambda_{1,\varepsilon}\},$$

so large that

$$\|u - \tilde{u}_N\|_{L^2(U,X;\gamma)} \leq \varepsilon^{\frac{1}{2} - \frac{p}{4(1-p)}}, \tag{6.31}$$

which is possible due to the $(\boldsymbol{b}, \xi, \delta, X)$-holomorphy of u (cp. Definition 4.1 (iii)). An appropriate value of ε depending on n will be chosen below. In the following for

$\boldsymbol{\nu} \in \mathcal{F}_1^N$ we denote by $\tilde{u}_{N,\boldsymbol{\nu}} \in X$ the PC coefficient of $\tilde{u}_N$ and for $\boldsymbol{\nu} \in \mathcal{F}$ as earlier $u_{\boldsymbol{\nu}} \in X$ is the PC coefficient of u.

Because

$$N \geq \max\{j \in \mathrm{supp}(\boldsymbol{\nu}) \, : \, \boldsymbol{\nu} \in \Lambda_{1,\varepsilon}\}$$

and $\chi_{0,0} = 0$, we have

$$\mathbf{I}_{\Lambda_{1,\varepsilon}} u = \mathbf{I}_{\Lambda_{1,\varepsilon}} \tilde{u}_N$$

(cp. Remark 4.4). Hence by (6.31)

$$\begin{aligned} \|u - \mathbf{I}_{\Lambda_{1,\varepsilon}} u\|_{L^2(U,X;\gamma)} &= \|u - \mathbf{I}_{\Lambda_{1,\varepsilon}} \tilde{u}_N\|_{L^2(U,X;\gamma)} \\ &\leq \varepsilon^{\frac{1}{2} - \frac{p}{4(1-p)}} + \|\tilde{u}_N - \mathbf{I}_{\Lambda_{1,\varepsilon}} \tilde{u}_N\|_{L^2(U,X;\gamma)}. \end{aligned} \tag{6.32}$$

We now give a bound of the second term on the right side of (6.32). By Lemma 6.11 we can write

$$\mathbf{I}_{\Lambda_{1,\varepsilon}} \tilde{u}_N = \sum_{\boldsymbol{\nu} \in \mathcal{F}_1^N} \tilde{u}_{N;\boldsymbol{\nu}} \mathbf{I}_{\Lambda_{1,\varepsilon}} H_{\boldsymbol{\nu}}$$

with unconditional convergence in $L^2(U, X; \gamma)$. Hence by Lemma 6.2 and (6.7) we have that

$$\begin{aligned} \|\tilde{u}_N - \mathbf{I}_{\Lambda_{1,\varepsilon}} \tilde{u}_N\|_{L^2(U,X;\gamma)} &= \left\| \sum_{\boldsymbol{\nu} \in \mathcal{F} \setminus \Lambda_{1,\varepsilon}} \tilde{u}_{N;\boldsymbol{\nu}} (H_{\boldsymbol{\nu}} - \mathbf{I}_{\Lambda_{1,\varepsilon}} H_{\boldsymbol{\nu}}) \right\|_{L^2(U,X;\gamma)} \\ &\leq \sum_{\boldsymbol{\nu} \in \mathcal{F} \setminus \Lambda_{1,\varepsilon}} \|\tilde{u}_{N;\boldsymbol{\nu}}\|_X \left(\|H_{\boldsymbol{\nu}}\|_{L^2(U;\gamma)} + \|\mathbf{I}_{\Lambda_{1,\varepsilon}} H_{\boldsymbol{\nu}}\|_{L^2(U;\gamma)} \right) \\ &\leq \sum_{\boldsymbol{\nu} \in \mathcal{F}_1^N \setminus \Lambda_{1,\varepsilon}} \|\tilde{u}_{N;\boldsymbol{\nu}}\|_X \, (1 + p_{\boldsymbol{\nu}}(3)) \\ &\leq 2 \sum_{\boldsymbol{\nu} \in \mathcal{F}_1^N \setminus \Lambda_{1,\varepsilon}} \|\tilde{u}_{N;\boldsymbol{\nu}}\|_X \, p_{\boldsymbol{\nu}}(3). \end{aligned}$$

Choosing $r > 4/p - 1$ ($q := p/(1-p)$, $\tau = 3$), according to Proposition 6.7, (6.14) and Theorem 4.8 (with $(\varrho_j^{-1})_{j \in \mathbb{N}} \in \ell^{p/(1-p)}(\mathbb{N})$ as in Theorem 4.8) the last sum is bounded by

$$C\left(\sum_{\nu\in\mathcal{F}_1^N} \beta_\nu(r,\varrho)\|\tilde{u}_{N,\nu}\|_X^2\right)\varepsilon^{\frac{1}{2}-\frac{q}{4}} \le C(\boldsymbol{b})\delta^2\varepsilon^{\frac{1}{2}-\frac{q}{4}} = C(\boldsymbol{b})\delta^2\varepsilon^{\frac{1}{2}-\frac{p}{4(1-p)}},$$

and the constant $C(\boldsymbol{b})$ from Theorem 4.8 does not depend on N and δ. Hence, by (6.32) we obtain

$$\|u - \mathbf{I}_{\Lambda_{1,\varepsilon}}u\|_{L^2(U,X;\gamma)} \le C_1\varepsilon^{\frac{1}{2}-\frac{p}{4(1-p)}}. \tag{6.33}$$

From (6.19) it follows that

$$|\mathrm{pts}(\Lambda_{1,\varepsilon})| \le C_2\varepsilon^{-\frac{q}{2}} = C_2\varepsilon^{-\frac{p}{2(1-p)}}.$$

For every $n \in \mathbb{N}$, we choose an $\varepsilon_n > 0$ satisfying the condition

$$n/2 \le C_2\varepsilon_n^{-\frac{p}{2(1-p)}} \le n.$$

Then due to (6.33), the claim holds true for the chosen ε_n. □

Remark 6.14 Comparing the best n-term convergence result in Remark 4.10 with the interpolation result of Theorem 6.13, we observe that the convergence rate is reduced by $1/2$, and moreover, rather than $p \in (0,1)$ as in Remark 4.10, Theorem 6.13 requires $p \in (0,2/3)$. This discrepancy can be explained as follows: Since $(H_\nu)_{\nu\in\mathcal{F}}$ forms an orthonormal basis of $L^2(U;\gamma)$, for the best n-term result we could resort to Parseval's identity, which merely requires ℓ^2-summability of the Hermite PC coefficients, i.e. $(\|u_\nu\|_X)_{\nu\in\mathcal{F}} \in \ell^2(\mathcal{F})$. Due to $(\|u_\nu\|_X)_{\nu\in\mathcal{F}} \in \ell^{\frac{2p}{2-p}}$ by Theorem 4.9, this is ensured as long as $p \in (0,1)$. On the other hand, for the interpolation result we had to use the triangle inequality, since the family $(\mathbf{I}_{\Lambda_{1,\varepsilon_n}}H_\nu)_{\nu\in\mathcal{F}}$ of interpolated multivariate Hermite polynomials does not form an orthonormal family of $L^2(U;\gamma)$. This argument requires the stronger condition $(\|u_\nu\|_X)_{\nu\in\mathcal{F}} \in \ell^1(\mathcal{F})$, resulting in the stronger assumption $p \in (0,2/3)$ of Theorem 6.13.

6.4 Quadrature Convergence Rate

We first prove a result on equality and unconditional convergence in the space X for quadrature operators, which is similar to that in Lemma 6.11. It is needed to establish the quadrature convergence rate.

Lemma 6.15 *Let u be $(\boldsymbol{b}, \xi, \delta, X)$-holomorphic for some $\boldsymbol{b} \in (0, \infty)^\infty$. Let $N \in \mathbb{N}$, and let $\tilde{u}_N : U \to X$ be as in Definition 4.1 and $\tilde{u}_{N,\boldsymbol{\nu}}$ as in Lemma 6.10. Let $\Lambda \subset \mathcal{F}_1$ be a finite downward closed set.*

Then we have

$$\boldsymbol{Q}_\Lambda \tilde{u}_N = \sum_{\boldsymbol{\nu} \in \mathcal{F}_1^N} \tilde{u}_{N;\boldsymbol{\nu}} \boldsymbol{Q}_\Lambda H_{\boldsymbol{\nu}} \tag{6.34}$$

with equality and unconditional convergence in X.

Proof For a function $v : U \to X$ by (6.8) and (6.24) we have

$$\mathbf{Q}_\Lambda v = \sum_{\boldsymbol{\nu} \in \Lambda} \sigma_{\Lambda;\boldsymbol{\nu}} \sum_{\boldsymbol{\mu} \in \mathcal{F}, \boldsymbol{\mu} \le \boldsymbol{\nu}} v(\chi_{\boldsymbol{\nu},\boldsymbol{\mu}}) \int_U L_{\boldsymbol{\nu},\boldsymbol{\mu}}(\boldsymbol{y}) \, \mathrm{d}\gamma(\boldsymbol{y}),$$

where $\chi_{\boldsymbol{\nu},\boldsymbol{\mu}} = (\chi_{\nu_j,\mu_j})_{j \in \mathbb{N}}$, $\sigma_{\Lambda;\boldsymbol{\nu}}$, $L_{\boldsymbol{\nu},\boldsymbol{\mu}}$ are defined in (6.5) and (6.25), respectively. By using this representation, we can prove the lemma in a way similar to the proof of Lemma 6.11 with some appropriate modifications. □

Analogous to Theorem 6.13 we obtain the following result for the quadrature convergence with an improved convergence rate compared to interpolation.

Theorem 6.16 *Let u be $(\boldsymbol{b}, \xi, \delta, X)$-holomorphic for some $\boldsymbol{b} \in \ell^p(\mathbb{N})$ and some $p \in (0, 4/5)$. Let $(c_{2,\boldsymbol{\nu}})_{\boldsymbol{\nu} \in \mathcal{F}}$ be as in Lemma 6.5 with ϱ as in Theorem 4.8. Then there exist $C > 0$ and, for every $N \in \mathbb{N}$, $\varepsilon_n > 0$ such that $|\mathrm{pts}(\Lambda_{2,\varepsilon_n})| \le n$ (with $\Lambda_{2,\varepsilon_n}$ as in (6.17)) and*

$$\left\| \int_U u(\boldsymbol{y}) \, \mathrm{d}\gamma(\boldsymbol{y}) - \boldsymbol{Q}_{\Lambda_{2,\varepsilon_n}} u \right\|_X \le C n^{-\frac{2}{p} + \frac{5}{2}}.$$

Proof For $\varepsilon > 0$ small enough and satisfying $|\Lambda_{2,\varepsilon}| > 0$, take $N \in \mathbb{N}$, $N \ge \max\{j \in \mathrm{supp}(\boldsymbol{\nu}) : \boldsymbol{\nu} \in \Lambda_{2,\varepsilon}\}$ so large that

$$\left\| \int_U [u(\boldsymbol{y}) - \tilde{u}_N(\boldsymbol{y})] \, \mathrm{d}\gamma(\boldsymbol{y}) \right\|_X \le \|u - \tilde{u}_N\|_{L^2(U,X;\gamma)} \le \varepsilon^{\frac{1}{2} - \frac{p}{8(1-p)}}, \tag{6.35}$$

which is possible due to the $(\boldsymbol{b}, \xi, \delta, X)$-holomorphy of u (cp. Definition 4.1 (iii)). An appropriate value of ε depending on n will be chosen below. In the following for $\boldsymbol{\nu} \in \mathcal{F}$ we denote by $\tilde{u}_{N,\boldsymbol{\nu}}$ the Wiener-Hermite PC expansion coefficient of $\tilde{u}_N$ and as earlier $u_{\boldsymbol{\nu}}$ is the Wiener-Hermite PC expansion coefficient of u.

Because

$$N \ge \max\{j \in \mathrm{supp}(\boldsymbol{\nu}) : \boldsymbol{\nu} \in \Lambda_{2,\varepsilon}\}$$

and $\chi_{0,0} = 0$, we have $\mathbf{Q}_{\Lambda_{2,\varepsilon}} u = \mathbf{Q}_{\Lambda_{2,\varepsilon}} \tilde{u}_N$ (cp. Remark. 4.4). Hence by (6.35)

$$\begin{aligned}\left\| \int_U u(\boldsymbol{y}) \,\mathrm{d}\gamma(\boldsymbol{y}) - \mathbf{Q}_{\Lambda_{2,\varepsilon}} u \right\|_X &= \left\| \int_U u(\boldsymbol{y}) \,\mathrm{d}\gamma(\boldsymbol{y}) - \mathbf{Q}_{\Lambda_{2,\varepsilon}} \tilde{u}_N \right\|_X \\ &\le \varepsilon^{\frac{1}{2} - \frac{p}{8(1-p)}} + \left\| \int_U \tilde{u}_N(\boldsymbol{y}) \,\mathrm{d}\gamma(\boldsymbol{y}) - \mathbf{Q}_{\Lambda_{2,\varepsilon}} \tilde{u}_N \right\|_X .\end{aligned} \tag{6.36}$$

By Lemma 6.15 we have

$$\mathbf{Q}_{\Lambda_{2,\varepsilon}} \tilde{u}_N = \sum_{\boldsymbol{\nu} \in \mathcal{F}_1^N} \tilde{u}_{N;\boldsymbol{\nu}} \mathbf{Q}_{\Lambda_{2,\varepsilon}} H_{\boldsymbol{\nu}} = \sum_{\boldsymbol{\nu} \in \mathcal{F}_2^N} \tilde{u}_{N;\boldsymbol{\nu}} \mathbf{Q}_{\Lambda_{2,\varepsilon}} H_{\boldsymbol{\nu}}$$

with the equality and unconditional convergence in the space X. Since $\Lambda_{2,\varepsilon}$ is nonempty and downward closed we have $\mathbf{0} \in \Lambda_{2,\varepsilon}$. Then, by Lemma 6.4, (6.10), and using

$$\int_U H_{\boldsymbol{\nu}}(\boldsymbol{y}) \,\mathrm{d}\gamma(\boldsymbol{y}) = 0$$

for all $\mathbf{0} \neq \boldsymbol{\nu} \in \mathcal{F} \backslash \mathcal{F}_2$, we have that

$$\begin{aligned}&\left\| \int_U \tilde{u}_N(\boldsymbol{y}) \,\mathrm{d}\gamma(\boldsymbol{y}) - \mathbf{Q}_{\Lambda_{2,\varepsilon}} \tilde{u}_N \right\|_X \\ &= \left\| \sum_{\boldsymbol{\nu} \in \mathcal{F}_2 \backslash \Lambda_{2,\varepsilon}} \tilde{u}_{N;\boldsymbol{\nu}} \left(\int_U H_{\boldsymbol{\nu}}(\boldsymbol{y}) \,\mathrm{d}\gamma(\boldsymbol{y}) - \mathbf{Q}_{\Lambda_{2,\varepsilon}} H_{\boldsymbol{\nu}} \right) \right\|_X \\ &\le \sum_{\boldsymbol{\nu} \in \mathcal{F}_2 \backslash \Lambda_{2,\varepsilon}} \|\tilde{u}_{N;\boldsymbol{\nu}}\|_X (\|H_{\boldsymbol{\nu}}\|_{L^2(U;\gamma)} + |\mathbf{Q}_{\Lambda_{2,\varepsilon}} H_{\boldsymbol{\nu}}|) \\ &\le \sum_{\boldsymbol{\nu} \in \mathcal{F}_2 \backslash \Lambda_{2,\varepsilon}} \|\tilde{u}_{N;\boldsymbol{\nu}}\|_X \, (1 + p_{\boldsymbol{\nu}}(3)) \\ &\le 2 \sum_{\boldsymbol{\nu} \in \mathcal{F}_2 \backslash \Lambda_{2,\varepsilon}} \|\tilde{u}_{N;\boldsymbol{\nu}}\|_X \, p_{\boldsymbol{\nu}}(3).\end{aligned}$$

Choosing $r > 8/p - 5$ ($q = \frac{p}{1-p}$, $\tau = 3$), according to Proposition 6.7, (6.14) and Theorem 4.8 (with $(\varrho_j^{-1})_{j \in \mathbb{N}} \in \ell^{p/(1-p)}(\mathbb{N})$ as in Theorem 4.8) the last sum is bounded by

$$C \left(\sum_{\boldsymbol{\nu} \in \mathcal{F}} \beta_{\boldsymbol{\nu}}(r, \boldsymbol{\varrho}) \|\tilde{u}_{N,\boldsymbol{\nu}}\|_X^2 \right) \varepsilon^{\frac{1}{2} - \frac{q}{8}} \le C(\boldsymbol{b}) \delta^2 \varepsilon^{\frac{1}{2} - \frac{q}{8}} = C(\boldsymbol{b}) \varepsilon^{\frac{1}{2} - \frac{p}{8(1-p)}},$$

and the constant $C(\boldsymbol{b})$ from Theorem 4.8 does not depend on N and δ. Hence, by (6.35) and (6.36) we obtain that

$$\left\| \int_U u(\boldsymbol{y})\,\mathrm{d}\gamma(\boldsymbol{y}) - \mathbf{Q}_{\Lambda_{2,\varepsilon}} u \right\|_X \le C_1 \varepsilon^{\frac{1}{2} - \frac{p}{8(1-p)}}. \tag{6.37}$$

From (6.19) it follows that

$$|\mathrm{pts}(\Lambda_{k,\epsilon})| \le C_2 \varepsilon^{-\frac{q}{4}} = C_2 \varepsilon^{-\frac{p}{4(1-p)}}.$$

For every $n \in \mathbb{N}$, we choose an $\varepsilon_n > 0$ satisfying the condition

$$n/2 \le C_2 \varepsilon_n^{-\frac{p}{4(1-p)}} \le n.$$

Then due to (6.37) the claim holds true for the chosen ε_n. □

Remark 6.17 Interpolation formulas based on index sets like

$$\Lambda(\xi) := \{\boldsymbol{\nu} \in \mathcal{F} : \beta_{\boldsymbol{\nu}}(r, \boldsymbol{\varrho}) \le \xi^{2/q}\},$$

(where $\xi > 0$ is a large parameter), have been proposed in [43, 52] for the parametric, elliptic divergence-form PDE (3.17) with log-Gaussian inputs (3.18) satisfying the assumptions of Theorem 3.38 with $i = 1$. There, dimension-independent convergence rates of sparse-grid interpolation were obtained. Based on the weighted ℓ^2-summability of the Wiener-Hermite PC expansion coefficients of the form

$$\sum_{\boldsymbol{\nu} \in \mathcal{F}} \beta_{\boldsymbol{\nu}}(r, \boldsymbol{\varrho}) \|u_{\boldsymbol{\nu}}\|_X^2 < \infty \quad \text{with} \quad \left(p_{\boldsymbol{\nu}}(\tau, \lambda) \beta_{\boldsymbol{\nu}}(r, \boldsymbol{\varrho})^{-1/2}\right)_{\boldsymbol{\nu} \in \mathcal{F}} \in \ell^q(\mathcal{F}) \ (0 < q < 2), \tag{6.38}$$

the rate established in [52] is $\frac{1}{2}(1/q - 1/2)$ which is lower than those obtained in the present analysis. The improved rate $1/q - 1/2$ has been established in [44]. This rate coincides with the rate in Theorem 6.13 for the choice $q = p/(1-p)$.

The existence of Smolyak type quadratures with a proof of dimension-independent convergence rates was shown first in [31] and then in [43]. In [31], the symmetry of the GM and corresponding cancellations were not exploited, and these quadrature formulas provide the convergence rate $\frac{1}{2}(1/q - 1/2)$ which is lower (albeit dimension-independent) in terms of the number of function evaluations compared to Theorems 6.13 and 6.16. By using this symmetry, for a given weighted ℓ^2-summability of the Wiener-Hermite PC expansion coefficients (3.43) with $\sigma_{\boldsymbol{\nu}} = \beta_{\boldsymbol{\nu}}(r, \boldsymbol{\varrho})^{1/2}$, the rate established in [43] (see also [45]) is $2/q - 1/2$ which coincides with the rate of convergence that was obtained in Theorem 6.16 for the choice $q = p/(1-p)$.

Chapter 7
Multilevel Smolyak Sparse-Grid Interpolation and Quadrature

In this section we introduce a multilevel interpolation and quadrature algorithm which are suitable for numerical implementation. The presentation and arguments follow mostly [114] and [111, Section 3.2], where multilevel algorithms for the uniform measure on the hypercube $[-1, 1]^\infty$ were analyzed (in contrast to the case of a product GM on U, which we consider here). In Sect. 7.1, we introduce the setting for the multilevel algorithms, in particular a notation of "work-measure" related to the discretization of a Wiener-Hermite PC expansion coefficient $u_{\boldsymbol{\nu}}$ for $\boldsymbol{\nu}$ in the set of multi-indices that are active in a given (interpolation or quadrature) approximation. Section 7.2 describes the general structure of the algorithms, Sect. 7.3 addresses algorithms for the determination of sets $\Lambda \subset \mathcal{F}$ of active multi-indices and a corresponding allocation of discretization levels in linear in $|\Lambda|$ work and memory. Section 7.4 addresses the error analysis of the Smolyak sparse-grid interpolation, and Sect. 7.5 contains the error analysis of the corresponding Smolyak sparse-grid quadrature algorithm. All algorithms are formulated and analyzed in terms of several abstract hypotheses. Section 7.6 verifies these abstract conditions for a concrete family of parametric, elliptic PDEs. Finally, Sect. 7.7 addresses convergence rates achieveable with the mentioned Smolyak sparse-grid interpolation and quadrature algorithms *assuming at hand optimal multi-index sets*. The major finding being that the corresponding rates differ only by logarithmic terms from the error bounds furnished by those realized by the algorithms in Sects. 7.2–7.5.

7.1 Setting and Notation

To approximate the solution u to a parametric PDE as in the examples of the preceding sections, the interpolation operator $\mathbf{I}_\Lambda$ introduced in Sect. 6.1.1 requires function values of u at different interpolation points in the parameter space U. For

D. Dũng et al., *Analyticity and Sparsity in Uncertainty Quantification for PDEs with Gaussian Random Field Inputs*, Lecture Notes in Mathematics 2334,
https://doi.org/10.1007/978-3-031-38384-7_7

a parameter $\boldsymbol{y} \in U$, typically the PDE solution $u(\boldsymbol{y})$, which is a function belonging to a Sobolev space over a *physical domain* D, is not given in closed form and has to be approximated. The idea of multilevel approximations is to combine interpolants of approximations to u at different spatial accuracies, in order to reduce the overall computational complexity. This will now be formalized.

First, we assume given a sequence $(\mathfrak{w}_l)_{l\in\mathbb{N}_0} \subset \mathbb{N}$, exhibiting the properties of the following assumption. Throughout $\mathfrak{w}_l$ will be interpreted as a measure for the computational complexity of evaluating an approximation $u^l : U \to X$ of $u : U \to X$ at a parameter $\boldsymbol{y} \in U$. Here we use a superscript l rather than a subscript for the approximation level, as the subscript is reserved for the dimension truncated version u_N of u as in Definition 4.1.

Assumption 7.1 *The sequence $(\mathfrak{w}_l)_{l\in\mathbb{N}_0} \subseteq \mathbb{N}_0$ is strictly monotonically increasing and $\mathfrak{w}_0 = 0$. There exists a constant $K_{\mathfrak{W}} \geq 1$ such that for all $l \in \mathbb{N}$*

(i) $\sum_{j=0}^{l} \mathfrak{w}_j \leq K_{\mathfrak{W}} \mathfrak{w}_l$,
(ii) $l \leq K_{\mathfrak{W}}(1 + \log(\mathfrak{w}_l))$,
(iii) $\mathfrak{w}_l \leq K_{\mathfrak{W}}(1 + \mathfrak{w}_{l-1})$,
(iv) for every $r > 0$ there exists $C = C(r) > 0$ independent of l such that

$$\sum_{j=l}^{\infty} \mathfrak{w}_j^{-r} \leq C(1 + \mathfrak{w}_l)^{-r}.$$

Assumption 7.1 is satisfied if $(\mathfrak{w}_l)_{l\in\mathbb{N}}$ is exponentially increasing, (for instance $\mathfrak{w}_l = 2^l, l \in \mathbb{N}$). In the following we write $\mathfrak{W} := \{\mathfrak{w}_l : l \in \mathbb{N}_0\}$ and

$$\lfloor x \rfloor_{\mathfrak{W}} := \max\{\mathfrak{w}_l : \mathfrak{w}_l \leq x\}.$$

We work under the following *hypothesis on the discretization errors in physical space*: we quantify the convergence of the discretization scheme with respect to the discretization level $l \in \mathbb{N}$. Specifically, we assume the approximation u^l to u to behave asymptotically as

$$\|u(\boldsymbol{y}) - u^l(\boldsymbol{y})\|_X \leq C(\boldsymbol{y})\mathfrak{w}_l^{-\alpha} \qquad \forall l \in \mathbb{N}, \tag{7.1}$$

for some fixed convergence rate $\alpha > 0$ of the "physical space discretization" and with constant $C(\boldsymbol{y}) > 0$ depending on the parameter sequence $\boldsymbol{y}$. We will make this assumption on u^l more precise shortly. If we think of $u^l(\boldsymbol{y}) \in H^1(D)$ for the moment as a FEM approximation to the exact solution $u(\boldsymbol{y}) \in H^1(D)$ of some $\boldsymbol{y}$-dependent elliptic PDE, then $\mathfrak{w}_l$ could stand for the number of degrees of freedom of the finite element space. In this case α corresponds to the FEM convergence rate. Assumption (7.1) will for instance be satisfied if for each consecutive level the meshwidth is cut in half. Examples are provided by the FE spaces discussed in Sect. 2.6.2, Proposition 2.31. As long as the computational cost of computing the

FEM solution is proportional to the dimension $\mathfrak{w}_l$ of the FEM space, $\mathfrak{w}_l^{-\alpha}$ is the error in terms of the work $\mathfrak{w}_l$. Such an assumption usually holds in one spatial dimension, where the resulting stiffness matrix is tridiagonal. For higher spatial dimensions solving the corresponding linear system is often times not of linear complexity, in which case the convergence rate $\alpha > 0$ has to be adjusted accordingly.

We now state our assumptions on the sequence of functions $(u^l)_{l\in\mathbb{N}}$ approximating u. Equation (7.1) will hold in the L^2 sense over all parameters $\boldsymbol{y} \in U$, cp. Assumption 7.2 (iii), and Definition 4.1 (ii).

Assumption 7.2 *Let X be a separable Hilbert space and let $(\mathfrak{w}_l)_{l\in\mathbb{N}_0}$ satisfy Assumption 7.1. Furthermore, $0 < p_1 \leq p_2 < \infty$, $\boldsymbol{b}_1 \in \ell^{p_1}(\mathbb{N})$, $\boldsymbol{b}_2 \in \ell^{p_2}(\mathbb{N})$, $\xi > 0$, $\delta > 0$ and there exist functions $u \in L^2(U, X; \gamma)$, $(u^l)_{l\in\mathbb{N}} \subseteq L^2(U, X; \gamma)$ such that*

(i) $u \in L^2(U, X; \gamma)$ is $(\boldsymbol{b}_1, \xi, \delta, X)$-holomorphic,
(ii) $(u - u^l) \in L^2(U, X; \gamma)$ is $(\boldsymbol{b}_1, \xi, \delta, X)$-holomorphic for every $l \in \mathbb{N}$,
(iii) $(u - u^l) \in L^2(U, X; \gamma)$ is $(\boldsymbol{b}_2, \xi, \delta\mathfrak{w}_l^{-\alpha}, X)$-holomorphic for every $l \in \mathbb{N}$.

Remark 7.3 Items (ii) and (iii) are two assumptions on the domain of holomorphic extension of the discretization error $e_l := u - u^l : U \to X$. As pointed out in Remark 4.2, the faster the sequence $\boldsymbol{b}$ decays the larger the size of holomorphic extension, and the smaller δ the smaller the upper bound of this extension.

Hence items (ii) and (iii) can be interpreted as follows: Item (ii) implies that e_l has a large domain of holomorphic extension. Item (iii) is related to the assumption (7.1). It yields that by considering the extension of e_l on a smaller domain, we can get a (l-dependent) smaller upper bound of the extension of e_l (in the sense of Definition 4.1 (ii)). Hence there is a tradeoff between choosing the size of the domain of the holomorphic extension and the upper bound of this extension.

7.2 Multilevel Smolyak Sparse-Grid Algorithms

Let $\mathbf{l} = (l_{\boldsymbol{\nu}})_{\boldsymbol{\nu}\in\mathcal{F}} \subseteq \mathbb{N}_0$ be a family of natural numbers associating with each multiindex $\boldsymbol{\nu} \in \mathcal{F}$ of a PC expansion a discretization level $l_{\boldsymbol{\nu}} \in \mathbb{N}_0$. Typically, this is a family of discretization levels for some *hierarchic, numerical* approximation of the PDE in the physical domain D, associating with each multiindex $\boldsymbol{\nu} \in \mathcal{F}$ of a PC expansion of the parametric solution in the parameter domain a possibly coefficient-dependent discretization level $l_{\boldsymbol{\nu}} \in \mathbb{N}_0$. With the sequence $l_{\boldsymbol{\nu}} \in \mathbb{N}_0$, we associate sets of multiindices via

$$\Gamma_j = \Gamma_j(\mathbf{l}) := \{\boldsymbol{\nu} \in \mathcal{F} : l_{\boldsymbol{\nu}} \geq j\} \qquad \forall j \in \mathbb{N}_0. \tag{7.2}$$

Throughout we will assume that

$$|\mathbf{l}| := \|\mathbf{l}\|_{\ell^1(\mathcal{F})} := \sum_{\boldsymbol{\nu}\in\mathcal{F}} l_{\boldsymbol{\nu}} < \infty$$

and that $\mathbf{l}$ is monotonically decreasing, meaning that $\boldsymbol{\nu} \le \boldsymbol{\mu}$ implies $l_{\boldsymbol{\nu}} \ge l_{\boldsymbol{\mu}}$. In this case each $\Gamma_j \subseteq \mathcal{F}$, $j \in \mathbb{N}$, is finite and downward closed. Moreover $\Gamma_0 = \mathcal{F}$, and the sets $(\Gamma_j)_{j\in\mathbb{N}_0}$ are nested according to

$$\mathcal{F} = \Gamma_0 \supseteq \Gamma_1 \supseteq \Gamma_2 \ldots .$$

With $(u^l)_{l\in\mathbb{N}}$ as in Assumption 7.2, we now define the multilevel sparse-grid interpolation algorithm

$$\mathbf{I}_{\mathbf{l}}^{\mathrm{ML}} u := \sum_{j\in\mathbb{N}} (\mathbf{I}_{\Gamma_j} - \mathbf{I}_{\Gamma_{j+1}}) u^j . \tag{7.3}$$

A few remarks are in order. First, the index $\mathbf{l}$ indicates that the sets $\Gamma_j = \Gamma_j(\mathbf{l})$ depend on the choice of $\mathbf{l}$, although we usually simply write Γ_j in order to keep the notation succinct. Secondly, due to $|\mathbf{l}| < \infty$ it holds

$$\max_{\boldsymbol{\nu}\in\mathcal{F}} l_{\boldsymbol{\nu}} =: L < \infty$$

and thus $\Gamma_j = \emptyset$ for all $j > L$. Defining $\mathbf{I}_{\emptyset}$ as the constant 0 operator, the infinite series (7.3) can also be written as the finite sum

$$\mathbf{I}_{\mathbf{l}}^{\mathrm{ML}} u = \sum_{j=1}^{L} (\mathbf{I}_{\Gamma_j} - \mathbf{I}_{\Gamma_{j+1}}) u^j = \mathbf{I}_{\Gamma_1} u^1 + \mathbf{I}_{\Gamma_2} (u^2 - u^1) + \cdots + \mathbf{I}_{\Gamma_L} (u^L - u^{L-1}),$$

where we used $\mathbf{I}_{\Gamma_{L+1}} = 0$. If we had $\Gamma_1 = \cdots = \Gamma_L$, this sum would reduce to $\mathbf{I}_{\Gamma_L} u^L$, which is the interpolant of the approximation u^L at the (highest) discretization level L. The main observation of multilevel analyses is that it is beneficial not to choose all Γ_j equal, but instead to balance out the accuracy of the interpolant $\mathbf{I}_{\Gamma_j}$ (in the parameter) and the accuracy of the approximation u^j of u.

A multilevel sparse-grid quadrature algorithm is defined analogously via

$$\mathbf{Q}_{\mathbf{l}}^{\mathrm{ML}} u := \sum_{j\in\mathbb{N}} (\mathbf{Q}_{\Gamma_j} - \mathbf{Q}_{\Gamma_{j+1}}) u^j , \tag{7.4}$$

with $\Gamma_j = \Gamma_j(\mathbf{l})$ as in (7.2). In the following we will prove algebraic convergence rates of multilevel interpolation and quadrature algorithms w.r.t. the $L^2(U, X; \gamma)$- and X-norm, respectively. The convergence rates will hold in terms of the *work* of computing $\mathbf{I}_{\mathbf{l}}^{\mathrm{ML}}$ and $\mathbf{Q}_{\mathbf{l}}^{\mathrm{ML}}$.

As mentioned above, for a level $l \in \mathbb{N}$, we interpret $\mathfrak{w}_l \in \mathbb{N}$ as a measure of the computational complexity of evaluating u^l at an arbitrary parameter $\boldsymbol{y} \in U$. As discussed in Sect. 6.2.1, computing $\mathbf{I}_{\Gamma_j} u$ or $\mathbf{Q}_{\Gamma_j} u$ requires to evaluate the function u at each parameter in the set $\mathrm{pts}(\Gamma_j) \subseteq U$ introduced in (6.11). We recall the bound

$$|\mathrm{pts}(\Gamma_j)| \leq \sum_{\boldsymbol{\nu}\in\Gamma_j} p_{\boldsymbol{\nu}}(1),$$

on the cardinality of this set obtained in (6.12). As an upper bound of the work corresponding to the evaluation of all functions required for the multilevel interpolant in (7.3), we obtain

$$\sum_{j\in\mathbb{N}} \mathfrak{w}_j \left(\sum_{\boldsymbol{\nu}\in\Gamma_j(\mathbf{l})} p_{\boldsymbol{\nu}}(1) + \sum_{\boldsymbol{\nu}\in\Gamma_{j+1}(\mathbf{l})} p_{\boldsymbol{\nu}}(1) \right). \tag{7.5}$$

Since $\Gamma_{j+1} \subseteq \Gamma_j$, up the factor 2 the work of a sequence $\mathbf{l}$ is defined by

$$\mathrm{work}(\mathbf{l}) := \sum_{j=1}^{L} \mathfrak{w}_j \sum_{\boldsymbol{\nu}\in\Gamma_j(\mathbf{l})} p_{\boldsymbol{\nu}}(1) = \sum_{\boldsymbol{\nu}\in\mathcal{F}(\mathbf{l})} p_{\boldsymbol{\nu}}(1) \sum_{j=1}^{l_{\boldsymbol{\nu}}} \mathfrak{w}_j, \tag{7.6}$$

where we used the definition of $\Gamma_j(\mathbf{l})$ in (7.2), $L := \max_{\boldsymbol{\nu}\in\mathcal{F}} l_{\boldsymbol{\nu}} < \infty$ and the finiteness of the set

$$\mathcal{F}(\mathbf{l}) := \{\boldsymbol{\nu} \in \mathcal{F} : l_{\boldsymbol{\nu}} > 0\}.$$

The efficiency of the multilevel interpolant critically relies on a suitable choice of levels $\mathbf{l} = (l_{\boldsymbol{\nu}})_{\boldsymbol{\nu}\in\mathcal{F}}$. This will be achieved with the following algorithm, which constructs $\mathbf{l}$ based on two collections of positive real numbers, $(c_{\boldsymbol{\nu}})_{\boldsymbol{\nu}\in\mathcal{F}} \in \ell^{q_1}(\mathcal{F})$ and $(d_{\boldsymbol{\nu}})_{\boldsymbol{\nu}\in\mathcal{F}} \in \ell^{q_2}(\mathcal{F})$. In the next sections we will analyze this construction of levels and provide details on suitable choices of $(c_{\boldsymbol{\nu}})_{\boldsymbol{\nu}\in\mathcal{F}}$ and $(d_{\boldsymbol{\nu}})_{\boldsymbol{\nu}\in\mathcal{F}}$.

Algorithm 2 $(l_\nu)_{\nu\in\mathcal{F}} = \text{ConstructLevels}((c_\nu)_{\nu\in\mathcal{F}}, (d_\nu)_{\nu\in\mathcal{F}}, q_1, \alpha, \varepsilon)$

1: $(l_\nu)_{\nu\in\mathcal{F}} \leftarrow (0)_{\nu\in\mathcal{F}}$
2: $\Lambda_\varepsilon \leftarrow \{\nu \in \mathcal{F} : c_\nu^{-1} \geq \varepsilon\}$
3: **for** $\nu \in \Lambda_\varepsilon$ **do**
4: $\quad \delta \leftarrow \varepsilon^{-\frac{1/2-q_1/4}{\alpha}} d_\nu^{\frac{-1}{1+2\alpha}} \left(\sum_{\mu\in\Lambda_\varepsilon} d_\mu^{\frac{-1}{1+2\alpha}}\right)^{\frac{1}{2\alpha}}$
5: $\quad l_\nu \leftarrow \max\{j \in \mathbb{N}_0 : \mathfrak{w}_j \leq \delta\}$
6: **return** $(l_\nu)_{\nu\in\mathcal{F}}$

Note that the determination of the sets Λ_ε in line 2 of Algorithm 2 can be done with Algorithm 1.

7.3 Construction of an Allocation of Discretization Levels

We discuss the construction of an allocation of discretization levels along the coefficients of Wiener-Hermite PC expansion. It is valid for collections $(u_\nu)_{\nu\in\mathcal{F}}$ of Wiener-Hermite PC expansion coefficients taking values in a separable Hilbert space, say X, with additional regularity, being $X^s \subset X$, allowing for weaker (weighted) summability of the V^s-norms $(\|u_\nu\|_{X^s})_{\nu\in\mathcal{F}}$. In the setting of elliptic BVPs with log-Gaussian diffusion coefficient, $X = V = H_0^1(D)$, and X^s is, for example, a weighted Kondrat'ev space in D as introduced in Sect. 3.8.1. We phrase the result and the construction in abstract terms so that the allocation is applicable to more general settings, such as the parabolic IBVP in Sect. 4.3.2.

For a given, dense sequence $(X_l)_{l\in\mathbb{N}_0} \subset X$ of nested, finite-dimensional subspaces and target accuracy $0 < \varepsilon \leq 1$, in the numerical approximation of Wiener-Hermite PC expansions of random fields u taking values in X, we consider approximating the Wiener-Hermite PC expansion coefficients u_ν in X from X_l. The assumed density of the sequence $(X_l)_{l\in\mathbb{N}_0} \subset X$ in X ensures that for $u \in L^2(U, X; \gamma)$ the coefficients $(u_\nu)_{\nu\in\mathcal{F}} \subset X$ are square summable, in the sense that $(\|u_\nu\|_X)_{\nu\in\mathcal{F}} \in \ell_2(\mathcal{F})$

The following lemma is a variation of [111, Lemma 3.2.7]. Its proof, is, with several minor modifications, taken from [111, Lemma 3.2.7]. We remark that the construction of the map $\mathbf{l}(\varepsilon, \nu)$, as described in the lemma, mimicks Algorithm 2. Again, a convergence rate is obtained that is not prone to the so-called "curse of dimensionality", being limited only by the available sparsity in the coefficients of Wiener-Hermite PC expansion for the parametric solution manifold.

Lemma 7.4 *Let* $\mathfrak{W} = \{\mathfrak{w}_l : l \in \mathbb{N}_0\}$ *satisfy Assumption 7.1. Let* $q_1 \in [0, 2)$, $q_2 \in [q_1, \infty)$ *and* $\alpha > 0$. *Let*

(i) $(a_{j,\nu})_{\nu\in\mathcal{F}} \subseteq [0, \infty)$ *for every* $j \in \mathbb{N}_0$,

(ii) $(c_{\boldsymbol{\nu}})_{\boldsymbol{\nu}\in\mathcal{F}} \subseteq (0,\infty)$ *and* $(d_{\boldsymbol{\nu}})_{\boldsymbol{\nu}\in\mathcal{F}} \subseteq (0,\infty)$ *be such that*

$$(c_{\boldsymbol{\nu}}^{-1/2})_{\boldsymbol{\nu}\in\mathcal{F}} \in \ell^{q_1}(\mathcal{F}) \text{ and } \left(d_{\boldsymbol{\nu}}^{-1/2} p_{\boldsymbol{\nu}}(1/2+\alpha)\right)_{\boldsymbol{\nu}\in\mathcal{F}} \in \ell^{q_2}(\mathcal{F}),$$

(iii)

$$\sup_{j\in\mathbb{N}_0} \left(\sum_{\boldsymbol{\nu}\in\mathcal{F}} a_{j,\boldsymbol{\nu}}^2 c_{\boldsymbol{\nu}}\right)^{1/2} =: C_1 < \infty,$$

$$\sup_{j\in\mathbb{N}_0} \left(\sum_{\boldsymbol{\nu}\in\mathcal{F}} (\mathfrak{w}_j^{\alpha} a_{j,\boldsymbol{\nu}})^2 d_{\boldsymbol{\nu}}\right)^{1/2} =: C_2 < \infty. \tag{7.7}$$

For every $\varepsilon > 0$ *define* $\Lambda_\varepsilon = \{\boldsymbol{\nu} \in \mathcal{F} : c_{\boldsymbol{\nu}}^{-1} \geq \varepsilon\}$, $\omega_{\varepsilon,\boldsymbol{\nu}} := 0$ *for all* $\boldsymbol{\nu} \in \mathcal{F}\backslash\Lambda_\varepsilon$, *and define*

$$\omega_{\varepsilon,\boldsymbol{\nu}} := \left\lceil \varepsilon^{-\frac{1/2-q_1/4}{\alpha}} d_{\boldsymbol{\nu}}^{\frac{-1}{1+2\alpha}} \left(\sum_{\boldsymbol{\mu}\in\Lambda_\varepsilon} d_{\boldsymbol{\mu}}^{\frac{-1}{1+2\alpha}}\right)^{\frac{1}{2\alpha}} \right\rceil_{\mathfrak{W}} \in \mathfrak{W} \qquad \forall \boldsymbol{\nu} \in \Lambda_\varepsilon.$$

Furthermore, for every $\varepsilon > 0$ *and* $\boldsymbol{\nu} \in \mathcal{F}$ *let* $l_{\varepsilon,\boldsymbol{\nu}} \in \mathbb{N}_0$ *be the corresponding discretization level, i.e.,* $\omega_{\varepsilon,\boldsymbol{\nu}} = \mathfrak{w}_{l_{\varepsilon,\boldsymbol{\nu}}}$, *and define the maximal discretization level*

$$L(\varepsilon) := \max\{l_{\varepsilon,\boldsymbol{\nu}} : \boldsymbol{\nu} \in \mathcal{F}\}.$$

Denote $\boldsymbol{l}_\varepsilon = (l_{\varepsilon,\boldsymbol{\nu}})_{\boldsymbol{\nu}\in\mathcal{F}}$.

Then there exists a constant $C > 0$ *and tolerances* $\varepsilon_n \in (0,1]$ *such that for every* $n \in \mathbb{N}$ *holds* $\mathrm{work}(\boldsymbol{l}_{\varepsilon_n}) \leq n$ *and*

$$\sum_{\boldsymbol{\nu}\in\mathcal{F}} \sum_{j=l_{\varepsilon_n,\boldsymbol{\nu}}}^{L(\varepsilon_n)} a_{j,\boldsymbol{\nu}} \leq C(1+\log n)n^{-R},$$

where the rate R *is given by*

$$R = \min\left\{\alpha, \frac{\alpha(q_1^{-1}-1/2)}{\alpha+q_1^{-1}-q_2^{-1}}\right\}.$$

Proof Throughout this proof denote $\delta := 1/2 - q_1/4 > 0$. In the following

$$\tilde{\omega}_{\varepsilon,\boldsymbol{\nu}} := \varepsilon^{-\frac{\delta}{\alpha}} d_{\boldsymbol{\nu}}^{\frac{-1}{1+2\alpha}} \left(\sum_{\boldsymbol{\mu}\in\Lambda_\varepsilon} d_{\boldsymbol{\mu}}^{\frac{-1}{1+2\alpha}}\right)^{\frac{1}{2\alpha}} \qquad \forall \boldsymbol{\nu} \in \Lambda_\varepsilon,$$

i.e. $\omega_{\varepsilon,\boldsymbol{\nu}} = \lfloor \tilde{\omega}_{\varepsilon,\boldsymbol{\nu}} \rfloor_{\mathfrak{W}}$. Note that $0 < \tilde{\omega}_{\varepsilon,\boldsymbol{\nu}}$ is well-defined for all $\boldsymbol{\nu} \in \Lambda_\varepsilon$ since $d_{\boldsymbol{\nu}} > 0$ for all $\boldsymbol{\nu} \in \mathcal{F}$ by assumption. Due to Assumption 7.1 (iii) it holds

$$\frac{\tilde{\omega}_{\varepsilon,\boldsymbol{\nu}}}{K_{\mathfrak{W}}} \le 1 + \omega_{\varepsilon,\boldsymbol{\nu}} \le 1 + \tilde{\omega}_{\varepsilon,\boldsymbol{\nu}} \qquad \forall \boldsymbol{\nu} \in \Lambda_\varepsilon. \tag{7.8}$$

Since $(c_{\boldsymbol{\nu}}^{-1/2})_{\boldsymbol{\nu}\in\mathcal{F}} \in \ell^{q_1}(\mathcal{F})$ and (7.7), we get

$$\begin{aligned}\sum_{\boldsymbol{\nu}\in\mathcal{F}\setminus\Lambda_\varepsilon} a_{j,\boldsymbol{\nu}} &\le \Bigg(\sum_{\boldsymbol{\nu}\in\mathcal{F}\setminus\Lambda_\varepsilon} a_{j,\boldsymbol{\nu}}^2 c_{\boldsymbol{\nu}}\Bigg)^{1/2} \Bigg(\sum_{\boldsymbol{\nu}\in\mathcal{F}\setminus\Lambda_\varepsilon} c_{\boldsymbol{\nu}}^{-1}\Bigg)^{1/2} \\ &\le C_1 \Bigg(\sum_{c_{\boldsymbol{\nu}}^{-1}\le\varepsilon} c_{\boldsymbol{\nu}}^{-\frac{q_1}{2}} c_{\boldsymbol{\nu}}^{\frac{q_1}{2}-1}\Bigg)^{1/2} \le C\varepsilon^{\delta}\end{aligned}$$

with the constant C independent of j and ε. Thus,

$$\sum_{\boldsymbol{\nu}\in\mathcal{F}\setminus\Lambda_\varepsilon} \sum_{j=0}^{L(\varepsilon)} a_{j,\boldsymbol{\nu}} = \sum_{j=0}^{L(\varepsilon)} \sum_{\boldsymbol{\nu}\in\mathcal{F}\setminus\Lambda_\varepsilon} a_{j,\boldsymbol{\nu}} \le C_1(1+L(\varepsilon))\varepsilon^{\delta}. \tag{7.9}$$

Next with C_2 as in (7.7),

$$\begin{aligned}\sum_{\boldsymbol{\nu}\in\Lambda_\varepsilon} \sum_{j=l_{\varepsilon,\boldsymbol{\nu}}}^{L(\varepsilon)} a_{j,\boldsymbol{\nu}} &= \sum_{\boldsymbol{\nu}\in\Lambda_\varepsilon} \sum_{j=l_{\varepsilon,\boldsymbol{\nu}}}^{L(\varepsilon)} a_{j,\boldsymbol{\nu}} \mathfrak{w}_j^{\alpha} \mathfrak{w}_j^{-\alpha} d_{\boldsymbol{\nu}}^{1/2} d_{\boldsymbol{\nu}}^{-1/2} \\ &\le \Bigg(\sum_{\boldsymbol{\nu}\in\Lambda_\varepsilon} \sum_{j=0}^{L(\varepsilon)} (a_{j,\boldsymbol{\nu}} \mathfrak{w}_j^{\alpha} d_{\boldsymbol{\nu}}^{1/2})^2\Bigg)^{\frac{1}{2}} \Bigg(\sum_{\boldsymbol{\nu}\in\Lambda_\varepsilon} \sum_{j\ge l_{\varepsilon,\boldsymbol{\nu}}} (d_{\boldsymbol{\nu}}^{-1/2} \mathfrak{w}_j^{-\alpha})^2\Bigg)^{\frac{1}{2}} \\ &\le C_2(1+L(\varepsilon)) \Bigg(\sum_{\boldsymbol{\nu}\in\Lambda_\varepsilon} \sum_{j\ge l_{\varepsilon,\boldsymbol{\nu}}} (d_{\boldsymbol{\nu}}^{-1/2} \mathfrak{w}_j^{-\alpha})^2\Bigg)^{\frac{1}{2}}.\end{aligned} \tag{7.10}$$

Assumption 7.1 (iv) implies for some C_3

$$\sum_{j\ge l_{\varepsilon,\boldsymbol{\nu}}} \mathfrak{w}_j^{-2\alpha} \le C_3^2 (1+\mathfrak{w}_{l_{\varepsilon,\boldsymbol{\nu}}})^{-2\alpha} = C_3^2 (1+\omega_{\varepsilon,\boldsymbol{\nu}})^{-2\alpha},$$

so that by (7.8) and (7.10)

$$\sum_{\nu\in\Lambda_\varepsilon}\sum_{j=l_{\varepsilon,\nu}}^{L(\varepsilon)} a_{j,\nu} \le C_3C_2(1+L(\varepsilon))\left(\sum_{\nu\in\Lambda_\varepsilon}\left(d_\nu^{-1/2}(1+\omega_{\varepsilon,\nu})^{-\alpha}\right)^2\right)^{\frac{1}{2}}$$

$$\le C_3C_2K_{\mathfrak{W}}^{\alpha}(1+L(\varepsilon))\left(\sum_{\nu\in\Lambda_\varepsilon}\left(d_\nu^{-1/2}\tilde{\omega}_{\varepsilon,\nu}^{-\alpha}\right)^2\right)^{\frac{1}{2}}. \tag{7.11}$$

Inserting the definition of $\tilde{\omega}_{\varepsilon,\nu}$, we have

$$\left(\sum_{\nu\in\Lambda_\varepsilon}\left(d_\nu^{-1/2}\tilde{\omega}_{\varepsilon,\nu}^{-\alpha}\right)^2\right)^{\frac{1}{2}} = \varepsilon^\delta\left(\sum_{\mu\in\Lambda_\varepsilon} d_\mu^{\frac{-1}{1+2\alpha}}\right)^{-\alpha\frac{1}{2\alpha}}\left(\sum_{\nu\in\Lambda_\varepsilon} d_\nu^{-1} d_\nu^{\frac{2\alpha}{1+2\alpha}}\right)^{\frac{1}{2}} = \varepsilon^\delta, \tag{7.12}$$

where we used

$$-1+\frac{2\alpha}{1+2\alpha} = \frac{-(1+2\alpha)+2\alpha}{1+2\alpha} = \frac{-1}{1+2\alpha}.$$

Using Assumption 7.1 (ii) and the definition of work($\mathbf{l}_\varepsilon$) in (7.6) we get

$$L(\varepsilon) \le \log(1+\max_{\nu\in\mathcal{F}}\omega_{\varepsilon,\nu}) \le \log(1+\mathrm{work}(\mathbf{l}_\varepsilon)). \tag{7.13}$$

Hence, (7.9), (7.11), (7.12) and (7.13) yield

$$\sum_{\nu\in\mathcal{F}}\sum_{j=l_{\varepsilon,\nu}}^{L(\varepsilon)} a_{j,\nu} = \sum_{\nu\in\Lambda_\varepsilon}\sum_{j=l_{\varepsilon,\nu}}^{L(\varepsilon)} a_{j,\nu} + \sum_{\nu\in\mathcal{F}\setminus\Lambda_\varepsilon}\sum_{j=0}^{L(\varepsilon)} a_{j,\nu} \le C\big(1+\log(\mathrm{work}(\mathbf{l}_\varepsilon))\big)\varepsilon^\delta. \tag{7.14}$$

Next, we compute an upper bound for work($\mathbf{l}_\varepsilon$). By definition of work($\mathbf{l}_\varepsilon$) in (7.6), and using Assumption 7.1 (i) as well as $\omega_{\varepsilon,\boldsymbol{\nu}} = \mathfrak{w}_{l_{\varepsilon,\boldsymbol{\nu}}}$,

$$\begin{aligned}\mathrm{work}(\mathbf{l}_\varepsilon) &= \sum_{\boldsymbol{\nu}\in\Lambda_\varepsilon} p_{\boldsymbol{\nu}}(1) \sum_{\{j\in\mathbb{N}:\, j\le l_{\varepsilon,\boldsymbol{\nu}}\}} \mathfrak{w}_j \le \sum_{\boldsymbol{\nu}\in\Lambda_\varepsilon} p_{\boldsymbol{\nu}}(1) K_{\mathfrak{W}} \omega_{\varepsilon,\boldsymbol{\nu}} \\ &\le K_{\mathfrak{W}} \sum_{\boldsymbol{\nu}\in\Lambda_\varepsilon} p_{\boldsymbol{\nu}}(1)\tilde{\omega}_{\varepsilon,\boldsymbol{\nu}} \le K_{\mathfrak{W}} \varepsilon^{-\frac{\delta}{\alpha}} \left(\sum_{\boldsymbol{\nu}\in\Lambda_\varepsilon} p_{\boldsymbol{\nu}}(1) d_{\boldsymbol{\nu}}^{\frac{-1}{1+2\alpha}}\right)^{\frac{1}{2\alpha}+1} \\ &= K_{\mathfrak{W}} \varepsilon^{-\frac{\delta}{\alpha}} \left(\sum_{\boldsymbol{\nu}\in\Lambda_\varepsilon} \left(p_{\boldsymbol{\nu}}(1/2+\alpha) d_{\boldsymbol{\nu}}^{-1/2}\right)^{\frac{2}{1+2\alpha}}\right)^{\frac{1}{2\alpha}+1}, \end{aligned} \tag{7.15}$$

where we used $p_{\boldsymbol{\nu}}(1) = p_{\boldsymbol{\nu}}(1/2+\alpha)^{2/(1+2\alpha)}$ and the fact that $p_{\boldsymbol{\nu}}(1) \ge 1$ for all $\boldsymbol{\nu}$.

We distinguish between the two cases

$$\frac{2}{1+2\alpha} \ge q_2 \quad \text{and} \quad \frac{2}{1+2\alpha} < q_2.$$

In the first case, since $(p_{\boldsymbol{\mu}}(1/2+\alpha) d_{\boldsymbol{\mu}}^{-1/2})_{\boldsymbol{\mu}\in\mathcal{F}} \in \ell^{q_2}(\mathcal{F})$, (7.15) implies

$$\mathrm{work}(\mathbf{l}_\varepsilon) \le C\varepsilon^{-\frac{\delta}{\alpha}} \tag{7.16}$$

and hence, $\log(\mathrm{work}(\mathbf{l}_\varepsilon)) \le \log\left(C\varepsilon^{-\frac{\delta}{\alpha}}\right)$. Then (7.14) together with (7.16) implies

$$\sum_{\boldsymbol{\nu}\in\mathcal{F}} \sum_{j=l_{\varepsilon,\boldsymbol{\nu}}}^{L(\varepsilon)} a_{j,\boldsymbol{\nu}} \le C(1+|\log(\varepsilon^{-1})|)\varepsilon^{\delta}.$$

For every $n\in\mathbb{N}$, we can find $\varepsilon_n > 0$ such that $\frac{n}{2} \le C\varepsilon_n^{-\frac{\delta}{\alpha}} \le n$. Then the claim of the corollary in the case $\frac{2}{1+2\alpha} \ge q_2$ holds true for the chosen ε_n.

Finally, let us address the case $\frac{2}{1+2\alpha} < q_2$. Then, by (7.15) and using Hölder's inequality with $q_2\frac{1+2\alpha}{2} > 1$ we get

$$\mathrm{work}(\mathbf{l}_\varepsilon) \le K_{\mathfrak{W}}\varepsilon^{-\frac{\delta}{\alpha}} \|(p_{\boldsymbol{\nu}}(1/2+\alpha)d_{\boldsymbol{\nu}}^{-1/2})_{\boldsymbol{\nu}\in\mathcal{F}}\|_{\ell^{q_2}(\mathcal{F})}^{\frac{1}{\alpha}} |\Lambda_\varepsilon|^{\left(1-\frac{2}{q_2(1+2\alpha)}\right)\frac{1+2\alpha}{2\alpha}}.$$

Since

$$|\Lambda_\varepsilon| = \sum_{\boldsymbol{\nu}\in\Lambda_\varepsilon} 1 = \sum_{c_{\boldsymbol{\nu}}^{-1}\ge\varepsilon} c_{\boldsymbol{\nu}}^{-\frac{q_1}{2}} c_{\boldsymbol{\nu}}^{\frac{q_1}{2}} \le C\varepsilon^{-\frac{q_1}{2}},$$

we obtain

$$\mathrm{work}(\mathbf{l}_\varepsilon) \le K_{\mathfrak{W}}\varepsilon^{-\frac{\delta}{\alpha}-\frac{q_1}{2}(1-\frac{2}{q_2(1+2\alpha)})\frac{1+2\alpha}{2\alpha}} \le C\varepsilon^{-\frac{q_1}{2\alpha}\left(\alpha-\frac{1}{q_2}+\frac{1}{q_1}\right)}.$$

For every $n \in \mathbb{N}$, we can find $\varepsilon_n > 0$ such that

$$\frac{n}{2} \le C\varepsilon_n^{-\frac{q_1}{2\alpha}\left(\alpha-\frac{1}{q_2}+\frac{1}{q_1}\right)} \le n.$$

Thus the claim also holds true in the case $\frac{2}{1+2\alpha} < q_2$. □

7.4 Multilevel Smolyak Sparse-Grid Interpolation Algorithm

We are now in position to formulate a multilevel Smolyak sparse-grid interpolation convergence theorem. To this end, we observe that our proofs of approximation rates have been constructive: rather than being based on a best N-term selection from the infinite set of Wiener-Hermite PC expansion coefficients, a constructive selection process of "significant" Wiener-Hermite PC expansion coefficients, subject to a given prescribed approximation tolerance, has been provided. In the present section, we turn this into a concrete, numerical selection process with complexity bounds. In particular, we provide an *a-priori allocation* of discretization levels to Wiener-Hermite PC expansion coefficients. This results on the one hand in an explicit, algorithmic definition of a family of multilevel interpolants which is parametrized by an approximation threshold $\varepsilon > 0$. On the other hand, it will result in *mathematical convergence rate bounds in terms of computational work* rather than in terms of, for example, number of active Wiener-Hermite PC expansion coefficients. These rate bounds are free from the curse of dimensionality.

The idea is as follows: let $\boldsymbol{b}_1 = (b_{1,j})_{j\in\mathbb{N}} \in \ell^{p_1}(\mathbb{N})$, $\boldsymbol{b}_2 = (b_{2,j})_{j\in\mathbb{N}} \in \ell^{p_2}(\mathbb{N})$, and ξ be the two sequences and constant from Assumption 7.2. For two constants $K > 0$ and $r > 3$ (which are still at our disposal and which will be specified below), set for all $j \in \mathbb{N}$

$$\varrho_{1,j} := b_{1,j}^{p_1-1}\frac{\xi}{4\|\boldsymbol{b}_1\|_{\ell^{p_1}}}, \qquad \varrho_{2,j} := b_{2,j}^{p_2-1}\frac{\xi}{4\|\boldsymbol{b}_2\|_{\ell^{p_2}}}. \tag{7.17}$$

We let for all $\boldsymbol{\nu} \in \mathcal{F}$ (as in Lemma 6.5 for $k = 1$ and with $\tau = 3$)

$$c_{\boldsymbol{\nu}} := \prod_{j\in\mathbb{N}} \max\{1, K\varrho_{1,j}\}^2 \nu_j^{r-3}, \qquad d_{\boldsymbol{\nu}} := \prod_{j\in\mathbb{N}} \max\{1, K\varrho_{2,j}\}^2 \nu_j^{r-3}. \tag{7.18}$$

Based on those two multi-index collections, Algorithm 2 provides a collection of discretization levels which sequence depends on $\varepsilon > 0$ and is indexed over $\mathcal{F}$.

We denote it by $\mathbf{l}_\varepsilon = (l_{\varepsilon,\boldsymbol{\nu}})_{\boldsymbol{\nu}\in\mathcal{F}}$. We now state an upper bound for the error of the corresponding multilevel interpolants in terms of the work measure in (7.6) as $\varepsilon \to 0$.

Theorem 7.5 *Let $u \in L^2(U, X; \gamma)$ and $u^l \in L^2(U, X; \gamma)$, $l \in \mathbb{N}$, satisfy Assumption 7.2 with some constants $\alpha > 0$ and $0 < p_1 < 2/3$ and $p_1 \le p_2 < 1$. Set $q_1 := p_1/(1-p_1)$. Assume that $r > 2(1+(\alpha+1)q_1)/q_1+3$ (for r as defined in (7.18)). There exist constants $K > 0$ (in (7.18)) and $C > 0$ such that the following holds.*

For every $n \in \mathbb{N}$, there are positive constants $\varepsilon_n \in (0, 1]$ such that work$(\boldsymbol{l}_{\varepsilon_n}) \le n$ *and with $\boldsymbol{l}_{\varepsilon_n} = (l_{\varepsilon_n,\boldsymbol{\nu}})_{\boldsymbol{\nu}\in\mathcal{F}}$ as defined in Lemma 7.4 (where $c_{\boldsymbol{\nu}}$, $d_{\boldsymbol{\nu}}$ as in* (7.18)*) it holds*

$$\|u - \boldsymbol{I}^{\mathrm{ML}}_{\boldsymbol{l}_{\varepsilon_n}} u\|_{L^2(U,X;\gamma)} \le C(1+\log n)n^{-R}$$

with the convergence rate

$$R := \min\left\{\alpha, \frac{\alpha(p_1^{-1} - 3/2)}{\alpha + p_1^{-1} - p_2^{-1}}\right\}. \tag{7.19}$$

Proof Throughout this proof we write $\boldsymbol{b}_1 = (b_{1,j})_{j\in\mathbb{N}}$ and $\boldsymbol{b}_2 = (b_{2,j})_{j\in\mathbb{N}}$ for the two sequences in Assumption 7.2. We observe that Γ_j defined in (7.2) is downward closed for all $j \in \mathbb{N}_0$. This can be easily deduced from the fact that the multi-index collections $(c_{\boldsymbol{\nu}})_{\boldsymbol{\nu}\in\mathcal{F}}$ and $(d_{\boldsymbol{\nu}})_{\boldsymbol{\nu}\in\mathcal{F}}$ are monotonically increasing (i.e., e.g., $\boldsymbol{\nu} \le \boldsymbol{\mu}$ implies $c_{\boldsymbol{\nu}} \le c_{\boldsymbol{\mu}}$) and the definition of Λ_ε and $l_{\varepsilon,\boldsymbol{\nu}}$ in Algorithm 2. We will use this fact throughout the proof, without mentioning it at every instance.

Step 1. Given $n \in \mathbb{N}$, we choose $\varepsilon := \varepsilon_n$ as in Lemma 7.4. Fix $N \in \mathbb{N}$ such that

$$N > \max\{j : j \in \mathrm{supp}(\boldsymbol{\nu}),\ l_{\varepsilon,\boldsymbol{\nu}} > 0\}$$

and so large that

$$\|u - \tilde{u}_N\|_{L^2(U,X;\gamma)} \le n^{-R}, \tag{7.20}$$

where $\tilde{u}_N : U \to X$ is as in Definition 4.1. This is possible due to

$$\lim_{N\to\infty} \|u - \tilde{u}_N\|_{L^2(U,X;\gamma)} = 0,$$

which holds by the $(\boldsymbol{b}_1, \xi, \delta, X)$-holomorphy of u. By Assumption 7.2, for every $j \in \mathbb{N}$ the function $e^j := u - u^j \in L^2(U, X; \gamma)$ is $(\boldsymbol{b}_1, \xi, \delta, X)$-holomorphic and $(\boldsymbol{b}_2, \xi, \delta\mathfrak{w}^\gamma_j, X)$-holomorphic. For notational convenience we set $e^0 := u - 0 = u \in L^2(U, X; \gamma)$, so that e^0 is $(\boldsymbol{b}_1, \xi, \delta, X)$-holomorphic and $(\boldsymbol{b}_2, \xi, \delta, X)$-

holomorphic. Hence, for every $j \in \mathbb{N}_0$ there exists a function $\tilde{e}_N^j = \tilde{u}_N - \tilde{u}_N^j$ as in Definition 4.1 (iii).
In the rest of the proof we use the following facts:

(i) By Lemma 6.10, for every $j \in \mathbb{N}_0$, with the Wiener-Hermite PC expansion coefficients

$$\tilde{e}_{N,\boldsymbol{\nu}}^j := \int_U H_{\boldsymbol{\nu}}(\boldsymbol{y}) \tilde{e}_N^j(\boldsymbol{y}) \, \mathrm{d}\gamma(\boldsymbol{y}),$$

it holds

$$\tilde{e}_N^j(\boldsymbol{y}) = \sum_{\boldsymbol{\nu} \in \mathcal{F}} \tilde{e}_{N,\boldsymbol{\nu}}^j H_{\boldsymbol{\nu}}(\boldsymbol{y}) \qquad \forall \boldsymbol{y} \in U,$$

with pointwise absolute convergence.

(ii) By Lemma 6.5, upon choosing $K > 0$ in (7.18) large enough, and because $r > 3$,

$$C_0 c_{\boldsymbol{\nu}} p_{\boldsymbol{\nu}}(3) \leq \beta_{\boldsymbol{\nu}}(r, \boldsymbol{\varrho}_1), \qquad C_0 d_{\boldsymbol{\nu}} p_{\boldsymbol{\nu}}(3) \leq \beta_{\boldsymbol{\nu}}(r, \boldsymbol{\varrho}_2) \qquad \forall \boldsymbol{\nu} \in \mathcal{F}_1.$$

We observe that by definition of $\boldsymbol{\varrho}_i$, $i \in \{1, 2\}$, in (7.17), it holds $\varrho_{i,j} \sim b_{i,j}^{-(1-p_i)}$ and therefore $(\varrho_{i,j}^{-1})_{j \in \mathbb{N}} \in \ell^{q_i}(\mathbb{N})$ with $q_i := p_i/(1 - p_i)$, $i \in \{1, 2\}$.

(iii) Due to $r > 2(1 + (\alpha + 1)q_1)/q_1 + 3$, the condition of Lemma 6.6 is satisfied (with $k = 1$, $\tau = 3$ and $\theta = (\alpha + 1)q_1$). Hence the lemma gives

$$\sum_{\boldsymbol{\nu} \in \mathcal{F}} p_{\boldsymbol{\nu}}((\alpha+1)q_1) c_{\boldsymbol{\nu}}^{-q_1/2} < \infty \qquad \Rightarrow \qquad (p_{\boldsymbol{\nu}}(\alpha+1) c_{\boldsymbol{\nu}}^{-1/2})_{\boldsymbol{\nu} \in \mathcal{F}} \in \ell^{q_1}(\mathcal{F})$$

and similarly

$$\sum_{\boldsymbol{\nu} \in \mathcal{F}} p_{\boldsymbol{\nu}}((\alpha + 1)q_2) d_{\boldsymbol{\nu}}^{-q_2/2} < \infty \quad \Rightarrow \quad (p_{\boldsymbol{\nu}}(\alpha + 1) d_{\boldsymbol{\nu}}^{-1/2})_{\boldsymbol{\nu} \in \mathcal{F}} \in \ell^{q_2}(\mathcal{F}).$$

(iv) By Theorem 4.8 and item (ii), for all $j \in \mathbb{N}_0$

$$C_0 \sum_{\boldsymbol{\nu} \in \mathcal{F}} c_{\boldsymbol{\nu}} \|\tilde{e}_{N,\boldsymbol{\nu}}^j\|_X^2 p_{\boldsymbol{\nu}}(3) \leq \sum_{\boldsymbol{\nu} \in \mathcal{F}} \beta_{\boldsymbol{\nu}}(r, \boldsymbol{\varrho}_1) \|\tilde{e}_{N,\boldsymbol{\nu}}^j\|_X^2 \leq C\delta^2$$

and

$$C_0 \sum_{\boldsymbol{\nu} \in \mathcal{F}} d_{\boldsymbol{\nu}} \|\tilde{e}_{N,\boldsymbol{\nu}}^j\|_X^2 p_{\boldsymbol{\nu}}(3) \leq \sum_{\boldsymbol{\nu} \in \mathcal{F}} \beta_{\boldsymbol{\nu}}(r, \boldsymbol{\varrho}_2) \|\tilde{e}_{N,\boldsymbol{\nu}}^j\|_X^2 \leq C \frac{\delta^2}{\mathfrak{w}_j^{2\alpha}},$$

with the constant C independent of j, $\mathfrak{w}_j$ and N.

(v) Because $N \geq \max\{j \in \mathrm{supp}(\boldsymbol{\nu}) : l_{\varepsilon,\boldsymbol{\nu}} \geq 0\}$ and $\chi_{0,0} = 0$ we have

$$\mathbf{I}_{\Gamma_j}(u - u^j) = \mathbf{I}_{\Gamma_j} e^j = \mathbf{I}_{\Gamma_j} \tilde{e}_N^j$$

for all $j \in \mathbb{N}$ (cp. Remark 4.4). Similarly $\mathbf{I}_{\Gamma_j} u = \mathbf{I}_{\Gamma_j} \tilde{u}_N$ for all $j \in \mathbb{N}$.

Step 2. Observe that $\Gamma_j = \emptyset$ for all $j > L(\varepsilon) := \max_{\boldsymbol{\nu} \in \mathcal{F}} l_{\varepsilon,\boldsymbol{\nu}}$ (cp. (7.2)), which is finite due to $|\mathbf{l}_\varepsilon| < \infty$. With the conventions $\mathbf{I}_{\Gamma_0} = \mathbf{I}_{\mathcal{F}} = \mathrm{Id}$ (i.e. $\mathbf{I}_{\Gamma_0}$ is the identity) and $\mathbf{I}_\emptyset \equiv 0$ this implies

$$u = \mathbf{I}_{\Gamma_0} u = \sum_{j=0}^{L(\varepsilon)} (\mathbf{I}_{\Gamma_j} - \mathbf{I}_{\Gamma_{j+1}}) u = (\mathbf{I}_{\Gamma_0} - \mathbf{I}_{\Gamma_1}) u + \cdots + (\mathbf{I}_{\Gamma_{L(\varepsilon)-1}} - \mathbf{I}_{\Gamma_{L(\varepsilon)}}) u + \mathbf{I}_{\Gamma_{L(\varepsilon)}} u.$$

By definition of the multilevel interpolant in (7.3)

$$\begin{aligned}
\mathbf{I}_{\mathbf{l}_\varepsilon}^{\mathrm{ML}} u &= \sum_{j=1}^{L(\varepsilon)} (\mathbf{I}_{\Gamma_j} - \mathbf{I}_{\Gamma_{j+1}}) u^j \\
&= (\mathbf{I}_{\Gamma_1} - \mathbf{I}_{\Gamma_2}) u^1 + \cdots + (\mathbf{I}_{\Gamma_{L(\varepsilon)-1}} - \mathbf{I}_{\Gamma_{L(\varepsilon)}}) u^{L(\varepsilon)-1} + \mathbf{I}_{\Gamma_{L(\varepsilon)}} u^{L(\varepsilon)}.
\end{aligned}$$

By item (v) of Step 1, we can write

$$\begin{aligned}
(\mathbf{I}_{\Gamma_0} - \mathbf{I}_{\Gamma_1}) u = u - \mathbf{I}_{\Gamma_1} u = u - \mathbf{I}_{\Gamma_1} \tilde{u}_N &= (u - \tilde{u}_N) + (\mathbf{I}_{\Gamma_0} - \mathbf{I}_{\Gamma_1}) \tilde{u}_N \\
&= (u - \tilde{u}_N) + (\mathbf{I}_{\Gamma_0} - \mathbf{I}_{\Gamma_1}) \tilde{e}_N^0,
\end{aligned}$$

where in the last equality we used $e_N^0 = u_N$, by definition of $e^0 = u$ (and $\tilde{e}_N^0 = \tilde{u}_N \in L^2(U, X; \gamma)$ as in Definition 4.1). Hence, again by item (v),

$$\begin{aligned}
u - \mathbf{I}_{\mathbf{l}_\varepsilon}^{\mathrm{ML}} u &= (\mathbf{I}_{\Gamma_0} - \mathbf{I}_{\Gamma_1}) u + \sum_{j=1}^{L(\varepsilon)} (\mathbf{I}_{\Gamma_j} - \mathbf{I}_{\Gamma_{j+1}})(u - u^j) \\
&= (u - \tilde{u}_N) + (\mathbf{I}_{\Gamma_0} - \mathbf{I}_{\Gamma_1}) \tilde{u}_N + \sum_{j=1}^{L(\varepsilon)} (\mathbf{I}_{\Gamma_j} - \mathbf{I}_{\Gamma_{j+1}}) \tilde{e}_N^j \\
&= (u - \tilde{u}_N) + \sum_{j=0}^{L(\varepsilon)} (\mathbf{I}_{\Gamma_j} - \mathbf{I}_{\Gamma_{j+1}}) \tilde{e}_N^j.
\end{aligned} \tag{7.21}$$

We will use this representation to bound the norm $\|u - \mathbf{I}^{\mathrm{ML}}_{\mathbf{l}_\varepsilon} u\|_{L^2(U,X;\gamma)}$. From item (i) of Step 1 it follows that for every $j \in \mathbb{N}_0$

$$\tilde{e}^j_N(\boldsymbol{y}) = \sum_{\boldsymbol{\nu}\in\mathcal{F}} \tilde{e}^j_{N,\boldsymbol{\nu}} H_{\boldsymbol{\nu}}(\boldsymbol{y}),$$

with the equality and unconditional convergence in the space X for all $\boldsymbol{y} \in U$. Therefore, by the same argument as in the proof of Lemma 6.11, we can prove that

$$(\mathbf{I}_{\Gamma_j} - \mathbf{I}_{\Gamma_{j+1}})\tilde{e}^j_N = \sum_{\boldsymbol{\nu}\in\mathcal{F}} \tilde{e}^j_{N,\boldsymbol{\nu}} (\mathbf{I}_{\Gamma_j} - \mathbf{I}_{\Gamma_{j+1}}) H_{\boldsymbol{\nu}} \tag{7.22}$$

with equality and unconditional convergence in the space $L^2(\mathbb{R}^N, X; \gamma_N)$. Using (7.21) and

$$(\mathbf{I}_{\Gamma_j} - \mathbf{I}_{\Gamma_{j+1}}) H_{\boldsymbol{\nu}} = 0$$

for all $\boldsymbol{\nu} \in \Gamma_{j+1} \subseteq \Gamma_j$ by Lemma 6.2, we get

$$\|u - \mathbf{I}^{\mathrm{ML}}_{\mathbf{l}_\varepsilon} u\|_{L^2(U,X;\gamma)} \le \|u - \tilde{u}_N\|_{L^2(U,X;\gamma)} + \sum_{\boldsymbol{\nu}\in\mathcal{F}} \sum_{j=l_{\varepsilon,\boldsymbol{\nu}}}^{L(\varepsilon)} \|\tilde{e}^j_{N,\boldsymbol{\nu}}\|_X \|(\mathbf{I}_{\Gamma_j} - \mathbf{I}_{\Gamma_{j+1}}) H_{\boldsymbol{\nu}}\|_{L^2(U;\gamma)}. \tag{7.23}$$

Step 3. We wish to apply Lemma 7.4 to the bound (7.23). By (6.7), we have for all $\boldsymbol{\nu} \in \mathcal{F}$

$$\|(\mathbf{I}_{\Gamma_j} - \mathbf{I}_{\Gamma_{j+1}}) H_{\boldsymbol{\nu}}\|_{L^2(U;\gamma)} \le \|\mathbf{I}_{\Gamma_j} H_{\boldsymbol{\nu}}\|_{L^2(U;\gamma)} + \|\mathbf{I}_{\Gamma_{j+1}} H_{\boldsymbol{\nu}}\|_{L^2(U;\gamma)} \le 2p_{\boldsymbol{\nu}}(3).$$

Note these inequalities also hold when $j = 0$, that is when $\mathbf{I}_{\Gamma_0} = \mathrm{Id}$. By items (iii) and (iv) of Step 1, the collections $(a_{j,\boldsymbol{\nu}})_{\boldsymbol{\nu}\in\mathcal{F}}$, $j \in \mathbb{N}_0$, and $(c_{\boldsymbol{\nu}})_{\boldsymbol{\nu}\in\mathcal{F}}$, $(d_{\boldsymbol{\nu}})_{\boldsymbol{\nu}\in\mathcal{F}}$, satisfy the assumptions of Lemma 7.4. Therefore, (7.23), (7.20) and Lemma 7.4 give

$$\|u - \mathbf{I}^{\mathrm{ML}}_{\mathbf{l}_{\varepsilon_n}} u\|_{L^2(U,X;\gamma)} \le n^{-R} + \sum_{\boldsymbol{\nu}\in\mathcal{F}} \sum_{j=l_{\varepsilon_n,\boldsymbol{\nu}}}^{L(\varepsilon_n)} a_{j,\boldsymbol{\nu}} \le C(1 + \log n) n^{-R},$$

with the convergence rate

$$R = \min\left\{\alpha, \frac{\alpha(q_1^{-1} - 1/2)}{\alpha + q_1^{-1} - q_2^{-1}}\right\} = \min\left\{\alpha, \frac{\alpha(p_1^{-1} - 3/2)}{\alpha + p_1^{-1} - p_2^{-1}}\right\},$$

where we used $q_1 = p_1/(1 - p_1)$ and $q_2 = p_2/(1 - p_2)$ as stated in item (ii) of Step 1. □

7.5 Multilevel Smolyak Sparse-Grid Quadrature Algorithm

We next formulate an analog of Theorem 7.5 for a multilevel Smolyak sparse-grid quadrature algorithm. First, the definition of the multi-index sets in (7.18) (which are used to construct the quadrature via Algorithm 2) has to be slightly adjusted. Then, we state and prove a convergence rate result for the corresponding algorithm. Its proof is along the lines of the proof of Theorem 7.5.

Let $\boldsymbol{b}_1 = (b_{1,j})_{j\in\mathbb{N}} \in \ell^{p_1}(\mathbb{N})$, $\boldsymbol{b}_2 = (b_{2,j})_{j\in\mathbb{N}} \in \ell^{p_2}(\mathbb{N})$, and ξ be the two sequences and the constant from Assumption 7.2. For two constants $K > 0$ and $r > 3$, which are still at our disposal and which will be defined below, we set for all $j \in \mathbb{N}$

$$\varrho_{1,j} := b_{1,j}^{p_1-1}\frac{\xi}{4\|\boldsymbol{b}_1\|_{\ell^{p_1}}}, \qquad \varrho_{2,j} := b_{2,j}^{p_2-1}\frac{\xi}{4\|\boldsymbol{b}_2\|_{\ell^{p_2}}}. \tag{7.24}$$

Furthermore, we let for all $\boldsymbol{\nu} \in \mathcal{F}$ (as in Lemma 6.5 for $k = 2$ and with $\tau = 3$)

$$c_{\boldsymbol{\nu}} := \prod_{j\in\mathbb{N}} \max\{1, K\varrho_{1,j}\}^4 \nu_j^{r-3}, \qquad d_{\boldsymbol{\nu}} := \prod_{j\in\mathbb{N}} \max\{1, K\varrho_{2,j}\}^4 \nu_j^{r-3}. \tag{7.25}$$

Theorem 7.6 *Let $u \in L^2(U, X; \gamma)$ and $u^l \in L^2(U, X; \gamma)$, $l \in \mathbb{N}$, satisfy Assumption 7.2 with some constants $\alpha > 0$ and $0 < p_1 < 4/5$ and $p_1 \le p_2 < 1$. Set $q_1 := p_1/(1-p_1)$. Assume that $r > 2(1+(\alpha+1)q_1/2)/q_1+3$ (for r in (7.25)). There exists constants $K > 0$ (in (7.25)) and $C > 0$ such that the following holds.*

There exist $C > 0$ and, for every $n \in \mathbb{N}$ there exists $\varepsilon_n \in (0, 1]$ such that $\mathrm{work}(\boldsymbol{l}_{\varepsilon_n}) \le n$ *and with $\boldsymbol{l}_{\varepsilon_n} = (l_{\varepsilon_n,\boldsymbol{\nu}})_{\boldsymbol{\nu}\in\mathcal{F}}$ as in Corollary 7.4 (with $c_{\boldsymbol{\nu}}$, $d_{\boldsymbol{\nu}}$ as in* (7.25)*) it holds*

$$\left\|\int_U u(\boldsymbol{y})\,\mathrm{d}\gamma(\boldsymbol{y}) - \boldsymbol{Q}^{\mathrm{ML}}_{\boldsymbol{l}_{\varepsilon_n}} u\right\|_X \le C(1 + \log n)n^{-R},$$

with the convergence rate

$$R := \min\left\{\alpha, \frac{\alpha(2p_1^{-1} - 5/2)}{\alpha + 2p_1^{-1} - 2p_2^{-1}}\right\}. \tag{7.26}$$

Proof Throughout this proof we write $\boldsymbol{b}_1 = (b_{1,j})_{j\in\mathbb{N}}$ and $\boldsymbol{b}_2 = (b_{2,j})_{j\in\mathbb{N}}$ for the two sequences in Assumption 7.2. As in the proof of Theorem 7.5 we highlight that the multi-index set Γ_j which was defined in (7.2) is downward closed for all $j \in \mathbb{N}_0$.

Step 1. Given $n \in \mathbb{N}$, we choose $\varepsilon := \varepsilon_n$ as in Lemma 7.4. Fix $N \in \mathbb{N}$ such that $N > \max\{j : j \in \mathrm{supp}(\boldsymbol{\nu}), l_{\varepsilon,\boldsymbol{\nu}} > 0\}$ and so large that

$$\left\| \int_U (u(\boldsymbol{y}) - \tilde{u}_N(\boldsymbol{y}))\,\mathrm{d}\gamma(\boldsymbol{y}) \right\|_X \le n^{-R}, \tag{7.27}$$

where $\tilde{u}_N : U \to X$ is as in Definition 4.1 (this is possible due $\lim_{N\to\infty} \|u - \tilde{u}_N\|_{L^2(U,X;\gamma)} = 0$ which holds by the $(\boldsymbol{b}_1, \xi, \delta, X)$-holomorphy of u).
By Assumption 7.2, for every $j \in \mathbb{N}$ the function $e^j := u - u^j \in L^2(U, X; \gamma)$ is $(\boldsymbol{b}_1, \xi, \delta, X)$-holomorphic and $(\boldsymbol{b}_2, \xi, \delta\mathfrak{w}_j^\alpha, X)$-holomorphic. For notational convenience we set $e^0 := u - 0 = u \in L^2(U, X; \gamma)$, so that e^0 is $(\boldsymbol{b}_1, \xi, \delta, X)$-holomorphic and $(\boldsymbol{b}_2, \xi, \delta, X)$-holomorphic. Hence for every $j \in \mathbb{N}_0$ there exists a function $\tilde{e}_N^j = \tilde{u}_N - \tilde{u}_N^j$ as in Definition 4.1 (iii).
The following assertions are identical to the ones in the proof of Theorem 7.5, except that we now admit different summability exponents q_1 and q_2.

(i) By Lemma 6.10, for every $j \in \mathbb{N}_0$, with the Wiener-Hermite PC expansion coefficients

$$\tilde{e}_{N,\boldsymbol{\nu}}^j := \int_U H_{\boldsymbol{\nu}}(\boldsymbol{y})\tilde{e}_N^j(\boldsymbol{y})\,\mathrm{d}\gamma(\boldsymbol{y}),$$

it holds

$$\tilde{e}_N^j(\boldsymbol{y}) = \sum_{\boldsymbol{\nu}\in\mathcal{F}} \tilde{e}_{N,\boldsymbol{\nu}}^j H_{\boldsymbol{\nu}}(\boldsymbol{y}) \qquad \forall \boldsymbol{y} \in U,$$

with pointwise absolute convergence.

(ii) By Lemma 6.5, upon choosing $K > 0$ in (7.18) large enough, and because $r > 3$,

$$C_0 c_{\boldsymbol{\nu}} p_{\boldsymbol{\nu}}(3) \le \beta_{\boldsymbol{\nu}}(r, \boldsymbol{\varrho}_1), \qquad C_0 d_{\boldsymbol{\nu}} p_{\boldsymbol{\nu}}(3) \le \beta_{\boldsymbol{\nu}}(r, \boldsymbol{\varrho}_2) \qquad \forall \boldsymbol{\nu} \in \mathcal{F}_2.$$

Remark that by definition of $\boldsymbol{\varrho}_i$, $i \in \{1, 2\}$, in (7.24), it holds $\varrho_{i,j} \sim b_{i,j}^{-(1-p_i)}$ and therefore $(\varrho_{i,j}^{-1})_{j\in\mathbb{N}} \in \ell^{q_i}(\mathbb{N})$ with $q_i := p_i/(1 - p_i)$, $i \in \{1, 2\}$.

(iii) Due to $r > 2(1+2(\alpha+1)q_1)/q_1+3$, the condition of Lemma 6.6 is satisfied (with $k = 2$, $\tau = 3$ and $\theta = (\alpha+1)q_1/2$). Hence the lemma gives

$$\sum_{\boldsymbol{\nu}\in\mathcal{F}} p_{\boldsymbol{\nu}}((\alpha+1)q_1/2)c_{\boldsymbol{\nu}}^{-q_1/4} < \infty \;\Rightarrow\; (p_{\boldsymbol{\nu}}(\alpha+1)c_{\boldsymbol{\nu}}^{-1/2})_{\boldsymbol{\nu}\in\mathcal{F}} \in \ell^{q_1/2}(\mathcal{F})$$

and similarly

$$\sum_{\boldsymbol{\nu}\in\mathcal{F}} p_{\boldsymbol{\nu}}((\alpha+1)q_2/2)d_{\boldsymbol{\nu}}^{-q_2/4} < \infty \;\Rightarrow\; (p_{\boldsymbol{\nu}}(\alpha+1)d_{\boldsymbol{\nu}}^{-1/2})_{\boldsymbol{\nu}\in\mathcal{F}} \in \ell^{q_2/2}(\mathcal{F}).$$

(iv) By Theorem 4.8 and item (ii), for all $j \in \mathbb{N}_0$

$$C_0 \sum_{\boldsymbol{\nu}\in\mathcal{F}_2} c_{\boldsymbol{\nu}} \|\tilde{e}^j_{N,\boldsymbol{\nu}}\|_X^2 p_{\boldsymbol{\nu}}(3) \le \sum_{\boldsymbol{\nu}\in\mathcal{F}_2} \beta_{\boldsymbol{\nu}}(r,\boldsymbol{\varrho}_1)\|\tilde{e}^j_{N,\boldsymbol{\nu}}\|_X^2 \le C\delta^2$$

and

$$C_0 \sum_{\boldsymbol{\nu}\in\mathcal{F}_2} d_{\boldsymbol{\nu}} \|\tilde{e}^j_{N,\boldsymbol{\nu}}\|_X^2 p_{\boldsymbol{\nu}}(3) \le \sum_{\boldsymbol{\nu}\in\mathcal{F}_2} \beta_{\boldsymbol{\nu}}(r,\boldsymbol{\varrho}_2)\|\tilde{e}^j_{N,\boldsymbol{\nu}}\|_X^2 \le C\frac{\delta^2}{\mathfrak{w}_j^{2\alpha}},$$

with the constant C independent of j, $\mathfrak{w}_j$ and N.

(v) Because $N \ge \max\{j \in \mathrm{supp}(\boldsymbol{\nu}) : l_{\varepsilon,\boldsymbol{\nu}} \ge 0\}$ and $\chi_{0,0} = 0$ we have $\mathbf{Q}_{\Gamma_j}(u - u^j) = \mathbf{Q}_{\Gamma_j} e^j = \mathbf{Q}_{\Gamma_j}\tilde{e}^j_N$ for all $j \in \mathbb{N}$ (cp. Remark 4.4). Similarly $\mathbf{Q}_{\Gamma_j} u = \mathbf{Q}_{\Gamma_j}\tilde{u}_N$ for all $j \in \mathbb{N}$.

Step 2. Observe that $\Gamma_j = \emptyset$ for all

$$j > L(\varepsilon) := \max_{\boldsymbol{\nu}\in\mathcal{F}} l_{\varepsilon,\boldsymbol{\nu}}$$

(cp. (7.2)), which is finite due to $|\mathbf{l}_\varepsilon| < \infty$. With the conventions

$$\mathbf{Q}_{\Gamma_0} = \mathbf{Q}_{\mathcal{F}} = \int_U \cdot \,\mathrm{d}\gamma(\boldsymbol{y})$$

(i.e. $\mathbf{Q}_{\Gamma_0}$ is the exact integral operator) and $\mathbf{Q}_\emptyset \equiv 0$ this implies

$$\begin{aligned}\int_U u(\boldsymbol{y})\,\mathrm{d}\gamma(\boldsymbol{y}) = \mathbf{Q}_{\Gamma_0} u &= \sum_{j=0}^{L(\varepsilon)} (\mathbf{Q}_{\Gamma_j} - \mathbf{Q}_{\Gamma_{j+1}})u \\ &= (\mathbf{Q}_{\Gamma_0} - \mathbf{Q}_{\Gamma_1})u + \ldots + (\mathbf{Q}_{\Gamma_{L(\varepsilon)-1}} - \mathbf{Q}_{\Gamma_{L(\varepsilon)}})u + \mathbf{Q}_{\Gamma_{L(\varepsilon)}} u.\end{aligned}$$

By definition of the multilevel quadrature in (7.4)

$$\begin{aligned}\mathbf{Q}^{\mathrm{ML}}_{\mathbf{l}_\varepsilon} u &= \sum_{j=1}^{L(\varepsilon)} (\mathbf{Q}_{\Gamma_j} - \mathbf{Q}_{\Gamma_{j+1}}) u^j \\ &= (\mathbf{Q}_{\Gamma_1} - \mathbf{Q}_{\Gamma_2}) u^1 + \ldots + (\mathbf{Q}_{\Gamma_{L(\varepsilon)-1}} - \mathbf{Q}_{\Gamma_{L(\varepsilon)}}) u^{L(\varepsilon)} + \mathbf{Q}_{\Gamma_{L(\varepsilon)}} u^{L(\varepsilon)}.\end{aligned}$$

By item (v) of Step 1, we can write

$$\begin{aligned}(\mathbf{Q}_{\Gamma_0} - \mathbf{Q}_{\Gamma_1}) u &= \int_U u(\boldsymbol{y}) \, \mathrm{d}\gamma(\boldsymbol{y}) - \mathbf{Q}_{\Gamma_1} u \\ &= \int_U u(\boldsymbol{y}) \, \mathrm{d}\gamma(\boldsymbol{y}) - \mathbf{Q}_{\Gamma_1} \tilde{u}_N \\ &= \int_U (u(\boldsymbol{y}) - \tilde{u}_N(\boldsymbol{y})) \, \mathrm{d}\gamma(\boldsymbol{y}) + (\mathbf{Q}_{\Gamma_0} - \mathbf{Q}_{\Gamma_1}) \tilde{u}_N \\ &= \int_U (u(\boldsymbol{y}) - \tilde{u}_N(\boldsymbol{y})) \, \mathrm{d}\gamma(\boldsymbol{y}) + (\mathbf{Q}_{\Gamma_0} - \mathbf{Q}_{\Gamma_1}) \tilde{e}^0_N,\end{aligned}$$

where in the last equality we used $e^0_N = u_N$, by definition of $e^0 = u$ (and $\tilde{e}^0_N = \tilde{u}_N \in L^2(U, X; \gamma)$ as in Definition 4.1). Hence, again by item (v),

$$\begin{aligned}\int_U u(\boldsymbol{y}) \, \mathrm{d}\gamma(\boldsymbol{y}) - \mathbf{Q}^{\mathrm{ML}}_{\mathbf{l}_\varepsilon} u &= (\mathbf{Q}_{\Gamma_0} - \mathbf{Q}_{\Gamma_1}) u + \sum_{j=1}^{L(\varepsilon)} (\mathbf{Q}_{\Gamma_j} - \mathbf{Q}_{\Gamma_{j+1}})(u - u^j) \\ &= \int_U (u(\boldsymbol{y}) - \tilde{u}_N(\boldsymbol{y})) \, \mathrm{d}\gamma(\boldsymbol{y}) + (\mathbf{Q}_{\Gamma_0} - \mathbf{Q}_{\Gamma_1}) \tilde{u}_N \\ &\quad + \sum_{j=1}^{L(\varepsilon)} (\mathbf{Q}_{\Gamma_j} - \mathbf{Q}_{\Gamma_{j+1}}) \tilde{e}^j_N \\ &= \int_U (u(\boldsymbol{y}) - \tilde{u}_N(\boldsymbol{y})) \, \mathrm{d}\gamma(\boldsymbol{y}) + \sum_{j=0}^{L(\varepsilon)} (\mathbf{Q}_{\Gamma_j} - \mathbf{Q}_{\Gamma_{j+1}}) \tilde{e}^j_N.\end{aligned}$$

Let us bound the norm. From item (i) of Step 1 it follows that for every $j \in \mathbb{N}_0$,

$$\tilde{e}^j_N(\boldsymbol{y}) = \sum_{\boldsymbol{\nu} \in \mathcal{F}^N_1} \tilde{e}^j_{N,\boldsymbol{\nu}} H_{\boldsymbol{\nu}}(\boldsymbol{y}),$$

with the equality and unconditional convergence in X for all $\boldsymbol{y} \in \mathcal{F}_1^N$. Hence similar to Lemma 6.15 we have

$$(\mathbf{Q}_{\Gamma_j} - \mathbf{Q}_{\Gamma_{j+1}})e_N^j = \sum_{\boldsymbol{\nu} \in \mathcal{F}_1^N} \tilde{e}_{N,\boldsymbol{\nu}}^j (\mathbf{Q}_{\Gamma_j} - \mathbf{Q}_{\Gamma_{j+1}}) H_{\boldsymbol{\nu}}$$

with the equality and unconditional convergence in X. Since $(\mathbf{Q}_{\Gamma_j} - \mathbf{Q}_{\Gamma_{j+1}}) H_{\boldsymbol{\nu}} = 0 \in X$ for all $\boldsymbol{\nu} \in \Gamma_{j+1} \subseteq \Gamma_j$ and all $\boldsymbol{\nu} \in \mathcal{F} \backslash \mathcal{F}_2$ by Lemma 6.2, we get

$$\left\| \int_U u(\boldsymbol{y}) \, \mathrm{d}\gamma(\boldsymbol{y}) - \mathbf{Q}_{\mathbf{l}_\varepsilon}^{\mathrm{ML}} u \right\|_X \le \left\| \int_U (u(\boldsymbol{y}) - \tilde{u}_N(\boldsymbol{y})) \, \mathrm{d}\gamma(\boldsymbol{y}) \right\|_X + \sum_{\boldsymbol{\nu} \in \mathcal{F}_2} \sum_{j=l_{\varepsilon,\boldsymbol{\nu}}}^{L(\varepsilon)} \|\tilde{e}_{N,\boldsymbol{\nu}}^j\|_X |(\mathbf{Q}_{\Gamma_j} - \mathbf{Q}_{\Gamma_{j+1}}) H_{\boldsymbol{\nu}}|. \tag{7.28}$$

Step 3. We wish to apply Lemma 7.4 to the bound (7.28). By (6.10), for all $\boldsymbol{\nu} \in \mathcal{F}$

$$|(\mathbf{Q}_{\Gamma_j} - \mathbf{Q}_{\Gamma_{j+1}}) H_{\boldsymbol{\nu}}| \le |\mathbf{Q}_{\Gamma_j} H_{\boldsymbol{\nu}}| + |\mathbf{Q}_{\Gamma_{j+1}} H_{\boldsymbol{\nu}}| \le 2 p_{\boldsymbol{\nu}}(3).$$

Define

$$a_{j,\boldsymbol{\nu}} := \|\tilde{e}_{N,\boldsymbol{\nu}}^j\|_X \, p_{\boldsymbol{\nu}}(3) \qquad \forall \boldsymbol{\nu} \in \mathcal{F}_2,$$

and $a_{j,\boldsymbol{\nu}} := 0$ for $\boldsymbol{\nu} \in \mathcal{F} \backslash \mathcal{F}_2$. By items (iii) and (iv) of Step 1, the collections $(a_{j,\boldsymbol{\nu}})_{\boldsymbol{\nu} \in \mathcal{F}}$, $j \in \mathbb{N}_0$, and $(c_{\boldsymbol{\nu}})_{\boldsymbol{\nu} \in \mathcal{F}}$, $(d_{\boldsymbol{\nu}})_{\boldsymbol{\nu} \in \mathcal{F}}$, satisfy the assumptions of Lemma 7.4 (with $\tilde{q}_1 := q_1/2$ and $\tilde{q}_2 := q_2/2$). Therefore (7.28), (7.27) and Lemma 7.4 give

$$\left\| \int_U u(\boldsymbol{y}) \, \mathrm{d}\gamma(\boldsymbol{y}) - \mathbf{Q}_{\mathbf{l}_\varepsilon}^{\mathrm{ML}} u \right\|_X \le n^{-R} + \sum_{\boldsymbol{\nu} \in \mathcal{F}} \sum_{j=l_{\varepsilon,\boldsymbol{\nu}}}^{L(\varepsilon)} a_{j,\boldsymbol{\nu}} \le C(1 + \log n) n^{-R},$$

with

$$R = \min \left\{ \alpha, \frac{\alpha(\tilde{q}_1^{-1} - 1/2)}{\alpha + \tilde{q}_1^{-1} - \tilde{q}_2^{-1}} \right\} = \min \left\{ \alpha, \frac{\alpha(2p_1^{-1} - 5/2)}{\alpha + 2p_1^{-1} - 2p_2^{-1}} \right\},$$

where we used $\tilde{q}_1 = q_1/2 = p_1/(2 - 2p_1)$ and $\tilde{q}_2 = q_2/2 = p_2/(2 - 2p_2)$ as stated in item (ii) of Step 1. □

7.6 Examples for Multilevel Interpolation and Quadrature

We revisit the examples in Chaps. 4 and 5, and demonstrate how to verify the assumptions required for the multilevel convergence rate results in Theorems 7.5 and 7.6.

7.6.1 *Parametric Diffusion Coefficient in Polygonal Domain*

Let $D \subseteq \mathbb{R}^2$ be a bounded polygonal domain, and consider once more the elliptic equation

$$-\operatorname{div}(a\nabla\mathcal{U}(a)) = f \quad \text{in } D, \qquad \mathcal{U}(a) = 0 \quad \text{on } \partial D, \tag{7.29}$$

as in Sect. 4.3.1.

For $s \in \mathbb{N}_0$ and $\varkappa \in \mathbb{R}$, recall the Kondrat'ev spaces $\mathcal{W}^s_\infty(D)$ and $\mathcal{K}^s_\varkappa(D)$ with norms

$$\|u\|_{\mathcal{K}^s_\varkappa} := \sum_{|\alpha|\le s} \|r_D^{|\alpha|-\varkappa} D^\alpha u\|_{L^2} \qquad \text{and} \qquad \|u\|_{\mathcal{W}^s_\infty} := \sum_{|\alpha|\le s} \|r_D^{|\alpha|} D^\alpha u\|_{L^\infty}$$

introduced in Sect. 3.8.1. Here, as earlier, $r_D : D \to [0, 1]$ denotes a fixed smooth function that coincides with the distance to the nearest corner, in a neighbourhood of each corner. According to Theorem 3.29, assuming $s \ge 2$, $f \in \mathcal{K}^{s-2}_{\varkappa-1}(D)$ and $a \in \mathcal{W}^{s-1}_\infty(D)$ the solution $\mathcal{U}(a)$ of (7.29) belongs to $\mathcal{K}^s_{\varkappa+1}(D)$ provided that with

$$\rho(a) := \operatorname*{ess\,inf}_{x\in D} \Re(a(x)) > 0,$$

$$|\varkappa| < \frac{\rho(a)}{\tau\|a\|_{L^\infty}}, \tag{7.30}$$

where τ is a constant depending on D and s. Our goal is to treat, in a unified manner, a family of diffusion coefficients $a(y)$, $y \in U$, where for certain $y \in U$ the diffusion coefficient $a(y)$ is such that the right-hand side of (7.30) might be arbitrarily small. This only leaves us with the choice $\varkappa = 0$, see Remark 3.31. On the other hand, the motivation of using Kondrat'ev spaces in the analysis of approximations to PDE solutions $\mathcal{U}(a(y))$, is that functions in $\mathcal{K}^s_{\varkappa+1}(D)$ on polygonal domains in $\mathbb{R}^2$ can be approximated with the optimal convergence rate $\frac{s-1}{2}$ w.r.t. the H^1-norm by suitable finite element spaces (on graded meshes; i.e. this analysis accounts for corner singularities which prevent optimal convergence rates on uniform meshes). Such results are well-known, see for example [25], however they require $\varkappa > 0$. For this reason we need a stronger regularity result, giving uniform $\mathcal{K}^s_{\varkappa+1}$-regularity with $\varkappa > 0$ independent of the parameter. This is the purpose of the next theorem.

For its proof we shall need the following lemma, which is shown in a similar way as in [112, Lemma C.2]. We recall that

$$\|f\|_{W^s_\infty} := \sum_{|\boldsymbol{\nu}|\le s} \|D^{\boldsymbol{\nu}} f\|_{L^\infty}.$$

Lemma 7.7 *Let $s \in \mathbb{N}_0$ and let $D \subseteq \mathbb{R}^2$ be a bounded polygonal domain, $d \in \mathbb{N}$. Then there exist C_s and $\tilde{C}_s$ such that for any two functions $f, g \in \mathcal{W}^s_\infty(D)$*

(i) $\|fg\|_{\mathcal{W}^s_\infty} \le C_s \|f\|_{\mathcal{W}^s_\infty} \|g\|_{\mathcal{W}^s_\infty}$,

(ii) $\|\frac{1}{f}\|_{\mathcal{W}^s_\infty} \le \tilde{C}_s \frac{\|f\|^s_{\mathcal{W}^s_\infty}}{\operatorname{ess\,inf}_{\boldsymbol{x}\in D} |f(\boldsymbol{x})|^{s+1}}$ *if* $\operatorname{ess\,inf}_{\boldsymbol{x}\in D} |f(\boldsymbol{x})| > 0$.

These statements remain true if $\mathcal{W}^s_\infty(D)$ is replaced by $W^s_\infty(D)$. Furthermore, if $\varkappa \in \mathbb{R}$, then for $f \in \mathcal{K}^s_\varkappa(D)$ and $a \in \mathcal{W}^s_\infty(D)$

(iii) $\|fa\|_{\mathcal{K}^s_\varkappa} \le C_s \|f\|_{\mathcal{K}^s_\varkappa} \|a\|_{\mathcal{W}^s_\infty}$,

(iv) $\|\nabla f \cdot \nabla a\|_{\mathcal{K}^{s-1}_{\varkappa-1}} \le C_{s-1} \|f\|_{\mathcal{K}^s_{\varkappa+1}} \|a\|_{\mathcal{W}^s_\infty}$ *if* $s \ge 1$.

Proof We will only prove (i) and (ii) for functions in $\mathcal{W}^s_\infty(D)$. The case of $W^s_\infty(D)$ is shown similarly (by omitting all occurring functions r_D in the following).

Step 1. We start with (i), and show a slightly more general bound: for $\tau \in \mathbb{R}$ introduce

$$\|f\|_{\mathcal{W}^s_{\tau,\infty}} := \sum_{|\boldsymbol{\nu}|\le s} \|r_D^{\tau+|\boldsymbol{\nu}|} D^{\boldsymbol{\nu}} f\|_{L^\infty},$$

i.e. $\mathcal{W}^s_{0,\infty}(D) = \mathcal{W}^s_\infty(D)$. We will show that for $\tau_1 + \tau_2 = \tau$

$$\|fg\|_{\mathcal{W}^s_{\tau,\infty}} \le C_s \|f\|_{\mathcal{W}^s_{\tau_1,\infty}} \|g\|_{\mathcal{W}^s_{\tau_2,\infty}}. \tag{7.31}$$

Item (i) then follows with $\tau = \tau_1 = \tau_2 = 0$.

Using the multivariate Leibniz rule for Lipschitz functions, for any multiindex $\boldsymbol{\nu} \in \mathbb{N}_0^d$ with $d \in \mathbb{N}$ fixed,

$$D^{\boldsymbol{\nu}}(fg) = \sum_{\boldsymbol{\mu}\le\boldsymbol{\nu}} \binom{\boldsymbol{\nu}}{\boldsymbol{\mu}} D^{\boldsymbol{\nu}-\boldsymbol{\mu}} f D^{\boldsymbol{\mu}} g. \tag{7.32}$$

Thus if $|\boldsymbol{\nu}| \le s$

$$\begin{aligned}\|r_D^{\tau+|\boldsymbol{\nu}|} D^{\boldsymbol{\nu}}(fg)\|_{L^\infty} &\le \sum_{\boldsymbol{\mu}\le\boldsymbol{\nu}} \binom{\boldsymbol{\nu}}{\boldsymbol{\mu}} \|r_D^{\tau_1+|\boldsymbol{\nu}-\boldsymbol{\mu}|} D^{\boldsymbol{\nu}-\boldsymbol{\mu}} f\|_{L^\infty} \|r_D^{\tau_2+|\boldsymbol{\mu}|} D^{\boldsymbol{\mu}} g\|_{L^\infty} \\ &\le 2^{|\boldsymbol{\nu}|} \|f\|_{\mathcal{W}^s_{\tau_1,\infty}} \|g\|_{\mathcal{W}^s_{\tau_2,\infty}},\end{aligned}$$

where we used $\binom{\boldsymbol{\nu}}{\boldsymbol{\mu}} = \prod_{j=1}^{d} \binom{\nu_j}{\mu_j}$ and $\sum_{i=0}^{\nu_j} \binom{\nu_j}{i} = 2^{\nu_j}$. We conclude

$$\|fg\|_{\mathcal{W}^s_{\tau,\infty}} \le C_s \|f\|_{\mathcal{W}^s_{\tau_1,\infty}} \|g\|_{\mathcal{W}^s_{\tau_2,\infty}}$$

with $C_s = \sum_{|\boldsymbol{\nu}|\le s} 2^{|\boldsymbol{\nu}|}$. Hence (i) holds.

Step 2. We show (ii), and claim that for all $|\boldsymbol{\nu}| \le s$ it holds

$$D^{\boldsymbol{\nu}}\left(\frac{1}{f}\right) = \frac{p_{\boldsymbol{\nu}}}{f^{|\boldsymbol{\nu}|+1}}, \tag{7.33}$$

where $p_{\boldsymbol{\nu}}$ satisfies

$$\|p_{\boldsymbol{\nu}}\|_{\mathcal{W}^{s-|\boldsymbol{\nu}|}_{|\boldsymbol{\nu}|,\infty}} \le \hat{C}_{|\boldsymbol{\nu}|} \|f\|^{|\boldsymbol{\nu}|}_{\mathcal{W}^s_\infty} \tag{7.34}$$

for some $\hat{C}_{|\boldsymbol{\nu}|}$ solely depending on $|\boldsymbol{\nu}|$. We proceed by induction over $|\boldsymbol{\nu}|$ and start with $|\boldsymbol{\nu}| = 1$, i.e., $\boldsymbol{\nu} = \boldsymbol{e}_j = (\delta_{ij})_{i=1}^d$ for some $j \in \{1, \ldots, d\}$. Then $D^{\boldsymbol{e}_j}\frac{1}{f} = \frac{-\partial_j f}{f^2}$ and $p_{\boldsymbol{e}_j} = -\partial_j f$ satisfies

$$\begin{aligned}
\|p_{\boldsymbol{e}_j}\|_{\mathcal{W}^{s-1}_{1,\infty}} &= \sum_{|\boldsymbol{\mu}|\le s-1} \|r_D^{1+|\boldsymbol{\mu}|} D^{\boldsymbol{\mu}} p_{\boldsymbol{e}_j}\|_{L^\infty} \\
&= \sum_{|\boldsymbol{\mu}|\le s-1} \|r_D^{|\boldsymbol{\mu}+\boldsymbol{e}_j|} D^{\boldsymbol{\mu}+\boldsymbol{e}_j} f\|_{L^\infty} \le \|f\|_{\mathcal{W}^s_\infty},
\end{aligned}$$

i.e. $\hat{C}_1 = 1$. For the induction step fix $\boldsymbol{\nu}$ with $1 < |\boldsymbol{\nu}| < s$ and $j \in \{1, \ldots, d\}$. Then by the induction hypothesis $D^{\boldsymbol{\nu}}\frac{1}{f} = \frac{p_{\boldsymbol{\nu}}}{f^{|\boldsymbol{\nu}|+1}}$ and

$$\begin{aligned}
D^{\boldsymbol{\nu}+\boldsymbol{e}_j}\frac{1}{f} = \partial_j\left(\frac{p_{\boldsymbol{\nu}}}{f^{|\boldsymbol{\nu}|+1}}\right) &= \frac{f^{|\boldsymbol{\nu}|+1}\partial_j p_{\boldsymbol{\nu}} - (|\boldsymbol{\nu}|+1) f^{|\boldsymbol{\nu}|} p_{\boldsymbol{\nu}} \partial_j f}{f^{2|\boldsymbol{\nu}|+2}} \\
&= \frac{f\partial_j p_{\boldsymbol{\nu}} - (|\boldsymbol{\nu}|+1) p_{\boldsymbol{\nu}} \partial_j f}{f^{|\boldsymbol{\nu}|+2}},
\end{aligned}$$

and thus

$$p_{\boldsymbol{\nu}+\boldsymbol{e}_j} := f\partial_j p_{\boldsymbol{\nu}} - (|\boldsymbol{\nu}|+1) p_{\boldsymbol{\nu}} \partial_j f.$$

Observe that

$$\begin{aligned}
\|\partial_j g\|_{\mathcal{W}^s_{\tau,\infty}} &= \sum_{|\boldsymbol{\mu}|\le s} \|r_D^{\tau+|\boldsymbol{\mu}|} D^{\boldsymbol{\mu}+\boldsymbol{e}_j} g\|_{L^\infty} \\
&\le \sum_{|\boldsymbol{\mu}|\le s+1} \|r_D^{\tau+|\boldsymbol{\mu}|-1} D^{\boldsymbol{\mu}} g\|_{L^\infty} = \|g\|_{\mathcal{W}^{s+1}_{\tau-1,\infty}}.
\end{aligned} \tag{7.35}$$

Using (7.31) and (7.35), we get with $\tau := |\boldsymbol{\nu}| + 1$

$$\begin{aligned}
\|p_{\boldsymbol{\nu}+\boldsymbol{e}_j}\|_{\mathcal{W}^{s-\tau}_{\tau,\infty}} &\leq \|f\partial_j p_{\boldsymbol{\nu}}\|_{\mathcal{W}^{s-\tau}_{\tau,\infty}} + (|\boldsymbol{\nu}|+1)\|p_{\boldsymbol{\nu}}\partial_j f\|_{\mathcal{W}^{s-\tau}_{\tau,\infty}} \\
&\leq C_{s-\tau}\|f\|_{\mathcal{W}^{s-\tau}_{0,\infty}}\|\partial_j p_{\boldsymbol{\nu}}\|_{\mathcal{W}^{s-\tau}_{\tau,\infty}} \\
&\quad + (|\boldsymbol{\nu}|+1)C_{s-\tau}\|p_{\boldsymbol{\nu}}\|_{\mathcal{W}^{s-\tau}_{\tau-1,\infty}}\|\partial_j f\|_{\mathcal{W}^{s-\tau}_{1,\infty}} \\
&\leq C_{s-\tau}\|f\|_{\mathcal{W}^{s-\tau}_{0,\infty}}\|p_{\boldsymbol{\nu}}\|_{\mathcal{W}^{s-\tau+1}_{\tau-1,\infty}} \\
&\quad + (|\boldsymbol{\nu}|+1)C_{s-\tau}\|p_{\boldsymbol{\nu}}\|_{\mathcal{W}^{s-\tau+1}_{\tau-1,\infty}}\|f\|_{\mathcal{W}^{s-\tau+1}_{0,\infty}}.
\end{aligned}$$

Due to $\tau - 1 = |\boldsymbol{\nu}|$ and the induction hypothesis (7.34) for $p_{\boldsymbol{\nu}}$,

$$\begin{aligned}
\|p_{\boldsymbol{\nu}+\boldsymbol{e}_j}\|_{\mathcal{W}^{s-(|\boldsymbol{\nu}|+1)}_{|\boldsymbol{\nu}|+1,\infty}} &\leq C_{s-(|\boldsymbol{\nu}|+1)}\Big(\hat{C}_{|\boldsymbol{\nu}|}\|f\|_{\mathcal{W}^{s-(|\boldsymbol{\nu}|+1)}_{\infty}}\|f\|^{|\boldsymbol{\nu}|}_{\mathcal{W}^{s}_{\infty}} \\
&\quad +(|\boldsymbol{\nu}|+1)\hat{C}_{|\boldsymbol{\nu}|}\|f\|^{|\boldsymbol{\nu}|}_{\mathcal{W}^{s}_{\infty}}\|f\|_{\mathcal{W}^{s-|\boldsymbol{\nu}|}_{\infty}}\Big) \\
&\leq C_{s-(|\boldsymbol{\nu}|+1)}\hat{C}_{|\boldsymbol{\nu}|}(|\boldsymbol{\nu}|+2)\|f\|^{|\boldsymbol{\nu}|+1}_{\mathcal{W}^{\infty}_{s}}.
\end{aligned}$$

In all this shows the claim with $\hat{C}_1 := 1$ and inductively for $1 < k \leq s$,

$$\hat{C}_k := C_{s-k}\hat{C}_{k-1}(k+1).$$

By (7.33) and (7.34), for every $|\boldsymbol{\nu}| \leq s$

$$\left\| r_D^{|\boldsymbol{\nu}|} D^{\boldsymbol{\nu}}\left(\frac{1}{f}\right)\right\|_{L^\infty} \leq \hat{C}_{|\boldsymbol{\nu}|}\frac{\|f\|^{|\boldsymbol{\nu}|}_{\mathcal{W}^{s}_{\infty}}}{\operatorname{ess\,inf}_{\boldsymbol{x}\in D}|f(\boldsymbol{x})|^{|\boldsymbol{\nu}|+1}}.$$

Due to

$$\|f\|_{\mathcal{W}^{s}_{\infty}} \geq \|f\|_{L^\infty} \geq \operatorname{ess\,inf}_{\boldsymbol{x}\in D}|f(\boldsymbol{x})|,$$

this implies

$$\left\|\frac{1}{f}\right\|_{\mathcal{W}^{s}_{\infty}} = \sum_{|\boldsymbol{\nu}|\leq s}\left\| r_D^{|\boldsymbol{\nu}|} D^{\boldsymbol{\nu}}\left(\frac{1}{f}\right)\right\|_{L^\infty} \leq \tilde{C}_s\frac{\|f\|^{s}_{\mathcal{W}^{s}_{\infty}}}{\operatorname{ess\,inf}_{\boldsymbol{x}\in D}|f(\boldsymbol{x})|^{s+1}}$$

with $\tilde{C}_s := \sum_{|\boldsymbol{\nu}|\leq s}\hat{C}_{|\boldsymbol{\nu}|}$.

Step 3. We show (iii) and (iv). If $f \in \mathcal{K}^{s}_{\varkappa}(D)$ and $a \in \mathcal{W}^{s}_{\infty}(D)$, then by (7.32) for Sobolev functions,

$$r_D^{\boldsymbol{\nu}-\varkappa}D^{\boldsymbol{\nu}}(fa) = \sum_{\boldsymbol{\mu}\leq\boldsymbol{\nu}}\binom{\boldsymbol{\nu}}{\boldsymbol{\mu}}(r_D^{|\boldsymbol{\nu}-\boldsymbol{\mu}|-\varkappa}D^{\boldsymbol{\nu}-\boldsymbol{\mu}}f)(r_D^{|\boldsymbol{\mu}|}D^{\boldsymbol{\mu}}a)$$

and hence

$$
\begin{aligned}
\|fa\|_{\mathcal{K}^s_\varkappa} &= \sum_{|\nu|\le s} \|r_D^{|\nu|-\varkappa} D^\nu (fa)\|_{L^2} \\
&\le \sum_{|\nu|\le s} \sum_{\mu\le\nu} \binom{\nu}{\mu} \|r_D^{|\nu-\mu|-\varkappa} D^{\nu-\mu} f\|_{L^2} \|r_D^{|\mu|} D^\mu a\|_{L^\infty} \\
&\le C_s \sum_{|\nu|\le s} \|r_D^{|\nu|-\varkappa} D^\nu f\|_{L^2} \sum_{|\mu|\le s} \|r_D^{|\mu|} D^\mu a\|_{L^\infty} \\
&= C_s \|f\|_{\mathcal{K}^s_\varkappa} \|a\|_{\mathcal{W}^s_\infty}.
\end{aligned}
$$

Finally if $s \ge 1$,

$$
\begin{aligned}
\|\nabla f \cdot \nabla a\|_{\mathcal{K}^{s-1}_{\varkappa-1}} &= \sum_{|\nu|\le s-1} \left\| r_D^{|\nu|-\varkappa+1} D^\nu \left(\sum_{j=1}^d \partial_j f \partial_j a \right) \right\|_{L^2} \\
&\le \sum_{|\nu|\le s-1} \sum_{\mu\le\nu} \binom{\nu}{\mu} \sum_{j=1}^d \|r_D^{|\nu-\mu|-\varkappa} D^{\nu-\mu+e_j} f\|_{L^2} \|r_D^{|\mu|+1} D^{\mu+e_j} a\|_{L^\infty} \\
&\le C_{s-1} d \sum_{|\nu|\le s} \|r_D^{|\nu|-\varkappa-1} D^\nu f\|_{L^2} \sum_{|\mu|\le s} \|r_D^{|\mu|} D^\mu a\|_{L^\infty} \\
&= C_{s-1} d \|f\|_{\mathcal{K}^s_{\varkappa+1}} \|a\|_{\mathcal{W}^s_\infty}.
\end{aligned}
$$

□

The proof of the next theorem is based on Theorem 3.29. In order to get regularity in $\mathcal{K}^s_{\varkappa+1}(D)$ with $\varkappa > 0$ independent of the diffusion coefficient a, we now assume $a \in W^1_\infty(D) \cap \mathcal{W}^{s-1}_\infty(D)$ in lieu of the weaker assumption $a \in \mathcal{W}^{s-1}_\infty$ that was required in Theorem 3.29.

Theorem 7.8 *Let $D \subseteq \mathbb{R}^2$ be a bounded polygonal domain and $s \in \mathbb{N}$, $s \ge 2$. Then there exist $\varkappa > 0$ and $C_s > 0$ depending on D and s (but independent of a) such that for all $a \in W^1_\infty(D) \cap \mathcal{W}^{s-1}_\infty(D)$ and all $f \in \mathcal{K}^{s-2}_{\varkappa-1}(D)$ the weak solution $\mathcal{U} \in H^1_0(D)$ of (7.29) satisfies with $N_s := \frac{s(s-1)}{2}$*

$$
\|\mathcal{U}\|_{\mathcal{K}^s_{\varkappa+1}} \le C_s \frac{1}{\rho(a)} \left(\frac{\|a\|_{\mathcal{W}^{s-1}_\infty} + \|a\|_{W^1_\infty}}{\rho(a)} \right)^{N_s} \|f\|_{\mathcal{K}^{s-2}_{\varkappa-1}}. \tag{7.36}
$$

Proof Throughout this proof let $\varkappa \in (0, 1)$ be a constant such that

$$-\Delta : \mathcal{K}^j_{\varkappa+1}(D) \cap H^1_0(D) \to \mathcal{K}^{j-2}_{\varkappa-1}(D) \tag{7.37}$$

is a boundedly invertible operator for all $j \in \{2, \dots, s\}$; such $\varkappa$ exists by Theorem 3.29, and $\varkappa$ merely depends on D and s.

Step 1. We prove the theorem for $s = 2$, in which case $a \in W^1_\infty \cap \mathcal{W}^1_\infty = W^1_\infty$. Applying Theorem 3.29 directly to (7.29) yields the existence of *some* $\tilde{\varkappa} \in (0, \varkappa)$ (depending on a) such that $\mathcal{U} \in \mathcal{K}^2_{\tilde{\varkappa}+1}$. Here we use

$$f \in \mathcal{K}^0_{\varkappa-1}(D) \hookrightarrow \mathcal{K}^0_{\tilde{\varkappa}-1}(D)$$

due to $\tilde{\varkappa} \in (0, \varkappa)$. By the Leibniz rule for Sobolev functions we can write

$$-\operatorname{div}(a\nabla\mathcal{U}) = -a\Delta\mathcal{U} - \nabla a \cdot \nabla\mathcal{U}$$

in the sense of $\mathcal{K}^0_{\tilde{\varkappa}-1}(D)$: (i) it holds $\Delta\mathcal{U} \in \mathcal{K}^0_{\tilde{\varkappa}-1}(D)$ and

$$a \in W^1_\infty(D) \hookrightarrow L^\infty(D)$$

which implies $a\Delta\mathcal{U} \in \mathcal{K}^0_{\tilde{\varkappa}-1}(D)$ (ii) it holds

$$\nabla\mathcal{U} \in \mathcal{K}^1_{\tilde{\varkappa}}(D) \hookrightarrow \mathcal{K}^0_{\tilde{\varkappa}}(D),$$

and $\nabla a \in L^\infty(D)$ which implies $\nabla a \cdot \nabla\mathcal{U} \in \mathcal{K}^0_{\tilde{\varkappa}-1}(D)$. Hence,

$$-\operatorname{div}(a\nabla\mathcal{U}) = -a\Delta\mathcal{U} - \nabla a \cdot \nabla\mathcal{U} = f,$$

and further

$$-\Delta\mathcal{U} = \frac{1}{a}\Big(f + \nabla a \cdot \nabla\mathcal{U}\Big) =: \tilde{f} \in \mathcal{K}^0_{\tilde{\varkappa}-1}(D)$$

since $\frac{1}{a} \in L^\infty(D)$ due to $\rho(a) > 0$. Our goal is to show that in fact $\tilde{f} \in \mathcal{K}^0_{\varkappa-1}(D)$. Because of $-\Delta\mathcal{U} = \tilde{f}$ and $\mathcal{U}|_{\partial D} \equiv 0$, Theorem 3.29 then implies

$$\|\mathcal{U}\|_{\mathcal{K}^2_{\varkappa+1}} \le C\|\tilde{f}\|_{\mathcal{K}^0_{\varkappa-1}} \tag{7.38}$$

for a constant C solely depending on D.
Denote by C_H a constant (solely depending on D) such that

$$\|r_D^{-1}v\|_{L^2} \le C_H\|\nabla v\|_{L^2} \qquad \forall v \in H^1_0(D).$$

This constant exists as a consequence of Hardy's inequality, see e.g. [64] and [82, 93] for the statement and proof of the inequality on bounded Lipschitz domains. Then due to

$$\begin{aligned}\rho(a)\|\nabla\mathcal{U}\|_{L^2}^2 &\leq \Re\left(\int_D a\nabla\mathcal{U}\cdot\overline{\nabla\mathcal{U}}\,\mathrm{d}\mathbf{x}\right) = \Re\left(\int_D f\overline{\mathcal{U}}\,\mathrm{d}\mathbf{x}\right)\\ &\leq \|r_D^{1-\varkappa} f\|_{L^2}\|r_D^{\varkappa-1}\mathcal{U}\|_{L^2} \leq \|f\|_{\mathcal{K}^0_{\varkappa-1}}\|r_D^{-1}\mathcal{U}\|_{L^2}\\ &\leq C_H\|f\|_{\mathcal{K}^0_{\varkappa-1}}\|\nabla\mathcal{U}\|_{L^2}\end{aligned}$$

it holds

$$\|\nabla\mathcal{U}\|_{L^2} \leq \frac{C_H\|f\|_{\mathcal{K}^0_{\varkappa-1}}}{\rho(a)}.$$

Hence, using $r_D^{1-\varkappa} \leq 1$, we have that

$$\begin{aligned}\|\tilde{f}\|_{\mathcal{K}^0_{\varkappa-1}} &= \left\|\frac{r_D^{1-\varkappa}}{a}\left(f + \nabla a\cdot\nabla\mathcal{U}\right)\right\|_{L^2}\\ &\leq \left\|\frac{1}{a}\right\|_{L^\infty}\left(\|r_D^{1-\varkappa} f\|_{L^2} + \|\nabla a\|_{L^\infty}\|\nabla\mathcal{U}\|_{L^2}\right)\\ &\leq \frac{1}{\rho(a)}\left(\|f\|_{\mathcal{K}^0_{\varkappa-1}} + \|a\|_{W^1_\infty}\frac{C_H\|f\|_{\mathcal{K}^0_{\varkappa-1}}}{\rho(a)}\right)\\ &= \frac{\|f\|_{\mathcal{K}^0_{\varkappa-1}}}{\rho(a)}\left(1 + \frac{C_H\|a\|_{W^1_\infty}}{\rho(a)}\right)\\ &\leq (1+C_H)\frac{1}{\rho(a)}\frac{\|a\|_{W^1_\infty}}{\rho(a)}\|f\|_{\mathcal{K}^0_{\varkappa-1}}.\end{aligned}$$

The statement follows by (7.38).

Step 2. For general $s \in \mathbb{N}$, $s \geq 2$, we proceed by induction. Assume the theorem holds for $s-1 \geq 2$. Then for

$$f \in \mathcal{K}^{s-2}_{\varkappa-1}(D) \hookrightarrow \mathcal{K}^{s-3}_{\varkappa-1}(D)$$

and

$$a \in W^1_\infty(D) \cap \mathcal{W}^{s-1}_\infty(D) \hookrightarrow W^1_\infty(D) \cap \mathcal{W}^{s-2}_\infty(D),$$

we get

$$\|\mathcal{U}\|_{\mathcal{K}^{s-1}_{\varkappa+1}} \leq \frac{C_{s-1}}{\rho(a)} \left(\frac{\|a\|_{W^1_\infty} + \|a\|_{\mathcal{W}^{s-2}_\infty}}{\rho(a)} \right)^{N_{s-1}} \|f\|_{\mathcal{K}^{s-3}_{\varkappa-1}}. \tag{7.39}$$

As in Step 1, it holds

$$-\Delta \mathcal{U} = \frac{1}{a}\Big(f + \nabla a \cdot \nabla \mathcal{U}\Big) =: \tilde{f}.$$

By Lemma 7.7 and (7.39), for some constant C (which can change in each line, but solely depends on D and s) we have that

$$\begin{aligned}
\|\tilde{f}\|_{\mathcal{K}^{s-2}_{\varkappa-1}} &\leq C \left\| \frac{1}{a} \right\|_{\mathcal{W}^{s-2,\infty}} \|f + \nabla a \cdot \nabla \mathcal{U}\|_{\mathcal{K}^{s-2}_{\varkappa-1}} \\
&\leq C \frac{\|a\|^{s-2}_{\mathcal{W}^{s-2}}}{\rho(a)^{s-1}} \left(\|f\|_{\mathcal{K}^{s-2}_{\varkappa-1}} + \|a\|_{\mathcal{W}^{s-1}_\infty} \|\mathcal{U}\|_{\mathcal{K}^{s-1}_{\varkappa+1}} \right) \\
&\leq C \frac{\|a\|^{s-2}_{\mathcal{W}^{s-2}}}{\rho(a)^{s-1}} \Bigg(\|f\|_{\mathcal{K}^{s-2}_{\varkappa-1}} \\
&\qquad + C_{s-1} \frac{\|a\|_{\mathcal{W}^{s-1}_\infty}}{\rho(a)} \left(\frac{\|a\|_{W^1_\infty} + \|a\|_{\mathcal{W}^{s-2}_\infty}}{\rho(a)} \right)^{N_{s-1}} \|f\|_{\mathcal{K}^{s-3}_{\varkappa-1}} \Bigg) \\
&\leq C \frac{1}{\rho(a)} \left(\frac{\|a\|_{W^1_\infty} + \|a\|_{\mathcal{W}^{s-1}_\infty}}{\rho(a)} \right)^{N_{s-1}+1+(s-2)} \|f\|_{\mathcal{K}^{s-2}_{\varkappa-1}}.
\end{aligned}$$

Note that

$$N_{s-1} + (s-1) = \frac{(s-1)(s-2)}{2} + (s-1) = \frac{s(s-1)}{2} = N_s.$$

We now use (7.39) and the fact that (7.37) is a boundedly invertible isomorphism to conclude that there exist C_s such that (7.36) holds. □

Throughout the rest of this section D is assumed a bounded polygonal domain and $\varkappa > 0$ the constant from Theorem 7.8.

Assumption 7.9 *For some fixed $s \in \mathbb{N}$, $s \geq 2$, there exist constants $C > 0$ and $\alpha > 0$, and a sequence $(X_l)_{l \in \mathbb{N}}$ of subspaces of $X = H^1_0(D; \mathbb{C}) =: H^1_0$, such that*

(i) $\mathfrak{w}_l := \dim(X_l)$, $l \in \mathbb{N}$, *satisfies Assumption 7.1 (for some $K_{\mathfrak{W}} > 0$),*

(ii) for all $l \in \mathbb{N}$

$$\sup_{0 \neq u \in \mathcal{K}^s_{\varkappa+1}} \frac{\inf_{v \in X_l} \|u - v\|_{H^1_0}}{\|u\|_{\mathcal{K}^s_{\varkappa+1}}} \leq C \mathfrak{w}_l^{-\alpha}. \tag{7.40}$$

The constant α in Assumption 7.9 can be interpreted as the convergence rate of the finite element method. For the Kondrat'ev space $\mathcal{K}^s_{\varkappa+1}(D)$, finite element spaces X_l of piecewise polynomials of degree $s - 1$ have been constructed in [25, Theorem 4.4], which achieve the optimal (in space dimension 2) convergence rate

$$\alpha = \frac{s-1}{2} \tag{7.41}$$

in (7.40). For these spaces, Assumption 7.9 holds with this α, which consequently allows us to retain optimal convergence rates. Nonetheless we keep the discussion general in the following, and assume arbitrary positive $\alpha > 0$.

We next introduce the finite element solutions of (7.29) in the spaces X_l, and provide the basic error estimate.

Lemma 7.10 *Let Assumption 7.9 be satisfied for some $s \geq 2$. Let $f \in \mathcal{K}^{s-2}_{\varkappa-1}(D)$ and*

$$a \in W^1_\infty(D) \cap \mathcal{W}^{s-1}_\infty(D) \subseteq L^\infty(D)$$

with $\rho(a) > 0$ and denote for $l \in \mathbb{N}$ by $\mathcal{U}^l(a) \in X_l$ the unique solution of

$$\int_D a(\nabla \mathcal{U}^l)^\top \overline{\nabla v} \, \mathrm{d}\boldsymbol{x} = \langle f, v \rangle \qquad \forall v \in X_l,$$

where the right hand side denotes the (sesquilinear) dual pairing between $H^{-1}(D)$ and $H^1_0(D)$. Then for the solution $\mathcal{U}(a) \in H^1_0(D)$ it holds with the constants N_s, C_s from Theorem. 7.8,

$$\begin{aligned} \|\mathcal{U}(a) - \mathcal{U}^l(a)\|_{H^1_0} &\leq \mathfrak{w}_l^{-\alpha} C \frac{\|a\|_{L^\infty}}{\rho(a)} \|\mathcal{U}(a)\|_{\mathcal{K}^s_{\varkappa+1}} \\ &\leq \mathfrak{w}_l^{-\alpha} C C_s \frac{(\|a\|_{W^1_\infty} + \|a\|_{\mathcal{W}^{s-1}_\infty})^{N_s+1}}{\rho(a)^{N_s+2}} \|f\|_{\mathcal{K}^{s-2}_{\varkappa-1}}. \end{aligned}$$

Here $C > 0$ is the constant from Assumption 7.9.

Proof By Céa's lemma in complex form we derive that

$$\|\mathcal{U}(a) - \mathcal{U}^l(a)\|_{H_0^1} \leq \frac{\|a\|_{L^\infty}}{\rho(a)} \inf_{v \in X_l} \|\mathcal{U}(a) - v\|_{H_0^1}.$$

Hence the assertion follows by Assumption 7.9 and (7.36). □

Throughout the rest of this section, as earlier we expand the logarithm of the diffusion coefficient

$$a(\boldsymbol{y}) = \exp\left(\sum_{j \in \mathbb{N}} y_j \psi_j\right)$$

in terms of a sequence $\psi_j \in W^1_\infty(D) \cap \mathcal{W}^{s-1}_\infty(D)$, $j \in \mathbb{N}$. Denote

$$b_{1,j} := \|\psi_j\|_{L^\infty}, \quad b_{2,j} := \max\left\{\|\psi_j\|_{W^1_\infty}, \|\psi_j\|_{\mathcal{W}^{s-1}_\infty}\right\} \tag{7.42}$$

and $\boldsymbol{b}_1 := (b_{1,j})_{j \in \mathbb{N}}$, $\boldsymbol{b}_2 := (b_{2,j})_{j \in \mathbb{N}}$.

Example 7.11 Let $D = [0, 1]$ and $\psi_j(x) = \sin(jx)j^{-r}$ for some $r > 2$. Then $\boldsymbol{b}_1 \in \ell^{p_1}(\mathbb{N})$ for every $p_1 > \frac{1}{r}$ and $\boldsymbol{b}_2 \in \ell^{p_2}(\mathbb{N})$ for every $p_2 > \frac{1}{r-(s-1)}$.

In the next proposition we verify Assumption 7.2. This will yield validity of the multilevel convergence rates proved in Theorems 7.5 and 7.6 in the present setting as we discuss subsequently.

Proposition 7.12 *Let Assumption 7.9 be satisfied for some $s \geq 2$ and $\alpha > 0$. Let $\boldsymbol{b}_1 \in \ell^{p_1}(\mathbb{N})$, $\boldsymbol{b}_2 \in \ell^{p_2}(\mathbb{N})$ with $p_1, p_2 \in (0, 1)$.*

Then there exist $\xi > 0$ and $\delta > 0$ such that

$$u(\boldsymbol{y}) := \mathcal{U}\left(\exp\left(\sum_{j \in \mathbb{N}} y_j \psi_j\right)\right) \tag{7.43}$$

is $(\boldsymbol{b}_1, \xi, \delta, H_0^1)$-holomorphic, and for every $l \in \mathbb{N}$

(i) $u^l(\boldsymbol{y}) := \mathcal{U}^l(\exp(\sum_{j \in \mathbb{N}} y_j \psi_j))$ is $(\boldsymbol{b}_1, \xi, \delta, H_0^1)$-holomorphic,
(ii) $u - u^l$ is $(\boldsymbol{b}_1, \xi, \delta, H_0^1)$-holomorphic,
(iii) $u - u^l$ is $(\boldsymbol{b}_2, \xi, \delta \mathfrak{w}_l^{-\alpha}, H_0^1)$-holomorphic.

Proof

Step 1. We show (i) and (ii). The argument to show that u^l is $(\boldsymbol{b}_1, \xi, \delta, H_0^1)$-holomorphic (for some constants $\xi > 0$, $\delta > 0$ independent of l) is essentially the same as in Sect. 4.3.1.

We wish to apply Theorem 4.11 with $E = L^\infty(D)$ and $X = H_0^1$. To this end let

$$O_1 = \{a \in L^\infty(D; \mathbb{C}) : \rho(a) > 0\} \subset L^\infty(D; \mathbb{C}).$$

By assumption, $b_{1,j} = \|\psi_j\|_{L^\infty}$ satisfies $\boldsymbol{b}_1 = (b_{1,j})_{j\in\mathbb{N}} \in \ell^{p_1}(\mathbb{N}) \subseteq \ell^1(\mathbb{N})$, which corresponds to assumption (iv) of Theorem 4.11. It remains to verify assumptions (i), (ii) and (iii) of Theorem 4.11:

(i) $\mathcal{U}^l : O_1 \to H_0^1$ is holomorphic: This is satisfied because the operation of inversion of linear operators is holomorphic on the set of boundedly invertible linear operators. Denote by $A_l : X_l \to X_l'$ the differential operator

$$A_l u = -\operatorname{div}(a\nabla u) \in X_l'$$

via

$$\langle A_l u, v\rangle = \int_D a\nabla u^\top \overline{\nabla v}\, \mathrm{d}\boldsymbol{x} \quad \forall v \in X_l.$$

Observe that A_l depends boundedly and linearly (thus holomorphically) on a, and therefore, the map $a \mapsto A_l(a)^{-1} f = \mathcal{U}^l(a)$ is a composition of holomorphic functions. We refer once more to [111, Example 1.2.38] for more details.

(ii) It holds for all $a \in O$

$$\|\mathcal{U}^l(a)\|_{H_0^1} \le \frac{\|f\|_{X_l'}}{\rho(a)} \le \frac{\|f\|_{H^{-1}}}{\rho(a)}.$$

The first inequality follows by the same calculation as (4.20) (but with X replaced by X_l), and the second inequality follows by the definition of the dual norm, viz

$$\|f\|_{X_l'} = \sup_{0\neq v\in X_l} \frac{|\langle f, v\rangle|}{\|v\|_{H_0^1}} \le \sup_{0\neq v\in H_0^1} \frac{|\langle f, v\rangle|}{\|v\|_{H_0^1}} = \|f\|_{H^{-1}}.$$

(iii) For all $a, b \in O$ we have

$$\|\mathcal{U}^l(a) - \mathcal{U}^l(b)\|_{H_0^1} \le \|f\|_{H^{-1}} \frac{1}{\min\{\rho(a), \rho(b)\}^2} \|a - b\|_{L^\infty},$$

which follows again by the same calculation as in the proof of (4.21).

According to Theorem 4.11 the map

$$\mathcal{U}^l \in L^2(U, X_l; \gamma) \subseteq L^2(U, H_0^1; \gamma)$$

is $(\boldsymbol{b}_1, \xi_1, \tilde{C}_1, H_0^1)$-holomorphic, for some fixed constants $\xi_1 > 0$ and $\tilde{C}_1 > 0$ depending on O_1 but independent of l. In fact the argument also works with H_0^1 instead of X_l, i.e. also u is $(\boldsymbol{b}_1, \xi_1, \tilde{C}_1, H_0^1)$-holomorphic (with the same constants ξ_1 and $\tilde{C}_1$).

Finally, it follows directly from the definition that the difference $u - u^l$ is $(\boldsymbol{b}_1, \xi, 2\delta, H_0^1)$-holomorphic.

Step 2. To show (iii), we set

$$O_2 = \{a \in W^1_\infty(D) \cap \mathcal{W}^{s-1}_\infty(D) : \rho(a) > 0\},$$

and verify again assumptions (i), (ii) and (iii) of Theorem 4.11, but now with "E" in this lemma being $W^1_\infty(D) \cap \mathcal{W}^{s-1}_\infty(D)$. First, observe that with

$$b_{2,j} := \max\left\{\|\psi_j\|_{\mathcal{W}^{s-1}_\infty}, \|\psi_j\|_{W^1_\infty}\right\},$$

by assumption

$$\boldsymbol{b}_2 = (b_{2,j})_{j\in\mathbb{N}} \in \ell^{p_2}(\mathbb{N}) \hookrightarrow \ell^1(\mathbb{N})$$

which corresponds to the assumption (iv) of Theorem 4.11.

For every $l \in \mathbb{N}$:

(i) $\mathcal{U} - \mathcal{U}^l : O_2 \to H_0^1(D)$ is holomorphic: Since O_2 can be considered a subset of O_1 (and O_2 is equipped with a stronger topology than O_1), Fréchet differentiability follows by Fréchet differentiability of

$$\mathcal{U} - \mathcal{U}^l : O_1 \to H_0^1(D),$$

which holds by Step 1.

(ii) For every $a \in O_2$

$$\|(\mathcal{U} - \mathcal{U}^l)(a)\|_{H_0^1} \le \underbrace{\mathfrak{w}_l^{-\alpha} C C_s \|f\|_{\mathcal{K}^{s-2}_{\varkappa-1}}}_{=:\delta_l} \frac{(\|a\|_{W^1_\infty} + \|a\|_{\mathcal{W}^{s-1}_\infty})^{N_s+1}}{\rho(a)^{N_s+2}}$$

by Lemma 7.10.

(iii) For every $a, b \in O_2 \subseteq O_1$, by Step 1 and (4.21),

$$\begin{aligned}
&\|(\mathcal{U} - \mathcal{U}^l)(a) - (\mathcal{U} - \mathcal{U}^l)(b)\|_{H_0^1} \\
&\le \|\mathcal{U}(a) - \mathcal{U}(b)\|_{H_0^1} + \|\mathcal{U}^l(a) - \mathcal{U}^l(b)\|_{H_0^1} \\
&\le \|f\|_{H^{-1}} \frac{2}{\min\{\rho(a), \rho(b)\}^2} \|a - b\|_{L^\infty}.
\end{aligned}$$

We conclude with Theorem 4.11 that there exist ξ_2 and $\tilde{C}_2$ depending on O_2, D but independent of l such that $u - u^l$ is $(\boldsymbol{b}_2, \xi_2, \tilde{C}_2\delta_l, H_0^1)$-holomorphic.

In all, the proposition holds with

$$\xi := \min\{\xi_1, \xi_2\} \qquad \text{and} \qquad \delta := \max\{\tilde{C}_1, \tilde{C}_2 C C_s \|f\|_{\mathcal{K}^{s-2}_{\varkappa-1}}\}. \qquad \square$$

Items (ii) and (iii) of Proposition 7.12 show that Assumption 7.9 implies validity of Assumption 7.2. This in turn allows us to apply Theorems 7.5 and 7.6. Specifically, assuming the optimal convergence rate $\alpha = \frac{s-1}{2}$ in (7.41), we obtain that for u in (7.43) and every $n \in \mathbb{N}$ there is $\varepsilon := \varepsilon_n > 0$ such that $\text{work}(\mathbf{l}_\varepsilon) \leq n$ and the multilevel interpolant $\mathbf{I}^{\text{ML}}_{\mathbf{l}}$ defined in (7.3) satisfies

$$\|u - \mathbf{I}^{\text{ML}}_{\mathbf{l}_\varepsilon} u\|_{L^2(U, H_0^1; \gamma)} \leq C(1 + \log n) n^{-R_I},$$

$$R_I = \min\left\{\frac{s-1}{2}, \frac{\frac{s-1}{2}(\frac{1}{p_1} - \frac{3}{2})}{\frac{s-1}{2} + \frac{1}{p_1} - \frac{1}{p_2}}\right\},$$

and the multilevel quadrature operator $\mathbf{Q}^{\text{ML}}_{\mathbf{l}}$ defined in (7.4) satisfies

$$\left\| \int_U u(y)\,\mathrm{d}\gamma(y) - \mathbf{Q}^{\text{ML}}_{\mathbf{l}_\varepsilon} u \right\|_{H_0^1} \leq C(1 + \log n) n^{-R_Q},$$

$$R_Q = \min\left\{\frac{s-1}{2}, \frac{\frac{s-1}{2}(\frac{2}{p_1} - \frac{5}{2})}{\frac{s-1}{2} + \frac{2}{p_1} - \frac{2}{p_2}}\right\}.$$

Let us consider these convergence rates in the case where the ψ_j are algebraically decreasing, with this decrease encoded by some $r > 1$: if for fixed but arbitrarily small $\varepsilon > 0$ holds $\|\psi_j\|_{L^\infty} \sim j^{-r-\varepsilon}$, and we assume (cp. Ex. 7.11)

$$\max\left\{\|\psi_j\|_{W^1_\infty}, \|\psi_j\|_{\mathcal{W}^{s-1}_\infty}\right\} \sim j^{-r+(s-1)-\varepsilon},$$

then setting $s := r$ we can choose $p_1 = \frac{1}{r}$ and $p_2 = 1$. Inserting those numbers, the convergence rates become

$$R_I = \min\left\{\frac{r-1}{2}, \frac{\frac{r-1}{2}(r - \frac{3}{2})}{\frac{r-1}{2} + r - 1}\right\} = \frac{r}{3} - \frac{1}{2} \quad \text{and}$$

$$R_Q = \min\left\{\frac{r-1}{2}, \frac{\frac{r-1}{2}(2r - \frac{5}{2})}{\frac{r-1}{2} + 2r - 2}\right\} = \frac{2r}{5} - \frac{1}{2}.$$

7.6.2 Parametric Holomorphy of the Posterior Density in Bayesian PDE Inversion

Throughout this section we assume that $D \subseteq \mathbb{R}^2$ is a polygonal Lipschitz domain and that $f \in \mathcal{K}^{s-2}_{\varkappa-1}(D)$ with $\varkappa$ as in Theorem 7.8.

As in Chap. 5, to treat the approximation of the (unnormalized) posterior density or its integral, we need an upper bound on $\|u(\boldsymbol{y})\|_{H_0^1}$ for all $\boldsymbol{y}$. This is achieved by considering (7.29) with diffusion coefficient $a_0 + a$ where

$$\rho(a_0) := \operatorname*{ess\,inf}_{x \in D} \Re(a_0) > 0.$$

The shift of the diffusion coefficient by a_0 ensures uniform ellipticity for all

$$a \in \{a \in L^\infty(D, \mathbb{C}) \,:\, \rho(a) \geq 0\}.$$

As a consequence, solutions $\mathcal{U}(a_0 + a) \in X = H_0^1(D; \mathbb{C}) =: H_0^1$ of (7.29) satisfy the apriori bound (cp. (4.20))

$$\|\mathcal{U}(a_0 + a)\|_{H_0^1} \leq \frac{\|f\|_{H^{-1}}}{\rho(a_0)}.$$

As before, for a sequence of subspaces $(X_l)_{l \in \mathbb{N}}$ of $H_0^1(D, \mathbb{C})$, for $a \in O$ we denote by $\mathcal{U}^l(a) \in X_l$ the finite element approximation to $\mathcal{U}(a)$. By the same calculation as for $\mathcal{U}$ it also holds

$$\|\mathcal{U}^l(a_0 + a)\|_{H_0^1} \leq \frac{\|f\|_{H^{-1}}}{\rho(a_0)}$$

independent of l.

Assuming that $b_j = \|\psi_j\|_{L^\infty}$ satisfies $(b_j)_{j \in \mathbb{N}} \in \ell^1(\mathbb{N})$, the function $u(\boldsymbol{y}) = \mathcal{U}(a_0 + a(\boldsymbol{y}))$ with

$$a(\boldsymbol{y}) = \exp\left(\sum_{j \in \mathbb{N}} y_j \psi_j\right),$$

is well-defined. For a fixed *observation* $\boldsymbol{\mathfrak{d}} \in \mathbb{R}^m$ consider again the (unnormalized) posterior density given in (5.4),

$$\tilde{\pi}(\boldsymbol{y}|\boldsymbol{\mathfrak{d}}) := \exp\left(-(\boldsymbol{\mathfrak{d}} - \mathcal{O}(u(\boldsymbol{y})))^\top \boldsymbol{\Gamma}^{-1}(\boldsymbol{\mathfrak{d}} - \mathcal{O}(u(\boldsymbol{y})))\right).$$

Recall that $\mathcal{O} : X \to \mathbb{C}^m$ (the observation operator) is assumed to be a bounded linear map, and $\boldsymbol{\Gamma} \in \mathbb{R}^{m \times m}$ (the noise covariance matrix) is symmetric positive definite. For $l \in \mathbb{N}$ (tagging discretization level of the PDE), and with $u^l(\boldsymbol{y}) =$

$\mathcal{U}^l(a_0 + a(\mathbf{y}))$, we introduce approximations

$$\tilde{\pi}^l(\mathbf{y}|\mathfrak{d}) := \exp\Big(-(\mathfrak{d} - \mathcal{O}(u^l(\mathbf{y})))^\top \boldsymbol{\Gamma}^{-1}(\mathfrak{d} - \mathcal{O}(u^l(\mathbf{y})))\Big)$$

to $\tilde{\pi}(\mathbf{y}|\mathfrak{d})$. In the following we show the analog of Proposition 7.12, that is we show validity of the assumptions required for the multilevel convergence results.

Lemma 7.13 *Let $\mathcal{O} : H_0^1(D;\mathbb{C}) \to \mathbb{C}^m$ be a bounded linear operator, $\mathfrak{d} \in \mathbb{C}^m$ and $\boldsymbol{\Gamma} \in \mathbb{R}^{m\times m}$ symmetric positive definite. Set*

$$\Phi := \begin{cases} H_0^1(D;\mathbb{C}) \to \mathbb{C} \\ u \mapsto \exp(-(\mathfrak{d} - \mathcal{O}(u))^\top \boldsymbol{\Gamma}^{-1}(\mathfrak{d} - \mathcal{O}(u))). \end{cases}$$

Then the function Φ is continuously differentiable and for every $r > 0$ has a Lipschitz constant K solely depending on $\|\boldsymbol{\Gamma}^{-1}\|$, $\|\mathcal{O}\|_{L(H_0^1;\mathbb{C}^m)}$, $\|\mathfrak{d}\|$ and r, on the set

$$\{u \in H_0^1(D;\mathbb{C}) \, : \, \|u\|_{H_0^1} < r\}.$$

Proof The function Φ is continuously differentiable as a composition of continuously differentiable functions. Hence for u, v with $w := u - v$ and with the derivative $D\Phi : H_0^1 \to L(H_0^1;\mathbb{C})$ of Φ,

$$\Phi(u) - \Phi(v) = \int_0^1 D\Phi(v + tw)w \, \mathrm{d}t. \tag{7.45}$$

Due to the symmetry of $\boldsymbol{\Gamma}$ it holds

$$\begin{aligned} D\Phi(u + tw)w &= 2\mathcal{O}(w)^\top \boldsymbol{\Gamma}^{-1}(\mathfrak{d} - \mathcal{O}(u + tw)) \\ &\quad \exp\Big(-(\mathfrak{d} - \mathcal{O}(u + tw))^\top \boldsymbol{\Gamma}^{-1}(\mathfrak{d} - \mathcal{O}(u + tw))\Big). \end{aligned}$$

If $\|u\|_{H_0^1}, \|v\|_{H_0^1} < r$ then also $\|u + tw\|_{H_0^1} < r$ for all $t \in [0, 1]$ and we can bound

$$|D\Phi(u + tw)w| \le K\|w\|_{H_0^1},$$

where

$$\begin{aligned} K := 2\|\mathcal{O}\|_{L(H_0^1;\mathbb{C}^m)}\|\boldsymbol{\Gamma}^{-1}\|\Big(\|\mathfrak{d}\| + r\|\mathcal{O}\|_{L(H_0^1;\mathbb{C}^m)}\Big) \\ \exp\Big(\|\boldsymbol{\Gamma}^{-1}\|\Big(\|\mathfrak{d}\| + \|\mathcal{O}\|_{L(H_0^1;\mathbb{C}^m)}r\Big)^2\Big). \end{aligned} \tag{7.46}$$

The statement follows by (7.45). □

Remark 7.14 The reason why we require the additional positive a_0 term in (7.44), is to guarantee boundedness of the solution $\mathcal{U}(a)$ and Lipschitz continuity of Φ.

Proposition 7.15 *Let Assumption 7.9 be satisfied for some $s \geq 2$ and $\alpha > 0$. Let a_0, $(\psi_j)_{j\in\mathbb{N}} \subseteq W^1_\infty(D) \cap \mathcal{W}^{s-1}_\infty(D)$ and $\boldsymbol{b}_1 \in \ell^{p_1}(\mathbb{N})$, $\boldsymbol{b}_2 \in \ell^{p_2}(\mathbb{N})$ with $p_1, p_2 \in (0,1)$ (see (7.42) for the definition of $\boldsymbol{b}_1, \boldsymbol{b}_2$). Fix $\mathfrak{d} \in \mathbb{C}^m$.*

Then there exist $\xi > 0$ and $\delta > 0$ such that $\tilde{\pi}(\boldsymbol{y}|\mathfrak{d})$ is $(\boldsymbol{b}_1, \xi, \delta, \mathbb{C})$-holomorphic, and for every $l \in \mathbb{N}$

(i) $\tilde{\pi}^l(\boldsymbol{y}|\mathfrak{d})$ is $(\boldsymbol{b}_1, \xi, \delta, \mathbb{C})$-holomorphic,
(ii) $\tilde{\pi}(\boldsymbol{y}|\mathfrak{d}) - \tilde{\pi}^l(\boldsymbol{y}|\mathfrak{d})$ is $(\boldsymbol{b}_1, \xi, \delta, H_0^1)$-holomorphic,
(iii) $\tilde{\pi}(\boldsymbol{y}|\mathfrak{d}) - \tilde{\pi}^l(\boldsymbol{y}|\mathfrak{d})$ is $(\boldsymbol{b}_2, \xi, \delta\mathfrak{w}_l^{-\alpha}, H_0^1)$-holomorphic.

Proof

Step 1. We show (i) and (ii). Set

$$O_1 := \{a \in L^\infty(D;\mathbb{C}) \,:\, \rho(a) > 0\}.$$

By (7.44) for all $a \in O_1$ and all $l \in \mathbb{N}$ with $r := \frac{\|f\|_{H^{-1}}}{\rho(a_0)}$

$$\|\mathcal{U}^l(a_0 + a)\|_{H_0^1} \leq r \qquad \text{and} \qquad \|\mathcal{U}(a_0 + a)\|_{H_0^1} \leq r. \tag{7.47}$$

As in Step 1 of the proof of Proposition 7.12, one can show that $u(\boldsymbol{y}) = \mathcal{U}(a_0 + a(\boldsymbol{y}))$ and $u^l(\boldsymbol{y}) = \mathcal{U}(a_0 + a(\boldsymbol{y}))$ where $a(\boldsymbol{y}) = \exp(\sum_{j\in\mathbb{N}} y_j\psi_j)$ are $(\boldsymbol{b}_1, \xi_1, \tilde{C}_1, H_0^1)$-holomorphic for certain $\xi_1 > 0$ and $\tilde{C}_1 > 0$ (the only difference to Proposition 7.12 is the affine offset a_0 in (7.29), which ensures a positive lower bound for $a + a_0$). In the following Φ is as in Lemma 7.13 and $T_{a_0}(a) := a_0 + a$ so that

$$\tilde{\pi}(\boldsymbol{y}|\mathfrak{d}) = \Phi(\mathcal{U}^l(T_{a_0}(a(\boldsymbol{y})))). \tag{7.48}$$

With $b_{1,j} = \|\psi_j\|_{L^\infty}$, by the assumption

$$\boldsymbol{b}_1 = (b_{1,j})_{j\in\mathbb{N}} \in \ell^{p_1}(\mathbb{N}) \hookrightarrow \ell^1(\mathbb{N})$$

which corresponds to assumption (iv) of Theorem 4.11. We now verify assumptions (i), (ii) and (iii) of Theorem 4.11 for (7.48).
For every $l \in \mathbb{N}$:

(i) The map

$$\Phi \circ \mathcal{U}^l \circ T_{a_0} : \begin{cases} O_1 \to \mathbb{C} \\ a \mapsto \Phi(\mathcal{U}(T_{a_0}(a))) \end{cases}$$

is holomorphic as a composition of holomorphic functions.

(ii) for all $a \in O_1$, since $\|\mathcal{U}^l(T_{a_0}(a))\|_{H_0^1} \leq r$

$$|\Phi(\mathcal{U}^l(T_{a_0}(a)))| \leq \exp((\|\mathfrak{d}\| + \|\mathcal{O}\|_{L(H_0^1(D;\mathbb{C});\mathbb{C}^m)} r)^2 \|\boldsymbol{\Gamma}^{-1}\|)$$

and thus assumption (ii) of Theorem 4.11 is trivially satisfied for some $\delta > 0$ independent of l,

(iii) for all $a, b \in O_1$ by Lemma 7.13 and the same calculation as in (4.21)

$$\begin{aligned} |\Phi(\mathcal{U}^l(T_{a_0}(a))) - \Phi(\mathcal{U}^l(T_{a_0}(b)))| &\leq K \|\mathcal{U}^l(T_{a_0}(a)) - \mathcal{U}^l(T_{a_0}(b))\|_{H_0^1} \\ &\leq K \frac{\|f\|_{H^{-1}}}{\rho(a_0)} \|a - b\|_{L^\infty}, \end{aligned} \tag{7.49}$$

where K is the constant given as in (7.46).

Now we can apply Theorem 4.11 to conclude that there exist ξ_1, δ_1 (independent of l) such that $\tilde{\pi}^l(\cdot|\mathfrak{d})$ is $(\boldsymbol{b}_1, \xi_1, \delta_1, H_0^1)$-holomorphic for every $l \in \mathbb{N}$. Similarly one shows that $\tilde{\pi}(\cdot|\mathfrak{d})$ is $(\boldsymbol{b}_1, \xi_1, \delta_1, H_0^1)$-holomorphic, and in particular $\tilde{\pi}(\cdot|\mathfrak{d}) - \tilde{\pi}^l(\cdot|\mathfrak{d})$ is $(\boldsymbol{b}_1, \xi_1, 2\delta_1, H_0^1)$-holomorphic.

Step 2. Set

$$O_2 = \{a \in W_\infty^1(D) \cap \mathcal{W}_\infty^{s-1}(D) \ : \ \rho(a) > 0\}.$$

We verify once more assumptions (i), (ii) and (iii) of Theorem 4.11 with "E" in this lemma being $W_\infty^1(D) \cap \mathcal{W}_\infty^{s-1}(D)$. With $b_{2,j} = \max\{\|\psi_j\|_{W_\infty^1}, \|\psi_j\|_{\mathcal{W}_\infty^{s-1}}\}$, by the assumption

$$\boldsymbol{b}_2 = (b_{2,j})_{j \in \mathbb{N}} \in \ell^{p_2}(\mathbb{N}) \hookrightarrow \ell^1(\mathbb{N}),$$

which corresponds to assumption (iv) of Theorem 4.11.
We will apply Theorem 4.11 with the function

$$\tilde{\pi}(\boldsymbol{y}|\mathfrak{d}) - \tilde{\pi}^l(\boldsymbol{y}|\mathfrak{d}) = \Phi(\mathcal{U}(T_{a_0}(a(\boldsymbol{y})))) - \Phi(\mathcal{U}^l(T_{a_0}(a(\boldsymbol{y})))). \tag{7.50}$$

For every $l \in \mathbb{N}$:

(i) By item (i) in Step 1 (and because $O_2 \subseteq O_1$)

$$\Phi \circ \mathcal{U} \circ T_{a_0} - \Phi \circ \mathcal{U}^l \circ T_{a_0} : \begin{cases} O_2 \to \mathbb{C} \\ a \mapsto \Phi(\mathcal{U}(T_{a_0}(a))) - \Phi(\mathcal{U}^l(T_{a_0}(a))) \end{cases}$$

is holomorphic,

(ii) for every $a \in O_2$, by Lemma 7.10

$$\begin{aligned}&\|\mathcal{U}(T_{a_0}(a)) - \mathcal{U}^l(T_{a_0}(a))\|_{H_0^1}\\&\le \mathfrak{w}_l^{-\alpha} C C_s \frac{(\|a_0 + a\|_{W^1_\infty} + \|a_0 + a\|_{\mathcal{W}^{s-1}_\infty})^{N_s+1}}{\rho(a_0 + a)^{N_s+2}} \|f\|_{\mathcal{K}^{s-2}_{\varkappa-1}}.\end{aligned}$$

Thus by (7.47) and Lemma 7.13

$$\begin{aligned}&|\Phi(\mathcal{U}(T_{a_0}(a))) - \Phi(\mathcal{U}^l(T_{a_0}(a)))|\\&\le \mathfrak{w}_l^{-\alpha} K C C_s \frac{(\|a_0 + a\|_{W^1_\infty} + \|a_0 + a\|_{\mathcal{W}^{s-1}_\infty})^{N_s+1}}{\rho(a_0 + a)^{N_s+2}} \|f\|_{\mathcal{K}^{s-2}_{\varkappa-1}},\end{aligned}$$

(iii) for all $a, b \in O_2 \subseteq O_1$ by (7.49) (which also holds for $\mathcal{U}^l$ replaced by $\mathcal{U}$):

$$\begin{aligned}&|\Phi(\mathcal{U}(T_{a_0}(a))) - \Phi(\mathcal{U}^l(T_{a_0}(a))) - (\Phi(\mathcal{U}(b)) - \Phi(\mathcal{U}^l(b)))|\\&\le 2K \frac{\|f\|_{H^{-1}}}{\rho(a_0)} \|a - b\|_{L^\infty}.\end{aligned}$$

By Theorem 4.11 and (7.50) we conclude that there exists $\delta > 0$ and ξ_2 independent of l such that $\tilde{\pi}(\boldsymbol{y}|\mathfrak{d}) - \tilde{\pi}^l(\boldsymbol{y}|\mathfrak{d})$ is $(\boldsymbol{b}_2, \delta \mathfrak{w}_l^{-\alpha}, \xi_2, H_0^1)$-holomorphic. □

Items (ii) and (iii) of Proposition 7.12 show that Assumption 7.9 implies validity of Assumption 7.2. This in turn allows us to apply Theorems 7.5 and 7.6. Specifically, assuming the optimal convergence rate (7.41), we obtain that for every $n \in \mathbb{N}$ there is $\varepsilon := \varepsilon_n > 0$ such that $\mathrm{work}(\mathbf{l}_\varepsilon) \le n$ and the multilevel interpolant $\mathbf{I}^{\mathrm{ML}}_{\mathbf{l}}$ defined in (7.3) satisfies

$$\|\tilde{\pi}(\cdot|\mathfrak{d}) - \mathbf{I}^{\mathrm{ML}}_{\mathbf{l}_\varepsilon} \tilde{\pi}(\cdot|\mathfrak{d})\|_{L^2(U, H_0^1; \gamma)} \le C(1 + \log n) n^{-R_I},$$

$$R_I = \min\left\{\frac{s-1}{2}, \frac{\frac{s-1}{2}(\frac{1}{p_1} - \frac{3}{2})}{\frac{s-1}{2} + \frac{1}{p_1} - \frac{1}{p_2}}\right\}.$$

Of higher practical interest is the application of the multilevel quadrature operator $\mathbf{Q}^{\mathrm{ML}}$ defined in (7.4). In case the prior is chosen as γ, then

$$\int_U \tilde{\pi}(\boldsymbol{y}|\mathfrak{d})\, \mathrm{d}\gamma(\boldsymbol{y})$$

equals the normalization constant in (5.2). It can be approximated with the error converging like

$$\left| \int_U \tilde{\pi}(\boldsymbol{y}|\mathfrak{d})\,\mathrm{d}\gamma(\boldsymbol{y}) - \mathbf{Q}^{\mathrm{ML}}_{\mathbf{l}_\varepsilon} u \right| \le C(1+\log n) n^{-R_Q},$$

$$R_Q := \min\left\{ \frac{s-1}{2}, \frac{\frac{s-1}{2}(\frac{2}{p_1}-\frac{5}{2})}{\frac{s-1}{2}+\frac{2}{p_1}-\frac{2}{p_2}} \right\}. \tag{7.51}$$

Typically, one is not merely interested in the normalization constant

$$Z = \int_U \tilde{\pi}(\boldsymbol{y}|\mathfrak{d})\,\mathrm{d}\gamma(\boldsymbol{y}),$$

but for example also in an estimate of the jth parameter y_j given as the conditional expectation, which up to multiplying with the normalization constant $\frac{1}{Z}$, corresponds to

$$\int_U y_j \tilde{\pi}(\boldsymbol{y}|\mathfrak{d})\,\mathrm{d}\gamma(\boldsymbol{y}).$$

Since $\boldsymbol{y} \mapsto y_j$ is analytic, one can show the same convergence rate as in (7.51) for the multilevel quadrature applied with the approximations $\boldsymbol{y} \mapsto y_j \tilde{\pi}^l(\boldsymbol{y}|\mathfrak{d})$ for $l \in \mathbb{N}$. Moreover, for example if $\phi : H_0^1(D;\mathbb{C}) \to \mathbb{C}$ is a bounded linear functional representing some quantity of interest, then we can show the same error convergence for the approximation of

$$\int_U \phi(u(\boldsymbol{y}))\tilde{\pi}(\boldsymbol{y}|\mathfrak{d})\,\mathrm{d}\gamma(\boldsymbol{y})$$

with the multilevel quadrature applied with the approximations $\phi(u^l(\boldsymbol{y}))\tilde{\pi}^l(\boldsymbol{y}|\mathfrak{d})\,\mathrm{d}\gamma(\boldsymbol{y})$ to the integrand for $l \in \mathbb{N}$.

7.7 Linear Multilevel Interpolation and Quadrature Approximation

In this section, we briefly recall some results from [43] (see also [45] for some corrections). The difference with Sects. 7.1–7.6 is, that the interpolation and quadrature operators presented in this section are *linear* operators; in contrast, the operators $\mathbf{I}^{\mathrm{ML}}$, $\mathbf{Q}^{\mathrm{ML}}$ in (7.3), (7.4) are in general nonlinear, since they build on the approximations u^n of u from Assumption 7.2. These approximations are not assumed to be linear (and, in general, are not linear) in u.

In this section we proceed similarly, but with $u^n := P_n u$ *for a linear operator* P_n; if u denotes the solution of an elliptic PDE in $H^1(D)$, P_n could for instance be the orthogonal projection from $H^1(D)$ into some fixed finite dimensional subspace. We emphasize, that such operators are not available in practice, and many widely used implementable algorithms (such as the finite element method, boundary element method, finite differences) realize projections that are *not of this type.* We will discuss this in more detail in Remark 7.31. Therefore the present results are mainly of theoretical rather than of practical importance. On a positive note, the convergence rates for both, Smolyak sparse-grid interpolation and quadrature obtained in this section via thresholding (see (7.54) ahead) improve the rates shown in the previous sections for the discretization levels allocated via Algorithm 2 by a logarithmic factor, cp. Theorems 7.5 and 7.6. Yet we emphasize that the latter are computable (in linear complexity, see Sec. 6.2.4).

7.7.1 Multilevel Smolyak Sparse-Grid Interpolation

In this section, we recall some results in [43] (see also, [45] for some corrections) on linear multilevel polynomial interpolation approximation in Bochner spaces.

In order to have a correct definition of interpolation operator let us impose some necessary restrictions on $v \in L^2(U, X; \gamma)$. Let $\mathcal{E}$ be a γ-measurable subset in U such that $\gamma(\mathcal{E}) = 1$ and $\mathcal{E}$ contains all $\boldsymbol{y} \in U$ with $|\boldsymbol{y}|_0 < \infty$, where $|\boldsymbol{y}|_0$ denotes the number of nonzero components y_j of $\boldsymbol{y}$. For a given $\mathcal{E}$ and separable Hilbert space X, let $C_{\mathcal{E}}(U)$ the set of all functions v on U taking values in X such that v are continuous on $\mathcal{E}$ w.r. to the local convex topology of $U := \mathbb{R}^\infty$ (see Example 2.5). We define $L^2_{\mathcal{E}}(U, X, \gamma) := L^2(U, X; \gamma) \cap C_{\mathcal{E}}(U)$. We will treat all elements $v \in L^2_{\mathcal{E}}(U, X, \gamma)$ as their representative belonging to $C_{\mathcal{E}}(U)$. Throughout this and next sections, we fix a set $\mathcal{E}$.

We define the univariate operator Δ^{I}_m for $m \in \mathbb{N}_0$ by

$$\Delta^{\mathrm{I}}_m := I_m - I_{m-1},$$

with the convention $I_{-1} = 0$, where I_m is defined in Sect. 6.1.1.

For $v \in L^2_{\mathcal{E}}(U, X; \gamma)$, we introduce the tensor product operator $\Delta^{\mathrm{I}}_{\boldsymbol{\nu}}$ for $\boldsymbol{\nu} \in \mathcal{F}$ by

$$\Delta^{\mathrm{I}}_{\boldsymbol{\nu}}(v) := \bigotimes_{j\in\mathbb{N}} \Delta^{\mathrm{I}}_{\nu_j}(v),$$

where the univariate operator $\Delta^{\mathrm{I}}_{\nu_j}$ is applied to the univariate function $\bigotimes_{j'=1}^{j-1} \Delta^{\mathrm{I}}_{\nu_{j'}}(v)$ by considering this function as a function of variable y_j with all remaining variables held fixed. From the definition of $L^2_{\mathcal{E}}(U, X; \gamma)$ one infers that the operators $\Delta^{\mathrm{I}}_{\boldsymbol{\nu}}$ are well-defined for all $\boldsymbol{\nu} \in \mathcal{F}$.

Let us recall a setting from [43] of linear fully discrete polynomial interpolation of functions in the Bochner space $L^2(U, X^2; \gamma)$, with the approximation error measured by the norm of the Bochner space $L^2(U, X^1; \gamma)$ for separable Hilbert spaces X^1 and X^2. To construct linear fully discrete methods of polynomial interpolation, besides weighted ℓ^2-summabilities with respect to X^1 and X^2 we need an approximation property on the spaces X^1 and X^2 combined in the following assumption.

Assumption 7.16 *For the Hilbert spaces X^1 and X^2 and $v \in L^2_{\mathcal{E}}(U, X^2; \gamma)$ represented by the series*

$$v = \sum_{\boldsymbol{\nu}\in\mathcal{F}} v_{\boldsymbol{\nu}} H_{\boldsymbol{\nu}}, \quad v_{\boldsymbol{\nu}} \in X^2, \tag{7.52}$$

there holds the following.

(i) *X^2 is a linear subspace of X^1 and $\|\cdot\|_{X^1} \le C\,\|\cdot\|_{X^2}$.*
(ii) *For $i = 1, 2$, there exist numbers q_i with $0 < q_1 \le q_2 < \infty$ and $q_1 < 2$, and families $(\sigma_{i;\boldsymbol{\nu}})_{\boldsymbol{\nu}\in\mathcal{F}}$ of numbers strictly larger than 1 such that $\sigma_{i;\boldsymbol{e}_{j'}} \le \sigma_{i;\boldsymbol{e}_j}$ if $j' < j$, and*

$$\sum_{\boldsymbol{\nu}\in\mathcal{F}} (\sigma_{i;\boldsymbol{\nu}} \|v_{\boldsymbol{\nu}}\|_{X^i})^2 \le M_i < \infty \quad \textit{and} \quad \left(p_{\boldsymbol{\nu}}(\tau,\lambda)\sigma_{i;\boldsymbol{\nu}}^{-1}\right)_{\boldsymbol{\nu}\in\mathcal{F}} \in \ell^{q_i}(\mathcal{F})$$

for every $\tau > \frac{17}{6}$ and $\lambda \ge 0$, where we recall that $(\boldsymbol{e}_j)_{j\in\mathbb{N}}$ is the standard basis of $\ell^2(\mathbb{N})$.
(iii) *There are a sequence $(V_n)_{n\in\mathbb{N}_0}$ of subspaces $V_n \subset X^1$ of dimension $\le n$, and a sequence $(P_n)_{n\in\mathbb{N}_0}$ of linear operators from X^1 into V_n, and a number $\alpha > 0$ such that*

$$\|P_n(v)\|_{X^1} \le C\|v\|_{X^1},$$
$$\|v - P_n(v)\|_{X^1} \le Cn^{-\alpha}\|v\|_{X^2}, \quad \forall n \in \mathbb{N}_0, \quad \forall v \in X^2. \tag{7.53}$$

Let Assumption 7.16 hold for Hilbert spaces X^1 and X^2 and $v \in L^2_{\mathcal{E}}(U, X^2; \gamma)$. Then we are able to construct a linear fully discrete polynomial interpolation approximation. We introduce the interpolation operator

$$\mathcal{I}_G : L^2_{\mathcal{E}}(U, X^2; \gamma) \to \mathcal{V}(G)$$

for a given finite set $G \subset \mathbb{N}_0 \times \mathcal{F}$ by

$$\mathcal{I}_G v := \sum_{(k,\boldsymbol{\nu})\in G} \delta_k \Delta^{\mathrm{I}}_{\boldsymbol{\nu}}(v),$$

where $\mathcal{V}(G)$ denotes the subspace in $L^2(U, X^1; \gamma)$ of all functions v of the form

$$v = \sum_{(k,\boldsymbol{\nu})\in G} v_k H_{\boldsymbol{\nu}}, \quad v_k \in V_{2^k}.$$

Notice that interpolation $v \mapsto \mathcal{I}_G v$ is a linear method of fully discrete polynomial interpolation approximation, which is the sum taken over the (finite) index set G, of mixed tensor products of dyadic scale successive differences of "spatial" approximations to v, and of successive differences of their parametric Lagrange interpolation polynomials.

Define for $\xi > 0$

$$G(\xi) := \begin{cases} \left\{(k, \boldsymbol{\nu}) \in \mathbb{N}_0 \times \mathcal{F} : 2^k \sigma_{2;\boldsymbol{\nu}}^{q_2} \le \xi\right\} & \text{if } \alpha \le 1/q_2 - 1/2; \\ \left\{(k, \boldsymbol{\nu}) \in \mathbb{N}_0 \times \mathcal{F} : \sigma_{1;\boldsymbol{\nu}}^{q_1} \le \xi,\ 2^{(\alpha+1/2)k} \sigma_{2;\boldsymbol{\nu}} \le \xi^{\vartheta}\right\} & \text{if } \alpha > 1/q_2 - 1/2, \end{cases} \tag{7.54}$$

where

$$\vartheta := \frac{1}{q_1} + \frac{1}{2\alpha}\left(\frac{1}{q_1} - \frac{1}{q_2}\right). \tag{7.55}$$

For any $\xi > 1$ we have that $G(\xi) \subset F(\xi)$ where

$$F(\xi) := \{(k, \boldsymbol{\nu}) \in \mathbb{N}_0 \times \mathcal{F} : k \le \log \xi,\ \boldsymbol{\nu} \in \Lambda(\xi)\}$$

and

$$\Lambda(\xi) := \begin{cases} \left\{\boldsymbol{\nu} \in \mathcal{F} : \sigma_{2;\boldsymbol{\nu}}^{q_2} \le \xi\right\} & \text{if } \alpha \le 1/q_2 - 1/2; \\ \left\{\boldsymbol{\nu} \in \mathcal{F} : \sigma_{1;\boldsymbol{\nu}}^{q_1} \le \xi\right\} & \text{if } \alpha > 1/q_2 - 1/2. \end{cases}$$

From [46, Lemma 3.3] it follows that

$$\bigcup_{\boldsymbol{\nu}\in\Lambda(\xi)} \operatorname{supp}(\boldsymbol{\nu}) \subset \{1, ..., \lfloor C\xi \rfloor\} \tag{7.56}$$

for some positive constant C that is independent of $\xi > 1$. Denote by $\Gamma_{\boldsymbol{\nu}}$ and $\Gamma(\Lambda)$, the set of interpolation points in the operators $\Delta_{\boldsymbol{\nu}}^{\mathrm{I}}$ and $\mathbf{I}_\Lambda$, respectively. We have that

$$\Gamma_{\boldsymbol{\nu}} = \{\boldsymbol{y}_{\boldsymbol{\nu}-\boldsymbol{e};\boldsymbol{m}} : \boldsymbol{e} \in \mathbb{E}_{\boldsymbol{\nu}};\ m_j = 0, \dots, s_j - e_j,\ j \in \mathbb{N}\},$$

and

$$\Gamma(\Lambda) = \bigcup_{\boldsymbol{\nu}\in\Lambda} \Gamma_{\boldsymbol{\nu}},$$

where $\mathbb{E}_{\boldsymbol{\nu}}$ is the subset in $\mathcal{F}$ of all $\boldsymbol{e}$ such that e_j is 1 or 0 if $\nu_j > 0$, and e_j is 0 if $\nu_j = 0$, and $\boldsymbol{y}_{\boldsymbol{\nu};\boldsymbol{m}} := (y_{\nu_j;m_j})_{j\in\mathbb{N}}$. Hence, by (7.56)

$$\Gamma(\Lambda(\xi)) \subset \mathbb{R}^{\lfloor C\xi \rfloor} \subset U,$$

and therefore, the operator $\mathcal{I}_{G(\xi)}$ is well-defined for any $v \in L^2_{\mathcal{E}}(U, X^2; \gamma)$ since v is continuous on $\mathbb{R}^{\lfloor C\xi \rfloor}$.

Theorem 7.17 *Let Assumption 7.16 hold for Hilbert spaces X^1 and X^2 and $v \in L^2_{\mathcal{E}}(U, X^2; \gamma)$. Then for each $n \in \mathbb{N}$, there exists a number ξ_n such that for the interpolation operator*

$$\mathcal{I}_{G(\xi_n)} : L^2_{\mathcal{E}}(U, X^2; \gamma) \to \mathcal{V}(G(\xi_n)),$$

we have $\dim \mathcal{V}(G(\xi_n)) \le n$ *and*

$$\|v - \mathcal{I}_{G(\xi_n)} v\|_{L^2(U,X^1;\gamma)} \le C n^{-\min(\alpha,\beta)}. \tag{7.57}$$

The rate α is as in (7.53) *and the rate β is given by*

$$\beta := \left(\frac{1}{q_1} - \frac{1}{2}\right)\frac{\alpha}{\alpha+\delta}, \quad \delta := \frac{1}{q_1} - \frac{1}{q_2}. \tag{7.58}$$

The constant C in (7.57) *is independent of v and n.*

Remark 7.18 Observe that the operator $\mathcal{I}_{G(\xi_n)}$ can be represented in the form of a multilevel Smolyak sparse-grid interpolation with k_n levels:

$$\mathcal{I}_{G(\xi_n)} = \sum_{k=0}^{k_n} \delta_k \mathbf{I}_{\Lambda_k(\xi_n)},$$

where $k_n := \lfloor \log_2 \xi_n \rfloor$, the operator $\mathbf{I}_\Lambda$ is defined as in (6.5), and for $k \in \mathbb{N}_0$ and $\xi > 1$,

$$\Lambda_k(\xi) := \begin{cases} \left\{ \boldsymbol{s} \in \mathcal{F} : \sigma_{2;\boldsymbol{s}}^{q_2} \le 2^{-k}\xi \right\} & \text{if } \alpha \le 1/q_2 - 1/2; \\ \left\{ \boldsymbol{s} \in \mathcal{F} : \sigma_{1;\boldsymbol{s}}^{q_1} \le \xi,\ \sigma_{2;\boldsymbol{s}} \le 2^{-(\alpha+1/2)k}\xi^{\vartheta} \right\} & \text{if } \alpha > 1/q_2 - 1/2. \end{cases}$$

In Theorem 7.17, the multilevel polynomial interpolation of $v \in L^2_{\mathcal{E}}(U, X^2; \gamma)$ by operators $\mathcal{I}_{G(\xi_n)}$ is a collocation method. It is based on the finite point-wise information in $\boldsymbol{y}$, more precisely, on $|\Gamma(\Lambda_0(\xi_n))| = \mathcal{O}(n)$ of particular values of v at the interpolation points $\boldsymbol{y} \in \Gamma(\Lambda_0(\xi_n))$ and the approximations of $v(\boldsymbol{y})$, $\boldsymbol{y} \in \Gamma(\Lambda_0(\xi_n))$, by $P_{2^k} v(\boldsymbol{y})$ for $k = 0, \dots, \lfloor \log_2 \xi_n \rfloor$ with $\lfloor \log_2 \xi_n \rfloor = \mathcal{O}(\log_2 n)$.

7.7.2 Multilevel Smolyak Sparse-Grid Quadrature

In this section, we recall results of [43] (see also [45]) on linear methods for numerical integration of functions from Bochner spaces as well as their linear functionals. We define the univariate operator Δ_m^{Q} for even $m \in \mathbb{N}_0$ by

$$\Delta_m^{\mathrm{Q}} := \mathcal{Q}_m - \mathcal{Q}_{m-2},$$

with the convention $Q_{-2} := 0$. We make use of the notation:

$$\mathcal{F}_{\mathrm{ev}} := \{\boldsymbol{\nu} \in \mathcal{F} : \nu_j \text{ even}, \ j \in \mathbb{N}\}.$$

For a function $v \in L^2_{\mathcal{E}}(U, X; \gamma)$, we introduce the operator $\Delta_{\boldsymbol{\nu}}^{\mathrm{Q}}$ defined for $\boldsymbol{\nu} \in \mathcal{F}_{\mathrm{ev}}$ by

$$\Delta_{\boldsymbol{\nu}}^{\mathrm{Q}}(v) := \bigotimes_{j \in \mathbb{N}} \Delta_{\nu_j}^{\mathrm{Q}}(v),$$

where the univariate operator $\Delta_{\nu_j}^{\mathrm{Q}}$ is applied to the univariate function $\bigotimes_{j'=1}^{j-1} \Delta_{\nu_{j'}}^{\mathrm{Q}}(v)$ by considering this function as a univariate function of y_j, with all other variables held fixed. As $\Delta_{\boldsymbol{\nu}}^{\mathrm{I}}$, the operators $\Delta_{\boldsymbol{\nu}}^{\mathrm{Q}}$ are well-defined for all $\boldsymbol{\nu} \in \mathcal{F}_{\mathrm{ev}}$.

Letting Assumption 7.16 hold for Hilbert spaces X^1 and X^2, we can construct linear fully discrete quadrature operators. For a finite set $G \subset \mathbb{N}_0 \times \mathcal{F}_{\mathrm{ev}}$, we introduce the quadrature operator $\mathcal{Q}_G$ which is defined for v by

$$\mathcal{Q}_G v := \sum_{(k,\boldsymbol{\nu}) \in G} \delta_k \Delta_{\boldsymbol{\nu}}^{\mathrm{Q}}(v). \tag{7.59}$$

If $\phi \in (X^1)'$ is a bounded linear functional on X^1, for a finite set $G \subset \mathbb{N}_0 \times \mathcal{F}_{\mathrm{ev}}$, the quadrature formula $\mathcal{Q}_G v$ generates the quadrature formula $\mathcal{Q}_G \langle \phi, v \rangle$ for integration of $\langle \phi, v \rangle$ by

$$\mathcal{Q}_G \langle \phi, v \rangle := \langle \phi, \mathcal{Q}_G v \rangle.$$

Define for $\xi > 0$,

$$G_{\text{ev}}(\xi) := \begin{cases} \{(k, \boldsymbol{\nu}) \in \mathbb{N}_0 \times \mathcal{F}_{\text{ev}} : 2^k \sigma_{2;\boldsymbol{\nu}}^{q_2} \le \xi\} & \text{if } \alpha \le 1/q_2 - 1/2; \\ \{(k, \boldsymbol{\nu}) \in \mathbb{N}_0 \times \mathcal{F}_{\text{ev}} : \sigma_{1;\boldsymbol{\nu}}^{q_1} \le \xi,\ 2^{(\alpha+1/2)k} \sigma_{2;\boldsymbol{\nu}} \le \xi^{\vartheta}\} & \text{if } \alpha > 1/q_2 - 1/2, \end{cases} \tag{7.60}$$

where ϑ is as in (7.55).

Theorem 7.19 *Let the hypothesis of Theorem 7.17 hold. Then we have the following.*

(i) *For each $n \in \mathbb{N}$ there exists a number ξ_n such that* $\dim \mathcal{V}(G_{\text{ev}}(\xi_n)) \le n$ *and*

$$\left\| \int_U v(\boldsymbol{y}) \, \mathrm{d}\gamma(\boldsymbol{y}) - \mathcal{Q}_{G_{\text{ev}}(\xi_n)} v \right\|_{X^1} \le C n^{-\min(\alpha,\beta)}. \tag{7.61}$$

(ii) *Let $\phi \in (X^1)'$ be a bounded linear functional on X^1. Then for each $n \in \mathbb{N}$ there exists a number ξ_n such that* $\dim \mathcal{V}(G_{\text{ev}}(\xi_n)) \le n$ *and*

$$\left| \int_U \langle \phi, v(\boldsymbol{y}) \rangle \, \mathrm{d}\gamma(\boldsymbol{y}) - \mathcal{Q}_{G_{\text{ev}}(\xi_n)} \langle \phi, v \rangle \right| \le C \|\phi\|_{(X^1)'} n^{-\min(\alpha,\beta)}. \tag{7.62}$$

The rate α is as in (7.53) *and the rate β is given by* (7.58). *The constants C in* (7.61) *and* (7.62) *are independent of v and n.*

The proof Theorem 7.19 are related to approximations in the norm of $L^1(U, X; \gamma)$ by special polynomial interpolation operators which generate the corresponding quadrature operators. Let us briefly describe this connection, for details see [43, 45].

Remark 7.20 We define the univariate interpolation operator $\Delta_m^{\mathrm{I}*}$ for even $m \in \mathbb{N}_0$ by

$$\Delta_m^{\mathrm{I}*} := I_m - I_{m-2},$$

with the convention $I_{-2} = 0$. The interpolation operators $\Delta_{\boldsymbol{\nu}}^{\mathrm{I}*}$ for $\boldsymbol{\nu} \in \mathcal{F}_{\text{ev}}$, I_Λ^* for a finite set $\Lambda \subset \mathcal{F}_{\text{ev}}$, and $\mathcal{I}_G^*$ for a finite set $G \subset \mathbb{N}_0 \times \mathcal{F}_{\text{ev}}$, are defined in a similar way as the corresponding quadrature operators $\Delta_{\boldsymbol{\nu}}^{\mathrm{Q}}$, $\mathcal{Q}_\Lambda$ and $\mathcal{Q}_G$ by replacing $\Delta_{\nu_j}^{\mathrm{Q}}$ with $\Delta_{\nu_j}^{\mathrm{I}*}$, $j \in \mathbb{N}$.

From the definitions it follows the equalities expressing the relationship between the interpolation and quadrature operators

$$\mathcal{Q}_\Lambda v = \int_U I_\Lambda^* v(\boldsymbol{y}) \, \mathrm{d}\gamma(\boldsymbol{y}), \quad \mathcal{Q}_\Lambda \langle \phi, v \rangle = \int_U \langle \phi, I_\Lambda^* v(\boldsymbol{y}) \rangle \, \mathrm{d}\gamma(\boldsymbol{y}),$$

and

$$\mathcal{Q}_G v = \int_U \mathcal{I}_G^* v(\boldsymbol{y}) \,\mathrm{d}\gamma(\boldsymbol{y}), \quad \mathcal{Q}_G \langle \phi, v \rangle = \int_U \langle \phi, \mathcal{I}_G^* v(\boldsymbol{y}) \rangle \,\mathrm{d}\gamma(\boldsymbol{y}).$$

Remark 7.21 Similarly to $\mathcal{I}_{G(\xi_n)}$, the operator $\mathcal{Q}_{G_{\mathrm{ev}}(\xi_n)}$ can be represented in the form of a multilevel Smolyak sparse-grid quadrature with k_n levels:

$$\mathcal{Q}_{G_{\mathrm{ev}}(\xi_n)} = \sum_{k=0}^{k_n} \delta_k \mathcal{Q}_{\Lambda_{\mathrm{ev},k}(\xi_n)},$$

where $k_n := \lfloor \log_2 \xi_n \rfloor$,

$$\mathcal{Q}_\Lambda := \sum_{\boldsymbol{\nu} \in \Lambda} \Delta_{\boldsymbol{\nu}}^{\mathrm{I}}, \quad \Lambda \subset \mathcal{F}_{\mathrm{ev}}, \tag{7.63}$$

and for $k \in \mathbb{N}_0$ and $\xi > 0$,

$$\Lambda_{\mathrm{ev},k}(\xi) := \begin{cases} \left\{ \boldsymbol{s} \in \mathcal{F}_{\mathrm{ev}} : \sigma_{2;\boldsymbol{s}}^{q_2} \le 2^{-k}\xi \right\} & \text{if } \alpha \le 1/q_2 - 1/2; \\ \left\{ \boldsymbol{s} \in \mathcal{F}_{\mathrm{ev}} : \sigma_{1;\boldsymbol{s}}^{q_1} \le \xi, \ \sigma_{2;\boldsymbol{s}} \le 2^{-(\alpha+1/2)k}\xi^{\vartheta} \right\} & \text{if } \alpha > 1/q_2 - 1/2. \end{cases}$$

Remark 7.22 The convergence rates established in Theorems 7.17 and 7.19 and in Theorems 7.5 and 7.6 are proven with respect to different parameters n as the dimension of the approximation space and the work (7.5), respectively. However, we could define the work of the operators $\mathcal{I}_{G(\xi_n)}$ and $\mathcal{Q}_{G_{\mathrm{ev}}(\xi_n)}$ similarly as

$$\sum_{k=0}^{k_n} 2^k |\Gamma(\Lambda_k(\xi_n))|,$$

and

$$\sum_{k=0}^{k_n} 2^k |\Gamma(\Lambda_{\mathrm{ev},k}(\xi_n))|,$$

respectively, and prove the same convergence rates with respect to this work measure as in Theorems 7.17 and 7.19.

7.7.3 Applications to Parametric Divergence-Form Elliptic PDEs

In this section, we apply the results in Sects. 7.7.1 and 7.7.2 to parametric divergence-form elliptic PDEs (3.17). The spaces V and W are as in Sect. 3.9.

Assumption 7.23 *There are a sequence $(V_n)_{n\in\mathbb{N}_0}$ of subspaces $V_n \subset V$ of dimension $\leq m$, and a sequence $(P_n)_{n\in\mathbb{N}_0}$ of linear operators from V into V_n, and a number $\alpha > 0$ such that*

$$\|P_n(v)\|_V \leq C\|v\|_V, \quad \|v - P_n(v)\|_V \leq Cn^{-\alpha}\|v\|_W, \quad \forall n \in \mathbb{N}_0, \quad \forall v \in W. \tag{7.64}$$

If Assumption 7.23 and the assumptions of Theorem 3.38 hold for the spaces $W^1 = V$ and $W^2 = W$ with some $0 < q_1 \leq q_2 < \infty$, then Assumption 7.16 holds for the spaces $X^i = W^i$, $i = 1, 2$, and the solution $u \in L^2(U, X^2; \gamma)$ to (3.17)–(3.18). Hence we obtain the following results on multilevel (fully discrete) approximations.

Theorem 7.24 *Let Assumption 7.23 hold. Let the hypothesis of Theorem 3.38 hold for the spaces $W^1 = V$ and $W^2 = W$ with some $0 < q_1 \leq q_2 < \infty$ and $q_1 < 2$. For $\xi > 0$, let $G(\xi)$ be the set defined by (7.54) for $\sigma_{i;\boldsymbol{\nu}}$ as in (3.59), $i = 1, 2$. Let α be as in (7.64). Then for every $n \in \mathbb{N}$ there exists a number ξ_n such that* $\dim \mathcal{V}(G(\xi_n)) \leq n$ *and*

$$\|u - \mathcal{I}_{G(\xi_n)}u\|_{L^2(U,V;\gamma)} \leq Cn^{-\min(\alpha,\beta)}, \tag{7.65}$$

where β is given by (7.58). The constant C in (7.65) is independent of u and n.

Theorem 7.25 *Let Assumption 7.23 hold. Let the assumptions of Theorem 3.38 hold for the spaces $W^1 = V$ and $W^2 = W$ for some $0 < q_1 \leq q_2 < \infty$ with $q_1 < 4$. Let α be the rate as given by (7.64). For $\xi > 0$, let $G_{\mathrm{ev}}(\xi)$ be the set defined by (7.60) for $\sigma_{i;\boldsymbol{\nu}}$ as in (3.59), $i = 1, 2$. Then we have the following.*

(i) *For each $n \in \mathbb{N}$ there exists a number ξ_n such that* $\dim \mathcal{V}(G_{\mathrm{ev}}(\xi_n)) \leq n$ *and*

$$\left\| \int_U v(\boldsymbol{y}) \, \mathrm{d}\gamma(\boldsymbol{y}) - \mathcal{Q}_{G_{\mathrm{ev}}(\xi_n)} v \right\|_V \leq Cn^{-\min(\alpha,\beta)}. \tag{7.66}$$

(ii) *Let $\phi \in V'$ be a bounded linear functional on V. Then for each $n \in \mathbb{N}$ there exists a number ξ_n such that* $\dim \mathcal{V}(G_{\mathrm{ev}}(\xi_n)) \leq n$ *and*

$$\left| \int_U \langle \phi, v(\boldsymbol{y}) \rangle \, \mathrm{d}\gamma(\boldsymbol{y}) - \mathcal{Q}_{G_{\mathrm{ev}}(\xi_n)} \langle \phi, v \rangle \right| \leq C\|\phi\|_{V'} n^{-\min(\alpha,\beta)}. \tag{7.67}$$

The rate β *is given by*

$$\beta := \left(\frac{2}{q_1} - \frac{1}{2}\right)\frac{\alpha}{\alpha+\delta}, \qquad \delta := \frac{2}{q_1} - \frac{2}{q_2}.$$

The constants C *in* (7.66) *and* (7.67) *are independent of* u *and* n.

Proof From Theorem 3.38, Lemma 3.39 and Assumption 7.23 we can see that the assumptions of Theorem 7.17 hold for $X^1 = V$ and $X^2 = W$ with $0 < q_1/2 \le q_2/2 < \infty$ and $q_1/2 < 2$. Hence, by applying Theorem 7.19 we prove the theorem. □

7.7.4 Applications to Holomorphic Functions

As noticed, the proof of the weighted ℓ_2-summability result formulated in Theorem 3.38 employs bootstrap arguments and induction on the differentiation order of derivatives with respect to the parametric variables, for details see [8, 9]. In the log-Gaussian case, this approach and technique are too complicated and difficult for extension to more general parametric PDE problems, in particular, of higher regularity. As it has been seen in the previous sections, the approach to a unified summability analysis of Wiener-Hermite PC expansions of various scales of function spaces based on parametric holomorphy, covers a wide range of parametric PDE problems. In this section, we apply the results in Sects. 7.7.1 and 7.7.2 on linear approximations and integration in Bochner spaces to approximation and numerical integration of parametric holomorphic functions based on weighted ℓ^2-summabilities of the coefficient sequences of the Wiener-Hermite PC expansion.

The following theorem on weighted ℓ_2-summability for $(\boldsymbol{b}, \xi, \delta, X)$-holomorphic functions can be derived from Theorem 4.9 and Lemma 3.39.

Theorem 7.26 *Let* v *be* $(\boldsymbol{b}, \xi, \delta, X)$*-holomorphic for some* $\boldsymbol{b} \in \ell^p(\mathbb{N})$ *with* $0 < p < 1$. *Let* $s = 1, 2$ *and* $\tau, \lambda \ge 0$. *Let further the sequence* $\boldsymbol{\varrho} = (\varrho_j)_{j\in\mathbb{N}}$ *be defined by*

$$\varrho_j := b_j^{p-1}\frac{\xi}{4\sqrt{r!}}\|\boldsymbol{b}\|_{\ell^p}.$$

Then, for any $r > \frac{2s(\tau+1)}{q}$,

$$\sum_{\boldsymbol{\nu}\in\mathcal{F}_s}(\sigma_{\boldsymbol{\nu}}\|v_{\boldsymbol{\nu}}\|_X)^2 \le M < \infty \quad \textit{and} \quad \left(p_{\boldsymbol{\nu}}(\tau,\lambda)\sigma_{\boldsymbol{\nu}}^{-1}\right)_{\boldsymbol{\nu}\in\mathcal{F}_s} \in \ell^{q/s}(\mathcal{F}_s),$$

where $q := \frac{p}{1-p}$, $M := \delta^2 C(\boldsymbol{b})$ *and* $(\sigma_{\boldsymbol{\nu}})_{\boldsymbol{\nu}\in\mathcal{F}}$ *with* $\sigma_{\boldsymbol{\nu}} := \beta_{\boldsymbol{\nu}}(r, \boldsymbol{\varrho})^{1/2}$.

To treat multilevel approximations and integration of parametric, holomorphic functions, it is appropriate to replace Assumption 7.16 by its modification.

Assumption 7.27 *Assumption 7.16 holds with item* (ii) *replaced with item*

(ii') For $i = 1, 2$, v is $(\boldsymbol{b}_i, \xi, \delta, X^i)$-holomorphic for some $\boldsymbol{b}_i \in \ell^{p_i}(\mathbb{N})$ with $0 < p_1 \le p_2 < 1$.

Assumption 7.27 is a condition for fully discrete approximation of $(\boldsymbol{b}, \xi, \delta, X)$-holomorphic functions. This is formalized in the following corollary of Theorem 7.26.

Corollary 7.28 *Assumption 7.27 implies Assumption 7.16 for $q_i := \frac{p_i}{1-p_i}$ and $(\sigma_{i;\boldsymbol{\nu}})_{\boldsymbol{\nu}\in\mathcal{F}}$, $i = 1, 2$, where*

$$\sigma_{i;\boldsymbol{\nu}} := \beta_{i;\boldsymbol{\nu}}(r, \boldsymbol{\varrho}_i)^{1/2}, \quad \varrho_{i;j} := b_{i;j}^{p_i - 1} \frac{\xi}{4\sqrt{r!}} \|\boldsymbol{b}_i\|_{\ell^{p_i}}.$$

We formulate results on multilevel quadrature of parametric holomorphic functions as consequences of Corollary 7.28 and Theorems 7.17 and 7.19.

Theorem 7.29 *Let Assumption 7.27 hold for the Hilbert spaces X^1 and X^2 with $p_1 < 2/3$, and $v \in L^2(U, X^2; \gamma)$. For $\xi > 0$, let $G(\xi)$ be the set defined by* (7.54) *for $\sigma_{i;\boldsymbol{\nu}}$, $i = 1, 2$ as given in Corollary 7.28. Then for every $n \in \mathbb{N}$ there exists a number ξ_n such that* $\dim \mathcal{V}(G(\xi_n)) \le n$ *and*

$$\|v - \mathcal{I}_{G(\xi_n)} v\|_{L^2(U,X^1;\gamma)} \le C n^{-R}, \tag{7.68}$$

where R is given by the formula (7.19) *and the constant C in*(7.68) *is independent of v and n.*

Theorem 7.30 *Let Assumption 7.27 hold for the Hilbert spaces X^1 and X^2 with $p_1 < 4/5$, and $v \in L^2(U, X^2; \gamma)$. For $\xi > 0$, let $G_{\mathrm{ev}}(\xi)$ be the set defined by* (7.60) *for $\sigma_{i;\boldsymbol{\nu}}$, $i = 1, 2$, as given in Corollary 7.28. Then we have the following.*

(i) *For each $n \in \mathbb{N}$ there exists a number ξ_n such that* $\dim \mathcal{V}(G_{\mathrm{ev}}(\xi_n)) \le n$ *and*

$$\left\| \int_U v(\boldsymbol{y}) \, \mathrm{d}\gamma(\boldsymbol{y}) - \mathcal{Q}_{G_{\mathrm{ev}}(\xi_n)} v \right\|_{X^1} \le C n^{-R}. \tag{7.69}$$

(ii) *Let $\phi \in (X^1)'$ be a bounded linear functional on X^1. Then for each $n \in \mathbb{N}$ there exists a number ξ_n such that* $\dim \mathcal{V}(G_{\mathrm{ev}}(\xi_n)) \le n$ *and*

$$\left| \int_U \langle \phi, v(\boldsymbol{y}) \rangle \, \mathrm{d}\gamma(\boldsymbol{y}) - \mathcal{Q}_{G_{\mathrm{ev}}(\xi_n)} \langle \phi, v \rangle \right| \le C \|\phi\|_{(X^1)'} n^{-R}, \tag{7.70}$$

where the convergence rate R is given by the formula (7.26) *and the constants C in* (7.69) *and* (7.70) *are independent of v and n.*

Remark 7.31 We comment on the relation of the results of Theorems 7.5 and 7.6 to the results of [43] which are presented in Theorems 7.24 and 7.25, on multilevel approximation of solutions to parametric divergence-form elliptic PDEs with log-Gaussian inputs.

Specifically, in [43], by combining spatial and parametric approximability in the spatial domain and weighted ℓ^2-summability of the $V := H_0^1(D)$ and W norms of Wiener-Hermite PC expansion coefficients obtained in [8, 9], the author constructed linear non-adaptive methods of fully discrete approximation by truncated Wiener-Hermite PC expansion and polynomial interpolation approximation as well as fully discrete weighted quadrature for parametric and stochastic elliptic PDEs with log-Gaussian inputs, and proved the convergence rates of approximation by them. The results in [43] are based on Assumption 7.23 that requires the *existence of a sequence* $(P_n)_{n\in\mathbb{N}_0}$ *of linear operators independent of* $\boldsymbol{y}$, from $H_0^1(D)$ into n-dimensional subspaces $V_n \subset H_0^1(D)$ such that

$$\|P_n(v)\|_{H_0^1} \le C_1\|v\|_{H_0^1} \text{ and } \|v - P_n(v)\|_{H_0^1} \le C_2 n^{-\alpha}\|v\|_W$$

for all $n \in \mathbb{N}_0$ and for all $v \in W$, where the constants C_1, C_2 are independent of n. *The assumption of* P_n *being independent of* $\boldsymbol{y}$ *is however typically not satisfied if* $P_n(u(\boldsymbol{y})) = u^n(\boldsymbol{y})$ *is a numerical approximation to* $u(\boldsymbol{y})$ (such as, e.g., a Finite-Element or a Finite-Difference discretization).

In contrast, the present approximation rate analysis is based on quantified, parametric holomorphy of the discrete approximations u^l to u as in Assumption 7.2. For example, assume that $u : U \to H_0^1(D)$ is the solution of the parametric PDE

$$-\operatorname{div}(a(\boldsymbol{y})\nabla u(\boldsymbol{y})) = f$$

for some $f \in L^2(D)$ and a parametric diffusion coefficient $a(\boldsymbol{y}) \in L^\infty(D)$ such that

$$\operatorname*{ess\,inf}_{\boldsymbol{x}\in D} a(\boldsymbol{y}, \boldsymbol{x}) > 0 \quad \forall \boldsymbol{y} \in U.$$

Then $u^l : U \to H_0^1(D)$ could be a numerical approximation to u, such as the FEM solution: for every $l \in \mathbb{N}$ there is a finite dimensional discretization space $X_l \subseteq H_0^1(D)$, and

$$\int_D \nabla u^l(\boldsymbol{y})^\top a(\boldsymbol{y})\nabla v \,\mathrm{d}\boldsymbol{x} = \int_D f v \,\mathrm{d}\boldsymbol{x}$$

for every $v \in X_l$ and for every $\boldsymbol{y} \in U$. Hence $u^l(\boldsymbol{y})$ is the orthogonal projection of $u(\boldsymbol{y})$ onto X_l w.r.t. the inner product

$$\langle v, w\rangle_{a(\boldsymbol{y})} := \int_D \nabla v^\top a(\boldsymbol{y})\nabla w \,\mathrm{d}\boldsymbol{x}$$

on $H_0^1(D)$. We may write this as $u^l(y) = P_l(y)u(y)$, for a y-dependent projector

$$P_l(y) : H_0^1(D) \to X_l.$$

This situation is covered by Assumption 7.2.

The preceding comments can be extended to the results on multilevel approximation of holomorphic functions in Theorems 7.5 and 7.6 to the results in Theorems 7.29 and 7.30. On the other hand, as noticed above, the convergence rates in Theorems 7.29 and 7.30 are slightly better than those obtained in Theorems 7.5 and 7.6.

Chapter 8
Conclusions

We established holomorphy of parameter-to-solution maps

$$E \ni a \mapsto u = \mathcal{U}(a) \in X$$

for linear, elliptic, parabolic, and other PDEs in various scales of function spaces E and X, including in particular standard and corner-weighted Sobolev spaces. Our discussion focused on non-compact parameter domains which arise from uncertain inputs from function spaces expressed in a suitable basis with Gaussian distributed coefficients. We introduced and used a form of quantified, parametric holomorphy in products of strips to show that this implies summability results of coefficients of the Wiener-Hermite PC expansion of such infinite parametric functions. Specifically, we proved weighted ℓ^2-summability and ℓ^p-summability results for Wiener-Hermite PC expansions of certain parametric, deterministic solution families $\{u(\boldsymbol{y}) : \boldsymbol{y} \in U\} \subset X$, for a given "log-affine" parametrization (3.18) of admissible random input data $a \in E$.

We introduced and analyzed constructive, deterministic, sparse-grid ("stochastic collocation") algorithms based on univariate Gauss-Hermite points, to efficiently sample the parametric, deterministic solutions in the possibly infinite-dimensional parameter domain $U = \mathbb{R}^\infty$. The sparsity of the coefficients of Wiener-Hermite PC expansion was shown to entail corresponding convergence rates of the presently developed sparse-grid sampling schemes. In combination with suitable Finite Element discretizations in the physical, space(-time) domain (which include proper mesh-refinements to account for singularities in the physical domain) we proved convergence rates for abstract, multilevel algorithms which employ different combinations of sparse-grid interpolants in the parametric domain with space(-time) discretizations at different levels of accuracy in the physical domain.

The presently developed, holomorphic setting was also shown to apply to the corresponding Bayesian inverse problems subject to PDE constraints: here, the density of the Bayesian posterior with respect to a Gaussian random field prior was

D. Dũng et al., *Analyticity and Sparsity in Uncertainty Quantification for PDEs with Gaussian Random Field Inputs*, Lecture Notes in Mathematics 2334,
https://doi.org/10.1007/978-3-031-38384-7_8

shown to generically inherit quantified holomorphy from the parametric forward problem, thereby facilitating the use of the developed sparse-grid collocation and integration algorithms also for the efficient deterministic computation of Bayesian estimates of PDEs with uncertain inputs, subject to noisy observation data.

Our approximation rate bounds are free from the curse-of-dimensionality and only limited by the PC coefficient summability. They will therefore also be relevant for convergence rate analyses of other approximation schemes, such as Gaussian process emulators or neural networks (see, e.g., [44, 46, 99, 104] and references there).

References

1. M. Abramowitz, I.A. Stegun, *Handbook of Mathematical Functions* (Dover, New York, 1965)
2. R.A. Adams, J.J.F. Fournier, *Sobolev Spaces*. Pure and Applied Mathematics, vol. 140, 2nd edn. (Elsevier/Academic Press, Amsterdam, 2003)
3. R.J. Adler, *The Geometry of Random Fields*. Wiley Series in Probability and Mathematical Statistics (John Wiley & Sons, Chichester, 1981)
4. R. Andreev, A. Lang, Kolmogorov-Chentsov theorem and differentiability of random fields on manifolds. Potential Anal. **41**(3), 761–769 (2014)
5. A. Ayache, M.S. Taqqu, Rate optimality of wavelet series approximations of fractional Brownian motion. J. Fourier Anal. Appl. **9**(5), 451–471 (2003)
6. I. Babuška, R.B. Kellogg, J. Pitkäranta, Direct and inverse error estimates for finite elements with mesh refinements. Numer. Math. **33**(4), 447–471 (1979)
7. I. Babuška, F. Nobile, R. Tempone, A stochastic collocation method for elliptic partial differential equations with random input data. SIAM J. Numer. Anal. **45**(3), 1005–1034 (2007)
8. M. Bachmayr, A. Cohen, D. Dũng, C. Schwab, Fully discrete approximation of parametric and stochastic elliptic PDEs. SIAM J. Numer. Anal. **55**(5), 2151–2186 (2017)
9. M. Bachmayr, A. Cohen, R. DeVore, G. Migliorati, Sparse polynomial approximation of parametric elliptic PDEs. Part II: Lognormal coefficients. ESAIM Math. Model. Numer. Anal. **51**(1), 341–363 (2017)
10. M. Bachmayr, A. Cohen, G. Migliorati, Sparse polynomial approximation of parametric elliptic PDEs. Part I: affine coefficients. ESAIM Math. Model. Numer. Anal. **51**. 321–339 (2017)
11. M. Bachmayr, A. Cohen, G. Migliorati, Sparse polynomial approximation of parametric elliptic PDEs. Part I: affine coefficients. ESAIM Math. Model. Numer. Anal. **51**(1), 321–339 (2017)
12. M. Bachmayr, A. Cohen, G. Migliorati, Representations of Gaussian random fields and approximation of elliptic PDEs with lognormal coefficients. J. Fourier Anal. Appl. **24**(3), 621–649 (2018)
13. M. Bachmayr, A. Djurdjevac, Multilevel representations of isotropic Gaussian random fields on the sphere (2022). https://doi.org/10.1093/imanum/drac034.
14. M. Bachmayr, I.G. Graham, V.K. Nguyen, R. Scheichl, Unified analysis of periodization-based sampling methods for Matérn covariances. SIAM J. Numer. Anal. **58**(5), 2953–2980 (2020)

D. Dũng et al., *Analyticity and Sparsity in Uncertainty Quantification for PDEs with Gaussian Random Field Inputs*, Lecture Notes in Mathematics 2334,
https://doi.org/10.1007/978-3-031-38384-7

15. M. Bachmayr, V.K. Nguyen, Identifiability of diffusion coefficients for source terms of non-uniform sign. Inverse Probl. Imaging **13**(5), 1007–1021 (2019)
16. J. Beck, R. Tempone, F. Nobile, L. Tamellini, On the optimal polynomial approximation of stochastic PDEs by Galerkin and collocation methods. Math. Models Methods Appl. Sci. **22**(9), 1250023, 33 (2012)
17. R. Bellman, *Dynamic Programming*. Princeton Landmarks in Mathematics (Princeton University Press, Princeton, 2010). Reprint of the 1957 edition, With a new introduction by Stuart Dreyfus.
18. J. Bergh, J. Löfström, *Interpolation Spaces: An Introduction*. Grundlehren der Mathematischen Wissenschaften, vol. 223 (Springer-Verlag, Berlin-New York, 1976)
19. M. Bieri, R. Andreev, C. Schwab, Sparse tensor discretization of elliptic SPDEs. SIAM J. Sci. Comput. **31**(6), 4281–4304 (2009/2010). MR 2566594
20. D. Boffi, F. Brezzi, M. Fortin, *Mixed Finite Element Methods and Applications*. Springer Series in Computational Mathematics, vol. 44 (Springer, Heidelberg, 2013)
21. V.I. Bogachev, *Gaussian Measures*. Mathematical Surveys and Monographs, vol. 62 (American Mathematical Society, Providence, 1998)
22. A. Bonito, A. Demlow, J. Owen, A priori error estimates for finite element approximations to eigenvalues and eigenfunctions of the Laplace-Beltrami operator. SIAM J. Numer. Anal. **56**(5), 2963–2988 (2018)
23. S.C. Brenner, L.R. Scott, *The Mathematical Theory of Finite Element Methods*. Texts in Applied Mathematics, vol. 15, 3rd edn. (Springer, New York, 2008)
24. C. Băcuţă, H. Li, V. Nistor, Differential operators on domains with conical points: precise uniform regularity estimates. Rev. Roum. Math. Pure Appl. **62**, 383–411 (2017)
25. C. Băcuţă, V. Nistor, L.T. Zikatanov, Improving the rate of convergence of 'high order finite elements' on polygons and domains with cusps. Numer. Math. **100**(2), 165–184 (2005)
26. C. Băcuţă, V. Nistor, L.T. Zikatanov, Improving the rate of convergence of high-order finite elements on polyhedra. I: a priori estimates. Numer. Funct. Anal. Optim. **26**(6), 613–639 (2005)
27. C. Băcuţă, V. Nistor, L.T. Zikatanov, Improving the rate of convergence of high-order finite elements on polyhedra. II: mesh refinements and interpolation. Numer. Funct. Anal. Optim. **28**(7–8), 775–824 (2007)
28. J. Charrier, Strong and weak error estimates for elliptic partial differential equations with random coefficients. SIAM J. Numer. Anal. **50**(1), 216–246 (2012)
29. J. Charrier, A. Debussche, Weak truncation error estimates for elliptic PDEs with lognormal coefficients. Stoch. Partial Differ. Equ. Anal. Comput. **1**(1), 63–93 (2013)
30. J. Charrier, R. Scheichl, A.L. Teckentrup, Finite element error analysis of elliptic PDEs with random coefficients and its application to multilevel Monte Carlo methods. SIAM J. Numer. Anal. **51**(1), 322–352 (2013). MR 3033013
31. P. Chen, Sparse quadrature for high-dimensional integration with Gaussian measure. ESAIM Math. Model. Numer. Anal. **52**(2), 631–657 (2018)
32. A. Chkifa, A. Cohen, C. Schwab, High-dimensional adaptive sparse polynomial interpolation and applications to parametric PDEs. Found. Comput. Math. **14**(4), 601–633 (2014)
33. P.G. Ciarlet, *The Finite Element Method for Elliptic Problems*, Studies in Mathematics and its Applications, vol. 4 (North-Holland Publishing, Amsterdam-New York-Oxford, 1978)
34. Z. Ciesielski, Hölder conditions for realizations of Gaussian processes. Trans. Am. Math. Soc. **99**, 403–413 (1961)
35. G. Cleanthous, A.G. Georgiadis, A. Lang, E. Porcu, Regularity, continuity and approximation of isotropic Gaussian random fields on compact two-point homogeneous spaces. Stochastic Process. Appl. **130**(8), 4873–4891 (2020)
36. K.A. Cliffe, M.B. Giles, R. Scheichl, A.L. Teckentrup, Multilevel Monte Carlo methods and applications to elliptic PDEs with random coefficients. Comput. Vis. Sci. **14**(1), 3–15 (2011). MR 2835612
37. A. Cohen, R. DeVore, Approximation of high-dimensional parametric PDEs. Acta Numer. **24**, 1–159 (2015)

38. A. Cohen, R. Devore, C. Schwab, Convergence rates of best N-term Galerkin approximations for a class of elliptic sPDEs. Found. Comput. Math. **10**(6), 615–646 (2010)
39. A. Cohen, R. Devore, C. Schwab, Analytic regularity and polynomial approximation of parametric and stochastic elliptic PDE's. Anal. Appl. (Singap.) **9**(1), 11–47 (2011)
40. A. Cohen, C. Schwab, J. Zech, Shape holomorphy of the stationary Navier-Stokes equations. SIAM J. Math. Anal. **50**(2), 1720–1752 (2018)
41. R.R. Coifman, M. Maggioni, Diffusion wavelets for multiscale analysis on graphs and manifolds. Wavelets and splines: Athens 2005, Mod. Methods Math. (Nashboro Press, Brentwood, 2006), pp. 164–188
42. D. Dũng, Linear collocation approximation for parametric and stochastic elliptic PDEs. Sb. Math. **210**(4), 565–588 (2019)
43. D. Dũng, Sparse-grid polynomial interpolation approximation and integration for parametric and stochastic elliptic PDEs with lognormal inputs. ESAIM Math. Model. Numer. Anal. **55**, 1163–1198 (2021)
44. D. Dũng, Collocation approximation by deep neural ReLU networks for parametric elliptic PDEs with lognormal inputs. Sb. Math. **214**(4) (2023)
45. D. Dũng, Erratum to: "Sparse-grid polynomial interpolation approximation and integration for parametric and stochastic elliptic PDEs with lognormal inputs". ESAIM Math. Model. Numer. Anal. **57**, 893–897 (2023). Erratum to: ESAIM: M2AN **55**, 1163–1198 (2021)
46. D. Dũng, V.K. Nguyen, D.T. Pham, Deep ReLU neural network approximation of parametric and stochastic elliptic PDEs with lognormal inputs J. Comp. **79** 101779 (2023).
47. M. Dashti, K.J.H. Law, A.M. Stuart, J. Voss, MAP estimators and their consistency in Bayesian nonparametric inverse problems. Inverse Probl. **29**(9), 095017, 27 (2013)
48. M. Dashti, A.M. Stuart, *The Bayesian Approach to Inverse Problems*. Handbook of uncertainty quantification, vol. 1, 2, 3 (Springer, Cham, 2017), pp. 311–428
49. A. Demlow, Higher-order finite element methods and pointwise error estimates for elliptic problems on surfaces. SIAM J. Numer. Anal. **47**(2), 805–827 (2009)
50. C.R. Dietrich, G.N. Newsam, Fast and exact simulation of stationary Gaussian processes through circulant embedding of the covariance matrix. SIAM J. Sci. Comput. **18**(4), 1088–1107 (1997)
51. J. Dölz, H. Harbrecht, C. Schwab, Covariance regularity and $\mathcal{H}$-matrix approximation for rough random fields. Numer. Math. **135**(4), 1045–1071 (2017)
52. O.G. Ernst, B. Sprungk, L. Tamellini, Convergence of sparse collocation for functions of countably many Gaussian random variables (with application to elliptic PDEs). SIAM J. Numer. Anal. **56**(2), 877–905 (2018)
53. L.C. Evans, *Partial Differential Equations*. Graduate Studies in Mathematics, vol. 19, 2nd edn. (American Mathematical Society, Providence, 2010)
54. F.D. Gaspoz, P. Morin, Convergence rates for adaptive finite elements. IMA J. Numer. Anal. **29**(4), 917–936 (2009)
55. D. Gilbarg, N.S. Trudinger, *Elliptic Partial Differential Equations of Second Order*. Classics in Mathematics (Springer-Verlag, Berlin, 2001). Reprint of the 1998 edition
56. E. Giné, R. Nickl, *Mathematical Foundations of Infinite-dimensional Statistical Models*. Cambridge Series in Statistical and Probabilistic Mathematics, vol. 40 (Cambridge University Press, New York, 2016)
57. C.J. Gittelson, Representation of Gaussian fields in series with independent coefficients. IMA J. Numer. Anal. **32**(1), 294–319 (2012)
58. C. Gittelson, J. Könnö, C. Schwab, R. Stenberg, The multi-level Monte Carlo finite element method for a stochastic Brinkman problem. Numer. Math. **125**(2), 347–386 (2013)
59. I.G. Graham, F.Y. Kuo, J.A. Nichols, R. Scheichl, C. Schwab, I.H. Sloan, Quasi-Monte Carlo finite element methods for elliptic PDEs with lognormal random coefficients. Numer. Math. **131**, 329–368 (2014)
60. I.G. Graham, F.Y. Kuo, D. Nuyens, R. Scheichl, I.H. Sloan, Analysis of circulant embedding methods for sampling stationary random fields. SIAM J. Numer. Anal. **56**, 1871–1895 (2018)

61. P. Grisvard, *Elliptic Problems in Nonsmooth Domains*. Classics in Applied Mathematics, vol. 69 (Society for Industrial and Applied Mathematics (SIAM), Philadelphia, 2011). Reprint of the 1985 original [MR0775683], With a foreword by Susanne C. Brenner
62. B.Q. Guo, I. Babuška, On the regularity of elasticity problems with piecewise analytic data. Adv. in Appl. Math. **14**(3), 307–347 (1993)
63. H. Harbrecht, M. Peters, M. Siebenmorgen, Multilevel accelerated quadrature for PDEs with log-normally distributed diffusion coefficient. SIAM-ASA J. Uncertain. Quantif. **4**(1), 520–551 (2016)
64. G.H. Hardy, Note on a theorem of Hilbert. Math. Z. **6**(3–4), 314–317 (1920)
65. C. Heil, *A Basis Theory Primer*. Applied and Numerical Harmonic Analysis, Expanded edn. (Birkhäuser/Springer, New York, 2011)
66. L. Herrmann, C. Schwab, Multilevel quasi-Monte Carlo integration with product weights for elliptic PDEs with lognormal coefficients. ESAIM Math. Model. Numer. Anal. **53**(5), 1507–1552 (2019)
67. L. Herrmann, C. Schwab, QMC integration for lognormal-parametric, elliptic PDEs: local supports and product weights. Numer. Math. **141**(1), 63–102 (2019)
68. L. Herrmann, M. Keller, C. Schwab, Quasi-Monte Carlo Bayesian estimation under Besov priors in elliptic inverse problems. Math. Comp. **90**(330), 1831–1860 (2021)
69. E. Hille, Contributions to the theory of Hermitian series. II. The representation problem. Trans. Am. Math. Soc. **47**, 80–94 (1940)
70. V.H. Hoang, C. Schwab, Sparse tensor Galerkin discretization of parametric and random parabolic PDEs—analytic regularity and generalized polynomial chaos approximation. SIAM J. Math. Anal. **45**(5), 3050–3083 (2013)
71. V.H. Hoang, C. Schwab, N-term Wiener chaos approximation rate for elliptic PDEs with lognormal Gaussian random inputs. Math. Models Methods Appl. Sci. **24**(4), 797–826 (2014)
72. V.H. Hoang, C. Schwab, Convergence rate analysis of MCMC-FEM for Bayesian inversion of log-normal diffusion problems. Technical Report 2016–19, Seminar for Applied Mathematics, ETH Zürich, Switzerland, 2016
73. V.H. Hoang, J.H. Quek, C. Schwab, Analysis of a multilevel Markov chain Monte Carlo finite element method for Bayesian inversion of log-normal diffusions. Inverse Probl. **36**(3), 035021, 46 (2020)
74. J. Indritz, An inequality for Hermite polynomials. Proc. Am. Math. Soc. **12**, 981–983 (1961)
75. S. Janson, *Gaussian Hilbert Spaces*. Cambridge Tracts in Mathematics, vol. 129 (Cambridge University Press, Cambridge, 1997)
76. D. Kamilis, N. Polydorides, Uncertainty quantification for low-frequency, time-harmonic Maxwell equations with stochastic conductivity models. SIAM-ASA J. Uncertain. Quantif. **6**(4), 1295–1334 (2018)
77. K. Karhunen, Zur Spektraltheorie stochastischer Prozesse. Ann. Acad. Sci. Fennicae Ser. A. I. Math.-Phys. **1946**(34), 7 (1946)
78. Y. Kazashi, Quasi-Monte Carlo integration with product weights for elliptic PDEs with log-normal coefficients. IMA J. Numer. Anal. **39**(3), 1563–1593 (2019)
79. G. Kerkyacharian, S. Ogawa, P. Petrushev, D. Picard, Regularity of Gaussian processes on Dirichlet spaces. Constr. Approx. **47**(2), 277–320 (2018)
80. M. Kohr, S. Labrunie, H. Mohsen, V. Nistor, Polynomial estimates for solutions of parametric elliptic equations on complete manifolds. Stud. Univ. Babeş-Bolyai Math. **67**(2), 369–382 (2022)
81. V.A. Kozlov, J. Rossmann, Asymptotics of solutions of second order parabolic equations near conical points and edges. Bound. Value Probl. **2014**(252), 37 (2014)
82. A. Kufner, *Weighted Sobolev Spaces*. Teubner-Texte zur Mathematik [Teubner Texts in Mathematics], vol. 31 (BSB B. G. Teubner Verlagsgesellschaft, Leipzig, 1980). With German, French and Russian summaries
83. F.Y. Kuo, R. Scheichl, C. Schwab, I.H. Sloan, E. Ullmann, Multilevel Quasi-Monte Carlo methods for lognormal diffusion problems. Math. Comp. **86**(308), 2827–2860 (2017)

84. A. Lang, C. Schwab, Isotropic Gaussian random fields on the sphere: regularity, fast simulation and stochastic partial differential equations. Ann. Appl. Probab. **25**(6), 3047–3094 (2015)
85. M. Lassas, E. Saksman, S. Siltanen, Discretization-invariant Bayesian inversion and Besov space priors. Inverse Probl. Imaging **3**(1), 87–122 (2009)
86. H. Li, An anisotropic finite element method on polyhedral domains: interpolation error analysis. Math. Comp. **87**(312), 1567–1600 (2018)
87. M.A. Lifshits, *Gaussian Random Functions*. Mathematics and Its Applications, vol. 322 (Kluwer Academic Publishers, Dordrecht, 1995)
88. H. Luschgy, G. Pagès, Expansions for Gaussian processes and Parseval frames. Electron. J. Probab. **14**(42), 1198–1221 (2009)
89. B. Matérn, *Spatial Variation*. Lecture Notes in Statistics, vol. 36, 2nd edn. (Springer-Verlag, Berlin, 1986). With a Swedish summary
90. V. Maz'ya, J. Rossmann, *Elliptic Equations in Polyhedral Domains*. Mathematical Surveys and Monographs, vol. 162 (American Mathematical Society, Providence, 2010)
91. G.A. Muñoz, Y. Sarantopoulos, A. Tonge, Complexifications of real Banach spaces, polynomials and multilinear maps. Studia Math. **134**(1), 1–33 (1999)
92. F.J. Narcowich, P. Petrushev, J.D. Ward, Localized tight frames on spheres. SIAM J. Math. Anal. **38**(2), 574–594 (2006)
93. J. Nečas, Sur une méthode pour résoudre les équations aux dérivées partielles du type elliptique, voisine de la variationnelle. Ann. Scuola Norm. Sup. Pisa Cl. Sci. (3) **16**, 305–326 (1962)
94. R. Nickl, *Bayesian non-linear statistical inverse problems*. Zurich Lectures in Advanced Mathematics (EMS Press, Berlin, 2023), pp. xi+159
95. S. Labrunie, H. Mohsen, V. Nistor, Polynomial bounds for the solutions of parametric transmission problems on smooth, bounded domains. (2023). arXiv:2308.06215
96. R. Scheichl, A.M. Stuart, A.L. Teckentrup, Quasi-Monte Carlo and multilevel Monte Carlo methods for computing posterior expectations in elliptic inverse problems. SIAM-ASA J. Uncertain. Quantif. **5**(1), 493–518 (2017)
97. C. Schwab, R. Stevenson, Space-time adaptive wavelet methods for parabolic evolution problems. Math. Comput. **78**, 1293–1318 (2009)
98. C. Schwab, A.M. Stuart, Sparse deterministic approximation of Bayesian inverse problems. Inverse Probl. **28**(4), 045003, 32 (2012)
99. C. Schwab, J. Zech, Deep learning in high dimension: Neural network expression rates for analytic functions in $L^2(\mathbb{R}^d, \gamma_d)$. SIAM-ASA J. Uncertain. Quantif. **11**(1), 199–234 (2023)
100. M.L. Stein, *Interpolation of Spatial Data*. Springer Series in Statistics (Springer-Verlag, New York, 1999). Some theory for Kriging
101. I. Steinwart, C. Scovel, Mercer's theorem on general domains: on the interaction between measures, kernels, and RKHSs. Constr. Approx. **35**(3), 363–417 (2012)
102. S.B. Stečkin, On absolute convergence of orthogonal series. Dokl. Akad. Nauk SSSR (N.S.) **102**, 37–40 (1955)
103. A.M. Stuart, A.L. Teckentrup, Posterior consistency for Gaussian process approximations of Bayesian posterior distributions. Math. Comp. **87**(310), 721–753 (2018)
104. A.M. Stuart, A.L. Teckentrup, Posterior consistency for Gaussian process approximations of Bayesian posterior distributions. Math. Comp. **87**(310), 721–753 (2018)
105. G. Szegő, *Orthogonal Polynomials*. American Mathematical Society, vol. XXIII, 4th edn. (American Mathematical Society/Colloquium Publications, Providence, 1975)
106. A.L. Teckentrup, R. Scheichl, M.B. Giles, E. Ullmann, Further analysis of multilevel Monte Carlo methods for elliptic PDEs with random coefficients. Numer. Math. **125**(3), 569–600 (2013). MR 3117512
107. H. Triebel, *Interpolation Theory, Function Spaces, Differential Operators*, 2nd edn. (Johann Ambrosius Barth, Heidelberg, 1995)
108. C. Truesdell, *A First Course in Rational Continuum Mechanics, Vol. 1*. Pure and Applied Mathematics, vol. 71, 2nd edn. (Academic Press, Boston, 1991). General concepts

109. N. Wiener, The Homogeneous Chaos. Amer. J. Math. **60**(4), 897–936 (1938)
110. J. Wloka, *Partial Differential Equations* (Cambridge University Press, Cambridge, 1987). Translated from the German by C. B. Thomas and M. J. Thomas
111. J. Zech, Sparse-Grid Approximation of High-Dimensional Parametric PDEs. Dissertation 25683, ETH Zürich, 2018. http://dx.doi.org/10.3929/ethz-b-000340651
112. J. Zech, Y. Marzouk, Sparse approximation of triangular transports on bounded domains (2020). arXiv:2006.06994
113. J. Zech, C. Schwab, Convergence rates of high-dimensional Smolyak quadrature. ESAIM: Math. Model. Numer. Anal. **54**(4), 1259–1307 (2020)
114. J. Zech, D. Dũng, C. Schwab, Multilevel approximation of parametric and stochastic PDEs. Math. Models Methods Appl. Sci. **29**, 1753–1817 (2019)

Index

D. Dũng et al., *Analyticity and Sparsity in Uncertainty Quantification for PDEs with Gaussian Random Field Inputs*, Lecture Notes in Mathematics 2334,
https://doi.org/10.1007/978-3-031-38384-7

LECTURE NOTES IN MATHEMATICS Springer

Editors in Chief: J.-M. Morel, B. Teissier;

Editorial Policy

1. Lecture Notes aim to report new developments in all areas of mathematics and their applications – quickly, informally and at a high level. Mathematical texts analysing new developments in modelling and numerical simulation are welcome.

   Manuscripts should be reasonably self-contained and rounded off. Thus they may, and often will, present not only results of the author but also related work by other people. They may be based on specialised lecture courses. Furthermore, the manuscripts should provide sufficient motivation, examples and applications. This clearly distinguishes Lecture Notes from journal articles or technical reports which normally are very concise. Articles intended for a journal but too long to be accepted by most journals, usually do not have this "lecture notes" character. For similar reasons it is unusual for doctoral theses to be accepted for the Lecture Notes series, though habilitation theses may be appropriate.

2. Besides monographs, multi-author manuscripts resulting from SUMMER SCHOOLS or similar INTENSIVE COURSES are welcome, provided their objective was held to present an active mathematical topic to an audience at the beginning or intermediate graduate level (a list of participants should be provided).

   The resulting manuscript should not be just a collection of course notes, but should require advance planning and coordination among the main lecturers. The subject matter should dictate the structure of the book. This structure should be motivated and explained in a scientific introduction, and the notation, references, index and formulation of results should be, if possible, unified by the editors. Each contribution should have an abstract and an introduction referring to the other contributions. In other words, more preparatory work must go into a multi-authored volume than simply assembling a disparate collection of papers, communicated at the event.

3. Manuscripts should be submitted either online at www.editorialmanager.com/lnm to Springer's mathematics editorial in Heidelberg, or electronically to one of the series editors. Authors should be aware that incomplete or insufficiently close-to-final manuscripts almost always result in longer refereeing times and nevertheless unclear referees' recommendations, making further refereeing of a final draft necessary. The strict minimum amount of material that will be considered should include a detailed outline describing the planned contents of each chapter, a bibliography and several sample chapters. Parallel submission of a manuscript to another publisher while under consideration for LNM is not acceptable and can lead to rejection.

4. In general, **monographs** will be sent out to at least 2 external referees for evaluation.

   A final decision to publish can be made only on the basis of the complete manuscript, however a refereeing process leading to a preliminary decision can be based on a pre-final or incomplete manuscript.

   Volume Editors of **multi-author works** are expected to arrange for the refereeing, to the usual scientific standards, of the individual contributions. If the resulting reports can be forwarded to the LNM Editorial Board, this is very helpful. If no reports are forwarded or if other questions remain unclear in respect of homogeneity etc, the series editors may wish to consult external referees for an overall evaluation of the volume.

5. Manuscripts should in general be submitted in English. Final manuscripts should contain at least 100 pages of mathematical text and should always include

   - a table of contents;
   - an informative introduction, with adequate motivation and perhaps some historical remarks: it should be accessible to a reader not intimately familiar with the topic treated;
   - a subject index: as a rule this is genuinely helpful for the reader.
   - For evaluation purposes, manuscripts should be submitted as pdf files.

6. Careful preparation of the manuscripts will help keep production time short besides ensuring satisfactory appearance of the finished book in print and online. After acceptance of the manuscript authors will be asked to prepare the final LaTeX source files (see LaTeX templates online: https://www.springer.com/gb/authors-editors/book-authors-editors/manuscriptpreparation/5636) plus the corresponding pdf- or zipped ps-file. The LaTeX source files are essential for producing the full-text online version of the book, see http://link.springer.com/bookseries/304 for the existing online volumes of LNM). The technical production of a Lecture Notes volume takes approximately 12 weeks. Additional instructions, if necessary, are available on request from lnm@springer.com.

7. Authors receive a total of 30 free copies of their volume and free access to their book on SpringerLink, but no royalties. They are entitled to a discount of 33.3 % on the price of Springer books purchased for their personal use, if ordering directly from Springer.

8. Commitment to publish is made by a *Publishing Agreement*; contributing authors of multiauthor books are requested to sign a *Consent to Publish form*. Springer-Verlag registers the copyright for each volume. Authors are free to reuse material contained in their LNM volumes in later publications: a brief written (or e-mail) request for formal permission is sufficient.

Addresses:

Professor Jean-Michel Morel, CMLA, École Normale Supérieure de Cachan, France
E-mail: moreljeanmichel@gmail.com

Professor Bernard Teissier, Equipe Géométrie et Dynamique,
Institut de Mathématiques de Jussieu – Paris Rive Gauche, Paris, France
E-mail: bernard.teissier@imj-prg.fr

Springer: Ute McCrory, Mathematics, Heidelberg, Germany,
E-mail: lnm@springer.com

Printed in the United States
by Baker & Taylor Publisher Services